Computer
Applications
of Numerical Methods

SHAN S. KUO, University of New Hampshire

Computer
Applications
of Numerical Methods

▲▼ **ADDISON-WESLEY PUBLISHING COMPANY**

Reading, Massachusetts · Menlo Park, California · London · Don Mills, Ontario

To my parents

Preface

During the past fourteen years the author has been giving a course in numerical methods and computers at Yale and Tufts Universities and the University of New Hampshire. This course is designed to acquaint students with the numerical methods used in solving problems by means of a high-speed digital computer. The interests of his audience have included such diverse subjects as biology, chemical engineering, chemistry, civil engineering, economics, electrical engineering, computer science, mechanical engineering, physics, psychology, and pure mathematics.

When I began teaching this course, I found that many excellent texts in numerical methods were available, but most of them were either not computer-based or were too specialized for use in a course of this nature. Furthermore, most books in programming languages touched little on numerical methods. Accordingly, it became necessary to prepare some mimeographed notes from which this book developed.

In preparing these lecture notes, I had two objectives in mind: first, to provide the student with the necessary fundamental knowledge of the computer-oriented numerical methods for basic problems in algebra and analysis which form the building blocks for more complicated problems; second, to acquaint him with a high-speed digital computer. The student was required to program and solve meaningful problems on a computer. Experience in the classroom has indicated that this approach develops an excellent comprehension of the successful application of computer-oriented numerical methods.

This text is divided into three parts. Part I deals with man-machine communication in some detail. Part II describes the various numerical methods that have been proved suitable for electronic computers. Part III is concerned with modern topics in digital computation, including linear programming and the Monte Carlo method.

The book is a revision of *Numerical Methods and Computers*, published in 1965. In preparing it the popular IBM System/360 and its Basic FORTRAN IV language

vii

are used for illustrative purposes. Programs written in this subset of the language can be implemented on the IBM System/360, System/370, and many other computers. This subset of FORTRAN language is compatible with and encompasses American Standard Association (ASA) Basic FORTRAN.

This book incorporates not only some new material, but also additional problems as well as a very substantial number of new references. Chapters 2, 4, and 5, as well as every FORTRAN program in the book, have been entirely rewritten. I have added detailed discussions of Romberg's method in Chapter 12, propagation of error from initial data in Chapter 13, and dumps in Appendix B. Two new sections have also been added to Chapter 14 in connection with the generation of pseudo random numbers by the multiplicative congruence method.

The book is suited for courses on the methods of modern computation which are included in the curricula of most universities. The prerequisite in methematics is a good knowledge of calculus and, preferably, elementary differential equations. It is hoped that this book forms a bridge between programming techniques and methods of numerical analysis. The deliberate emphasis on the flow chart and the presentation of a *tested* FORTRAN program for each numerical method should provide the student with real insight into the techniques of computer problem solving, and scientists and engineers with a guide to the solution of advanced problems.

It is with pleasure that I gratefully acknowledge the help and encouragement of my friends, colleagues, and students. In particular, my thanks go to D. T. Chin, C. S. Chu, W. D'Avanzo, R. Desmaison, D. S. Fine, Y. W. Hsu, L. P. Kuan, J. M. Lee, G. H. Room, R. O. Sather, J. Stephenson, E. Vaines, L. A. Walsh, and C. Wolfe for their comments, criticisms, and help in eliminating errors. I also wish to express my appreciation to Drs. A. Wang and S. Chao for helpful comments on Chapter 2. Numerous ideas were drawn from the internal publications of Computation Centers at Dartmouth College, Massachusetts Institute of Technology, University of Michigan, University of New Hampshire, Tufts University, and Yale University.

I am also grateful for the expert typing by Jane Kelfer and Mary Lambert. Finally, I wish to thank the staff of Addison-Wesley Publishing Company for their continuous cooperation.

Durham, New Hampshire
December 1971   S.S.K.

Contents

COMMUNICATION WITH DIGITAL COMPUTERS

The subject matter of Part I is concerned with an introduction of computer components and the general problem of man-machine communication. It is divided into five chapters. Chapter 1 is introductory in aim and content while Chapter 2 deals with the basic units of digital computers. Chapter 3 represents logical flow charting, a graphical representation of a sequence of commands. Chapter 4 is concerned with the floating-point method of arithmetic. Chapter 5 presents the FORTRAN coding system together with many detailed examples.

Introduction

1.1 CHARACTERISTICS OF DIGITAL COMPUTERS

High-speed electronic computers are triggering a revolution in the solving of engineering and science problems. This revolution will free men from many mentally stultifying computational tasks, and more important, it will make possible the solution of engineering and science problems of greater complexity.

There are two classes of computers: analog and digital. An analog computer, like a slide rule, solves problems by converting numbers into physical quantities such as distance or electrical resistance. On the other hand, a high-speed digital computer is like a desk calculator in that it uses numbers to express all the variables and quantities of a problem. However, it differs from a desk calculator in one important respect: It can automatically perform a long and complete sequence of operations without intervention from the human operator.

This book deals exclusively with digital computations; references to analog computations are listed at the end of the chapter.

We shall now discuss some marked characteristics of digital computers, the first of which is their high speed. In a matter of seconds or minutes, vast quantities of computations can be automatically performed. Computations avoided because of their impracticality prior to the use of digital computers can now be handled as a matter of simple routine. Consider the case of comparative engineering design. The designer, without computers, would either make a "guestimate" from his experience or at best estimate only one or two alternatives. When a digital computer is used, a complete comparative study can be obtained to show the effect of numerous parameters in a short time.

The digital computer is not merely a glorified slide rule; it has the ability to store, or remember, various information for future use. Such information includes the

3

original data, commands of operation, and intermediate results. In addition, digital computers are able to change or modify the commands of operations internally, and frequently the intermediate results obtained are used internally to dictate the path of subsequent computations.

For example, at a point of computation, the answer is tested for zero. Depending on whether or not the answer is zero, the machine will automatically take one of two entirely different courses for subsequent computations. Thus the digital computer is also noted for its ability to make a logical decision.

1.2 A BRIEF HISTORY AND SURVEY OF THE APPLICATION OF DIGITAL COMPUTERS

The first all-electronic digital computer was completed in 1946 at the University of Pennsylvania and was named the Electrical Numerical Integrator and Automatic Computer (ENIAC). Vacuum tubes were used for most of its functions. This was a great improvement over the Mark I digital computer built at Harvard University in 1944, which made use of electromechanical relays.

In the period following the completion of the ENIAC, two significant developments made possible the present-day family of computers. One was the development of a memory device for holding a few hundred to several thousand numbers. The other was the realization that commands could also be stored in this memory device in a manner similar to that in which the numbers are stored. This knowledge made it possible to instruct the computer to follow these commands from each memory space as required. In the meantime, the representation of numbers in the binary system was put in use, and since that time the number of basic commands has been steadily increased. The vacuum tube computers constituted the first generation, from 1947 to about 1959.

Since 1951 there have been continuous advances in the design and components of digital computers. Among these were the introduction of solid-state or transistorized design, increased number of available memory spaces, faster speed of operation, and a further diversification of commands. The second generation of computers, characterized by the solid-state design, began to appear in the market in 1959.

Many super computers have been built since 1951, including LARC, Stretch, CDC 6600, 7600, and ILLIAC IV. They have been designed to reach to the limit of existing practice in terms of speed, size, and logic design.

Most computers built since 1965 belong to a third generation. These computers are characterized by the microelectronics technology (electronic chips) in hardware and multiprocessor multiprogramming systems in software.

Digital computers are presently being used to store and retrieve information quickly and economically, simulate complex business operations, create a "model" river system, help determine who wrote the Federalist papers whose authorship is disputed, chart the complex interrelations among the hundreds of electric signals reproduced by the living brain, formulate and prove mathematical theorems, take the first step toward translation of languages, forecast the weather, and analyze the decay tracks left by strange particles in bubble chambers.

Computers are also used to calculate spacecraft orbits, process payrolls, update transactions, control chemical blending processes, design structures and machine components, and assist in medical diagnoses. They are simulating aircraft flight characteristics, automating airline reservation procedures, and controlling inventories. Two-way communication between the brain of a chimpanzee and a computer has also been established.

A procedure called *time-sharing* which brings the user closer to the computer is rapidly gaining recognition. In this procedure, a number of consoles scattered in different locations are connected to one central computer so that a number of users can take over control, successively, as they need its services. While the large time-sharing systems are subject to continuing research, the use of conventional computers by means of on-line remote consoles may well be an important consideration in future development.

Another interesting development is in the area of engineering graphics. By using a light pen on a cathode-ray oscilloscope, one can draw two projections of a given object and ask the computer to straighten out lines or rectify angles. A perspective view can then be produced by the computer and, if desired, recorded on microfilm. Similarly, the equation of a surface may be read in and a contour plot can readily be made by the computer.

BIBLIOGRAPHY

General

Alt, F. L. (ed.), *Advances in Computers*. Academic Press, New York. This is intended to be a continuing publication, the first volume of which appeared in 1960.

Bowden, B. V., *Faster than Thought*. Pitman, London, 1964.

Fano, R. M., and F. J. Corbato, "Time-sharing on computers," *Scient. Am.*, **215**, 128–143 (Sept. 1966).

Feigenbaum, E. A., and J. Feldman (eds.), *Computers and Thought*. McGraw-Hill, New York, 1963.

Greenberger, M. (ed.), *Computers and the World of the Future*. The M.I.T. Press, Cambridge, Mass., 1962.

McCarthy, J., "Information," *Scient. Am.*, **215**, 64–73 (Sept. 1966).

Rosen, S., "Electronic computers, a historical survey," *Computer Surveys*, **1**, 7–36 (1969).

Thompson, R. E., "Computer plotting of graphs," *Engng. Educ.*, 470–471 (Feb. 1970).

Von Neumann, J., *The Computer and the Brain*. Yale Univ. Press, New Haven, Conn., 1958.

Analog Computer Techniques

Johnson, C. L., *Analog Computer Techniques*, 2nd ed. McGraw-Hill, New York, 1963.

Korn, G. A., and T. M. Korn, *Electronic Analog and Hybrid Computers*. McGraw-Hill, New York, 1964.

Rogers, A. E., and T. W. Connolly, *Analog Computation in Engineering Design*. McGraw-Hill, New York, 1960.

Warfield, J. N., *Introduction to Electronic Analog Computers*. Prentice-Hall, Englewood Cliffs, N.J., 1959.

Zulauf, E. C., and J. R. Burnett, *Introductory Analog Computation with Graphic Solutions*. McGraw-Hill, New York, 1966.

CHAPTER 2

Main Computer Components

2.1 FUNCTIONS OF COMPUTER COMPONENTS

A basic computer system consists of the five components shown in Fig. 2.1: input devices, a main memory, an arithmetic and logic unit (ALU), a control, and output devices. Physically, they may be either combined into a single unit or separated into several distinct parts. The main memory, arithmetic and logic, and control are usually housed in the same unit, known as central processing unit (CPU).†

The main function of input devices is to feed both data and commands into the memory unit. This is accomplished through units such as a keyboard, a punched card reader, a device to read magnetic or paper tape, a telephone, a teletype line, a disk, or an optical reader.

After the commands and data are stored in the memory unit, the commands are then used to instruct the computer to proceed with the computation or data processing. It is important to note that intermediate or final answers can also be stored in the memory unit for future use and output.

The actual work of computation is carried out in the arithmetic and logic unit. It performs all arithmetic operations, makes logical decisions, and carries out other manipulations such as transformations of data from one representation to another. However, the sequence of operations is supervised by the control unit. This is the "mastermind" of the entire system, as it controls the activities of each of the other units.

Finally, there are output devices which serve to carry the resultant data outside the central processor. These devices include card punch units, printers, and magnetic or paper tape units. An oscilloscope resembling a television set or a plotter may also be utilized to display the results in pictorial form.

---

† ALU and control unit together are also often called CPU.

7

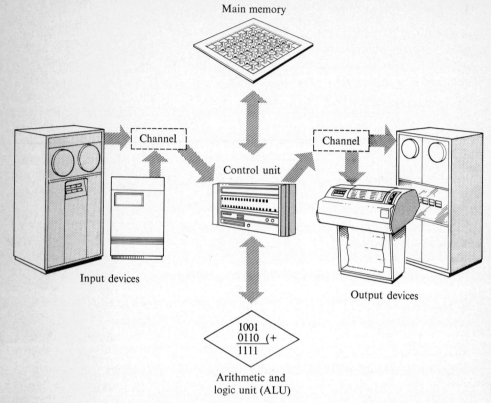

Fig. 2.1. Basic computer components. Channels are used in large systems.

2.2 MEMORY UNITS

There are two classes of memory units; namely, main memory and auxiliary memory units. The outstanding difference between these two classes lies in the *access time*, the time needed to locate a particular item of information and to transmit it to and from the memory unit. A memory unit is often called a storage unit.

The main memory unit is the seat of immediate "memory" of a computer system. It is composed of either monolithic "chips" or tiny magnetic cores. For a large system, the access time of a core memory is in the vicinity of 1 microsecond, and that of a monolithic memory is even shorter.

The main memory units have relatively limited storage capacity and are very expensive. The auxiliary memory units afford a larger capacity at the expense of a somewhat longer access time. In ascending order of the access time the major types of auxiliary memories are magnetic drums, direct access devices such as magnetic disk packs, data cells, and magnetic tapes. Thus the access time of a magnetic drum is faster than that of a disk pack, but by far slower than that of cores.

Main Memory Units

As mentioned in the last section, the memory unit serves to retain (a) initial data or instructions, (b) intermediate results, and (c) final answers. It is convenient to compare the memory unit to an array of post office boxes, with each box storing either a command or a datum. However, there is a basic difference. At any particular time, two or more communications can generally be stored in a single ordinary mail box, while only one command or one datum can be stored in a given memory location. As a second command or datum enters into a given memory location, the existing content is automatically erased. Also, unlike a mailbox, when the content in a given memory location A is transmitted to some other location B, the content in A is not disturbed; that is, it exists in both location A and location B. It cannot be erased from the original location until a new command or datum is fed into this original storage location. Memory location and storage location are synonymous.

We shall first discuss the *magnetic core*. Each core is a doughnut-shaped piece of ferromagnetic material about 0.02 to 0.1 in. in diameter. A wire is inserted through its hollow portion, as shown in Fig. 2.2(a), and when an electric current is passed along the wire, it sets up a magnetic field around the wire. The core is magnetized by the field and remains magnetized even if the current is removed (Fig. 2.2b). Now, if a current is passed along the wire in the opposite direction, the field around the wire is reversed, as shown in Fig. 2.2(c). This change of the magnetic state of a core is called *flipping*. Another wire, a sense wire, is also inserted through the hollow portion, and, when a magnetic core flips, a low voltage is induced along this sense wire. This voltage is then amplified and used in the computer (Fig. 2.3).

Another property of a magnetic core is that it does not change the direction of magnetization unless an electric current reaches a certain value, say, I. A practical application of this property is that cores can be arranged in the rectangular array of many columns and rows, and wires are passed through each row and each column (see Fig. 2.4). If electric current of $0.5I$ is transmitted in a particular row wire, and same current in a given column wire, then *the* core which contains both wires is magnetized in a direction. As shown in Fig. 2.4, the array of cores is known as a *core memory plane*. A third wire is also inserted through the hollow portion of a core. This sense wire helps read information from a memory core.

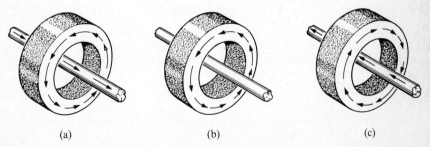

(a)   (b)   (c)

Fig. 2.2. Magnetic core. (a) Core is magnetized. (b) Core remains magnetized. (c) Core is *flipped*.

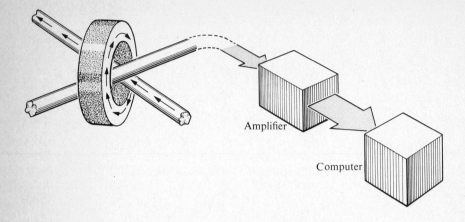

Fig. 2.3. Amplification of core flipping.

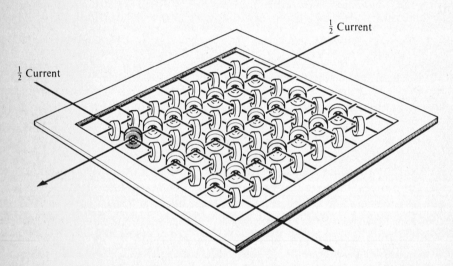

Fig. 2.4. Core memory plane.

A core storage unit is made of many such magnetic cores arranged in planes or sheets, as shown in Fig. 2.5. The corresponding cores in all sheets can be considered one distinct group. Such a group is usually identified with a unique number commonly known as an *address*, or a *location*.

In the core storage unit shown in Fig. 2.5, therefore, there are 49 addresses, each of which extends downward through the eight planes. In a digital computer containing, say, 8192 memory locations, the addresses would be numbered consecutively from 0000 to 8191.

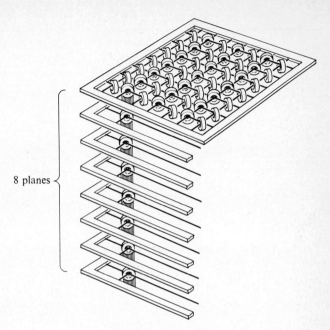

8 planes

Fig. 2.5. Core array.

It should be emphasized that the location number is totally different from the information stored in this location. For example, in location 7614, which has eight cores, the information stored may be

10110110

This information may be a piece of data or an instruction for future use, but it is *not* the number 7614 in binary notation. The use of the zeros and ones for the representation of information will be explained later in Sections 2.4 and 2.5.

The monolithic memory has been developed only very recently. Several models of IBM System/370 are already designed to use monolithic circuitry in all memory functions. The memory's basic building block is a silicon "chip" less than $\frac{1}{8}$ in. square. Each chip contains about 1400 microscopic circuit elements, such as resistors, transistors, and diodes, interconnected to form about 170 complete memory and support circuits. The high density of the monolithic circuits makes it possible to provide faster main memories to users. For example, the internal operating speed of System/370 Model 145, which has the monolithic main memory, is up to five times that of comparable System/360 Model 40, which has the conventional core memory unit.

Auxiliary Memory Units

Physically external to the main memory unit, the auxiliary storage units extend the storage capacity of a computer system at relatively low cost. They are used mainly for temporary storage, or for programs or data to which immediate access is not required. All auxiliary memories have the ability to retain the recorded information even if the power supplied to the computer is cut off.

The magnetic drum is a metal cylinder with a magnetic coating which can be demagnetized and magnetized readily, or left magnetized for a period of time.

Under operating conditions, the drum is rotating at high speed (in the vicinity of 10,000 rpm). When a magnetic-coated surface is passed directly under an active recording head (Fig. 2.6), the surface becomes magnetized as in a household tape recorder. It is evident that information stored on the drum's surface exists in a mixed pattern of magnetized and nonmagnetized areas. Each area represents a binary digit, that is, either a "0" or a "1." When reading from the drum, the process is just reversed. As the drum surface is passed under the recording head, the magnetized area induces small voltage in the head (Fig. 2.7) which can be amplified and transmitted to the computer for use. It is important to note that a piece of input data or an instruction can be stored in any specified location of a drum.

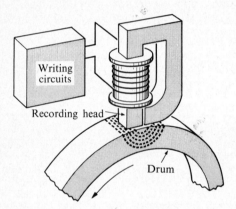

Fig. 2.6. Writing on a magnetic drum.

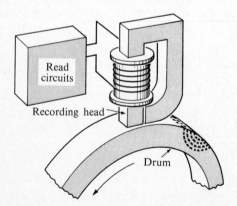

Fig. 2.7. Reading from a magnetic drum.

Magnetic drums are permanently fixed on their axes and are not interchangeable with other drums. The removable disk pack resembles a stack of records, and allows a user to easily change packs as he wishes. As the disks spin at about 2500 rpm, data are magnetically recorded on or retrieved from the surfaces of disks by the read-write heads. The density of data recording is about 200,000 bits per square inch of disk surface. Access time to data on disks can be as fast as 1/30,000 sec. A disk pack with access mechanism in a disk storage drive is shown in Fig. 2.8.

Data cells contain magazines of magnetic strips. The drive unit can address one of the strips for reading or writing. Finally, the magnetic tape is the storage medium which costs the least but has the slowest access. The magnetic tape used in a computer system is of the same type used for home tape recorders. Across the width (about 0.5 in.) of the tape, several small areas can be either magnetized or nonmagnetized. In Fig. 2.9 are shown 9-bit characters located at each position along the tape. Since a *density* of 800 characters per inch of tape is considered as standard, millions of characters can be stored on one reel of tape. The principal drawback of magnetic tape is that data on tape can be retrived only sequentially, and the resulting delay, from seconds to minutes, hampers computer operation.

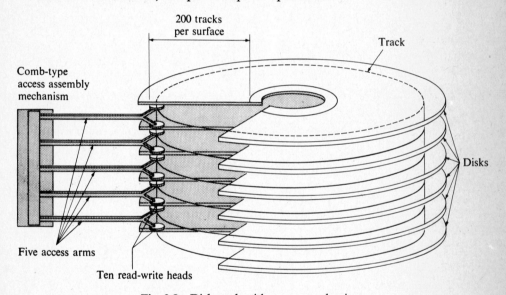

Fig. 2.8. Disk pack with access mechanism.

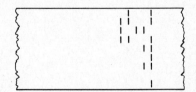

Fig. 2.9. Information recorded on a magnetic tape (not visible).

2.3 CPU, INPUT/OUTPUT DEVICES, AND CHANNELS

Aside from the main memory unit, the central processing unit (CPU) contains also the arithmetic and logic unit (ALU), and the control section. The ALU performs arithmetic when information is received on (1) the required operation and (2) the operand(s) to be operated on. The basic arithmetic operations are addition and subtraction. Multiplication and division are carried out respectively by repeated addition and repeated subtraction. The ALU can also compare two operands and indicate its result. Many large computer systems are designed to handle different data formats such as integer, decimal, and floating-point representations. Their ALU must also be capable of carrying out the operations on each of these data formats.

The control unit supervises the fetching, decoding and execution of each instruction. The desired instruction is first fetched from the main memory unit and then decoded during the so-called *instruction cycle*. In computers, the device which serves to locate the proper instruction for execution is called the *instruction address register*. This device is also responsible for setting up the conditions for the next instruction. For example, the next instruction to be executed may depend on the sign of the number in the arithmetic unit; if it is minus, the next instruction will be obtained from location 1678, for example. Conversely, if it is positive, then the next instruction might be obtained from another location, say 1296. The instruction is actually carried out during the *execution cycle*—which is then followed by a new set of instruction cycle and execution cycle, during which time another instruction will be fetched, decoded and executed, and so on.

The central processing unit is often referred to as the *main frame*. All units outside the main frame are collectively known as *peripheral equipment*. They include the auxiliary memories and a wide assortment of input/output devices: card readers, printers and plotters, optical scanning devices, and terminals with keyboards and TV-like screens.   More sophisticated input/output devices are also commercially available: printers that record output on microfilm at high speed (about 100,000 characters per second) and portable terminals which allow a user direct contact with a remote computer through a keyboard and telephone. Also, graphic devices of great versatility are growing in number and variety.

The continuing demand for new peripherals stems from the fact that a CPU can process data in the nanosecond range, while the electromechanical pheripheral devices still work relatively slowly. Specifically, the speed ratio between the fastest card reader and the slowest core memory is about 1/500. Much CPU time would be wasted if a card reader were attached directly to the CPU because the CPU has to wait when a card is being read into the main memory. In order to enable the CPU to do more computing, *channels* are installed between the CPU and peripheral equipment.

Each channel is in reality a mini-computer. In the IBM System/360 or System/370, two types of channels are in use: selector channels and multiplexor channels. The former is used between the CPU and high-speed devices such as drum or disc drives, while the latter is used primarily between the CPU and low-speed devices such as card readers or printers.

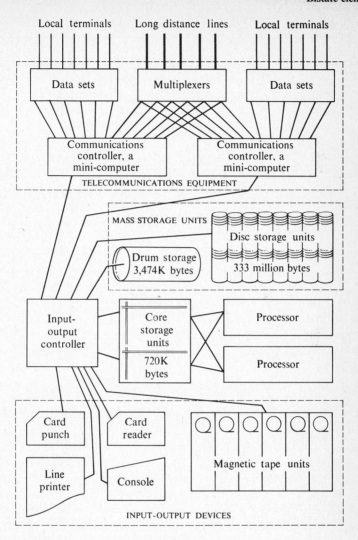

Fig. 2.10. Hardware used in a time-sharing computer system. $K = 2^{10} = 1024$.

In a time-sharing system, specially designed mini-computers often serve as communications controllers. Figure 2.10 shows schematically such a system in which two mini-computers are used to handle a total of about 170 terminals of all kinds. They are capable also of connecting with terminals of other types such as TV-like display units, and of communicating with other similiar mini-computers located in remote areas.

2.4 BISTATE ELEMENTS

A distinguishing characteristic of all computer elements is their ability to represent two (and only two) distinct states. The magnetic core mentioned in Section 2.2 is a typical example. The core may be in either of two magnetized directions. These two distinct conditions can be used to represent the information which the core holds. For instance, one condition may be interpreted as the digit 0; the other, as 1. Similarly, various computer elements may be used to represent yes or no, positive or negative, on or off, etc. Figure 2.11 shows some of the bistate computer elements.

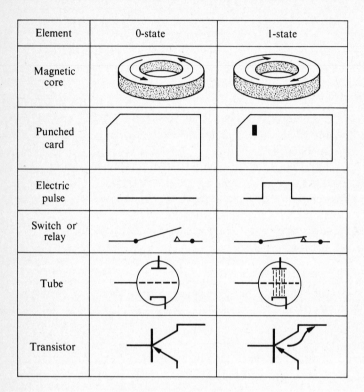

Element	0-state	1-state
Magnetic core		
Punched card		
Electric pulse		
Switch or relay		
Tube		
Transistor		

Fig. 2.11. Bistate elements.

This two-state character is important in that it makes it possible to denote the digits 0 or 1 in a particular storage location. The binary digits 0 and 1 are commonly called *bits*, which is a contraction of two words, "binary" and "digit." These *bits* are the basic units of information stored in digital computers.

There are two types of digital computers available commercially, binary computers and decimal computers. In a binary computer such as the IBM System/360 or the System/370, an internal machine code is represented simply by a string of 0's and 1's. For example, an instruction for adding two numbers is shown in Fig. 2.12. One number is stored in register 7; and the other in register 11.†

---

† A general register is a high-speed storage device. It is a part of the CPU, not part of main storage. There are 16 general registers in IBM System/360, each containing 32 bits.

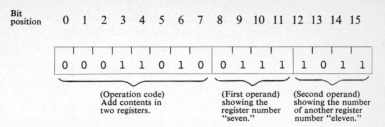

Fig. 2.12. The internal instruction for adding two numbers.

2.5 BYTES, HALFWORDS, FULLWORDS, AND DOUBLEWORDS

In Section 2.4, we stated that a bit is a single binary digit with a value of either 0 or 1. In the IBM System/360, or the System/370 computer, a sequence of eight adjacent bits is designed to be operated upon as a unit, known as a *byte*. Each byte has its own address. In Fig. 2.13, it is shown that a sequence of two adjacent bytes (16 bits) is defined as a *halfword*. Similarly, a sequence of four adjacent bytes (32 bits) is called a *word* (or *fullword*); and eight consecutive bytes (64 bits), a *doubleword*. Consider, for example, the instruction shown in Fig. 2.12. It occupies two bytes; thus it is a halfword instruction. Instructions may also occupy either four bytes (a fullword), or six bytes (three halfwords).

Byte	Byte	Byte	Byte	Byte	Byte	Byte	Byte	Byte
Halfword		Halfword		Halfword		Halfword		Halfword
Word				Word				
Doubleword								

Fig. 2.13. Byte: the basic addressable unit.

2.6 DATA FORMATS

In the previous section we learned that an instruction may be two, four, or six bytes long—each known as a fixed-length field. In contrast with an instruction, a piece of data may occupy either a fixed-length or a variable-length field, depending on the data format. There are four basic data formats in both the IBM System/360 and the System/370, namely, a fixed-point (or binary integer) format, two versions of decimal formats, and a floating-point format.

In this section we shall discuss briefly the first three formats, leaving the floating-point data format to the next chapter.

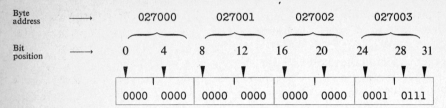

Fig. 2.14. Representation of the decimal number $+23$ in a fullword.

Table 2.1 EBCDIC character/numeral sets

Graphic	Bit Configuration	Graphic	Bit Configuration
(Blank)	0100 0000	F	1100 0110
¢	0100 1010	G	1100 0111
.	0100 1011	H	1100 1000
<	0100 1100	I	1100 1001
(	0100 1101	J	1101 0001
+	0100 1110	K	1101 0010
|	0100 1111	L	1101 0011
&	0101 0000	M	1101 0100
!	0101 1010	N	1101 0101
$	0101 1011	O	1101 0110
*	0101 1100	P	1101 0111
)	0101 1101	Q	1101 1000
;	0101 1110	R	1101 1001
→	0101 1111	S	1110 0010
–	0110 0000	T	1110 0011
/	0110 0001	U	1110 0100
,	0110 1011	V	1110 0101
%	0110 1100	W	1110 0110
–	0110 1101	X	1110 0111
>	0110 1110	Y	1110 1000
?	0110 1111	Z	1110 1001
:	0111 1010	0	1111 0000
#	0111 1011	1	1111 0001
@	0111 1100	2	1111 0010
'	0111 1101	3	1111 0011
=	0111 1110	4	1111 0100
"	0111 1111	5	1111 0101
A	1100 0001	6	1111 0110
B	1100 0010	7	1111 0111
C	1100 0011	8	1111 1000
D	1100 0100	9	1111 1001
E	1100 0101		

First, let us consider the fixed-point format. The binary integer equivalent to a decimal value, say, 23, represented in a fullword or 4 bytes, is shown in Fig. 2.14. Each fixed-length field is addressed by the leftmost byte of the field. The address of the word is shown in the address of the leftmost byte, or 027000.

Second, a decimal digit (0–9), alphabet (A through Z), and special characters such as $, @, or % are each represented internally as a fixed group of 8 binary digits, or a byte, according to a given code—known as EBCDIC (Extended Binary Coded Decimal Interchange Code). A rather complete table of character/numeral sets in EBCDIC is listed in Table 2.1. For example, the information, S. KUO-6/7, can be represented in this code by 10 bytes, each for a character, as illustrated in Fig. 2.15. Such character representation of decimal digits is also known as zoned decimal. Decimal digits can also be represented in the packed-decimal form, two digits per byte; with the sign represented in the rightmost four bits. For example, the decimal number –167 can be represented in a string of 0's and 1's as illustrated in Fig. 2.16.

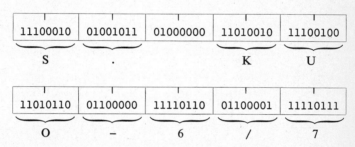

Fig. 2.15. Representation of alphabetic information in a computer. It takes 10 bytes.

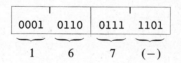

Fig. 2.16. Representation of a decimal number − 167 in packed decimal form. It takes 2 bytes.

BIBLIOGRAPHY

Bell, C. G., and A. Newell, *Computer Structures: readings and examples.* McGraw-Hill, New York, 1971.

Buckholz, W., *Planning a Computer System . . . Project Stretch.* McGraw-Hill, New York, 1962.

Chu, Y., *Digital Computer Design Fundamentals.* McGraw-Hill, New York, 1962.

Evans, D. C., "Computer logic and memory," *Scient. Am.,* **215,** 74–85 (Sept. 1966).

Hartree, D. R., *Calculating Instruments and Machines.* University of Illinois Press, Urbana, 1949.

Heath, F. G., "Large-scale integration in electronics," *Scient. Am.*, **222**, 22 (Feb. 1970).

Huskey, H. D., and G. A. Korn, *Computer Handbook.* McGraw-Hill, New York, 1962.

Liveslay, R. K., *An Introduction to Automatic Digital Computers*, 2nd ed. University Press, Cambridge, England, 1960.

Phister, M., Jr., *Logical Design of Digital Computers.* Wiley, New York, 1958.

Pierce, J. R., "The transmission of computer data," *Scient. Am.*, **215**, 144–159 (Sept. 1966).

Sutherland, I. E., "Computer inputs and outputs," *Scient. Am.*, **215**, 86–111 (Sept. 1966).

CHAPTER 3

Logical Flow Charting

3.1 THE SOLUTION OF A SCIENTIFIC OR ENGINEERING PROBLEM ON COMPUTERS

The steps involved in solving a scientific or engineering problem by high-speed digital computers include:

1. Converting the physical system to an idealized mathematical model, and, according to this model, formulating the mathematical equations.
2. Selecting a numerical procedure suitable for a digital computer. (Numerical methods are discussed in Part II.)
3. Drawing a detailed flow chart, i.e., a graphical representation of a sequence of operations, such as "read, compute, compare, write, etc."
4. Based on the flow chart, writing these operations in a language which a specific machine will recognize and accept. The process of writing computer instructions is called "coding."
5. Making a "test run" on the computer. If the machine yields the incorrect answers, or if the machine operates in a manner not planned in the program (such as a permanent loop), the coding should be checked or "debugged."
6. The corrected coding can be used at any time to make a "production run."

If the mathematical model is found to be a poor selection, one should then repeat step 1 after step 5 a few times, until a better model is found.

3.2 FLOW CHART

As mentioned in the previous section, a flow chart is a graphic representation of the course of solution to a given problem. It provides an overall picture of the algebraic and logical processes. The completion of a detailed flow chart is important since coding systems have not yet been standardized, although much effort has been made toward the formulation of a universal algebraic computer language acceptable to all computers. Even if such a language were available, the flow chart would still be a desirable way of representing the procedure for solving a problem. In discussing a complicated program, a flow chart is practically indispensable.

Before presenting a simple example of a flow chart, let us recall that digital computers are at best capable of executing three types of instructions.

The first type is the transfer of data. This type is exemplified by such instructions as "move the content of location 1520 to location 324," "read the constant C_1 into its proper storage location, 1079," and "write out the result which is stored presently in the location 1622."

The second type is arithmetic operations. For example, a machine is capable of performing the following operations: $(A + B \cdot C)/(D - C) \cdot A$.

The third type of machine operation is making logical decisions. For example: Let the answer obtained at a point of computation be A. Then, depending on whether A is positive or negative, the subsequent computations will take one of two entirely different paths.

We shall now illustrate flow charting by a simple example. It is required to evaluate the following polynomial:

$$x = \sum_{i=0}^{8} c_i y^i,$$

where c_i $(i = 0, 1, \ldots, 8)$ are known constants. We arbitrarily require that the summation will be made starting from the first term, that is, $c_0 y^0$.

The problem, as solved with a computer, may consist of the following steps:

1. Read in all known values, $c_0, c_1, \ldots, c_8$, and y (transfer of data).

2. Set $x = 0$; that is, store zero in the location for x (transfer of data).

3. Set $i = 0$; that is, store the constant 0 in the location for i (transfer of data).

4. Compute $x_i = c_i y^i$. Note that x_i has its own location (arithmetic operation).

5. Add x and x_i; store the sum in the location for x. This automatically erases the old value of x (arithmetic operation).

6. Increase i by 1 (arithmetic operation).

7. Check whether the present value i is equal to 9 (making a logical decision). If it is not equal to 9, go to step 4 and continue; if i is equal to 9, the required answer, x, has been obtained. The i-value has taken on the ten values $0, 1, \ldots, 9$, but the last one was not used. So we have been through the loop nine times. Finally, print out this answer.

Figure 3.1 shows the corresponding flow chart for the example. It is seen that a rectangular box is used in connection with an arithmetic operation, while an oval is used to indicate a logical decision or a branch instruction. Furthermore, arrows are used in two different senses; one serves to indicate the direction of computer operation;

the other (used *inside* a rectangular box) serves to express "replaced by." For example, in Fig. 3.1,

$$\boxed{x \leftarrow 0}$$

means that the content of x is replaced by zero. Similarly, the step

$$\boxed{x \leftarrow x + x_i}$$

means that the content of x is replaced by the result of the addition of the content of x to that of x_i. It is important to note that this step is often expressed in a flow chart as $x = x + x_i$. The equals sign, as used here, must be interpreted as "replaced by," not as "equal to."

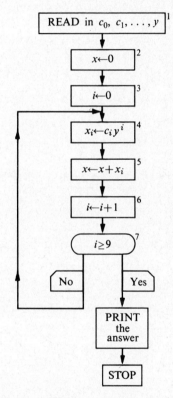

Fig. 3.1. Flow chart for computing $x = \sum_{i=0}^{8} c_i y^i$.

In Fig. 3.1, step 7 is sometimes expressed in the following way:

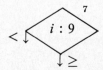

Step 7 is important in that advantage has been taken of the computer's ability to make a logical decision by comparing one quantity with another. Based on this comparison, there are two possible paths of subsequent operations, and the two arrows coming out of the diamond (or oval) serve to indicate these two possibilities.

Figure 3.2 shows another version of the flow chart given in Fig. 3.1.

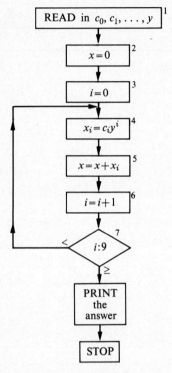

Fig. 3.2. Another version of a flow chart for computing $x = \sum_{i=0}^{8} c_i y^i$.

3.3 LOOP

In the previous section, a sample flow chart was discussed to evaluate the following polynomial:

$$x = \sum_{i=0}^{8} c_i y^i.$$

Steps 4 through 7 are said to be in a loop since the computation will follow the sequence of steps 4, 5, 6, 7, 4, 5, 6, 7, 4, ... as long as the condition $i < 9$ is satisfied. (There is a total of 36 steps.) Hence a loop is a repetition of a group of instructions in a program.

A fundamental property of digital computers is their ability to repeat a group of instructions at high speed. A loop in a flow chart serves to reflect this important characteristic. It is very infrequent that a flow chart does not contain one or more loops. Many flow charts even have loops within a loop; this idea will be discussed in the next section.

Returning to Fig. 3.2, we note that we set $x = 0$ and $i = 0$ before executing the loop. This line of reasoning is clear if we examine our procedure closely. In block 6, for example, each i is determined by the one preceding it; however, in the first pass through the loop, the computer needs an initial value of i to work with. It is thus seen that i must be defined (initialized) before the loop is entered, and the equation with which we are working demands that the first value of i be zero. The reasoning here is similar to that used in summations. Thus

$$\sum_{i=0}^{m} x_i \quad \text{and} \quad \sum_{i=1}^{m} x_i$$

have different meanings, and hence the initial value for i must be defined. Similarly, each value for x is defined in terms of the x preceding it, and therefore the initial value of x must be defined as in step 2.

Referring again to Fig. 3.2, we find that step 6 is used to update the quantity i, and hence this step is called updating or incrementing.

So far we have discussed a possible flow chart for evaluating the polynomial

$$x = \sum_{i=0}^{N} c_i y^i,$$

where $N = 8$.

We shall now consider the case for $N = 500$. Making a change in step 7 (Fig. 3.2) of

$$\boxed{i : 9}$$

to

$$\boxed{i : 501}$$

would certainly meet the need. But by doing so, we are required to read in the 501 constants for c_i and the one value for y. This may not be desirable if the storage of a given computer is limited to, say, only 500 locations. To avoid this difficulty, one must not read in all known values $c_0, c_1, \ldots, c_{500}$, and y in the first step. Figure 3.3 shows a possible flow chart for computing the sum of the 501 terms. It is noted that, in the first step, the y-value alone is read in. Furthermore, a new step, 4(a), is inserted at the beginning of the loop to read in the c_i-values. Thus we need only one storage location for the c_i instead of 501.

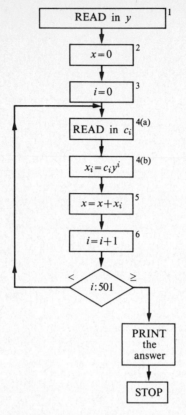

Fig. 3.3. Flow chart for computing $x = \sum_{i=0}^{500} c_i y^i$.

In the above discussion, the coefficients c_i are assumed to take on random values. If, one the other hand, they follow a certain pattern, it is desirable to *generate* these 501 constants and to avoid reading in a large amount of data. For example, if

$$c_i = i^{1/2}, \quad i = 0, 1, 2, \ldots, 250,$$

$$c_i = 3i^2, \quad i = 251, 252, \ldots, 500,$$

then the c_i can be generated as shown in Fig. 3.4. Generating numbers in a computer is an arithmetic operation. It is faster than reading in numbers, which is a data transfer operation.

3.4 LOOP WITHIN A LOOP

In the flow chart shown in Fig. 3.4 it was assumed that the value for y was a single constant. Suppose now that we had to repeat the process for n different values of y. One method of solving the problem is to run the program n times, but an easier solution is given in Fig. 3.5. Here we expand the flow chart shown in Fig. 3.4 so that

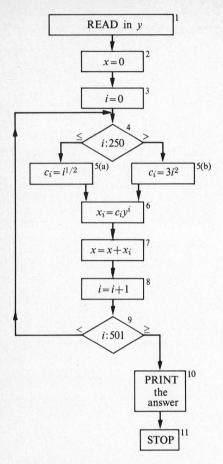

Fig. 3.4. Flow chart for computing
$x = \sum_{i=0}^{500} c_i y^i$:
$c_i = \sqrt{i}, \quad 0 \le i \le 250;$
$c_i = 3i^2, \quad 251 \le i \le 500.$

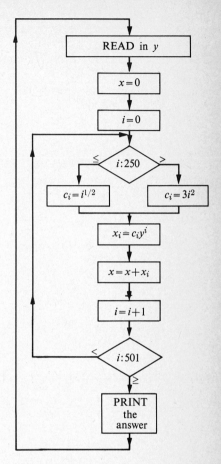

Fig. 3.5. Flow chart for computing
$x = \sum_{i=0}^{500} c_i y^i$ for several y-values:
$c_i = \sqrt{i}, \quad 0 \le i \le 250;$
$c_i = 3i^2, \quad 251 \le i \le 500.$

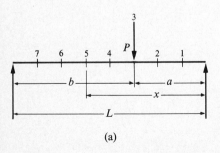

(a)

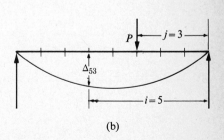

(b)

Fig. 3.6. Deflection of a beam due to load P.

the necessary steps for the evaluation of the polynomial are executed n times, with different values of y. This illustrates the concept of a loop within a loop. A possible extension to three or more loops is obvious.

As a second example for a loop within a loop we shall consider a problem in the strength of materials. Let us draw a flow chart for computing the deflections of the simply supported beam shown in Fig. 3.6. The beam is divided into eight equal segments. We denote by Δ_{ij} the deflection of the beam at station i due to a concentrated load P applied at point j. Thus the deflection Δ_{53}, shown in Fig. 3.6(b), represents the deflection at station 5 due to a load P acting at station 3. Since both i and j range from 1 to 7, it is clear that a total of 49 deflections are to be computed. The problem is to draw a simple flow chart, using loops, for computing the 49 deflections of the 7 different stations.

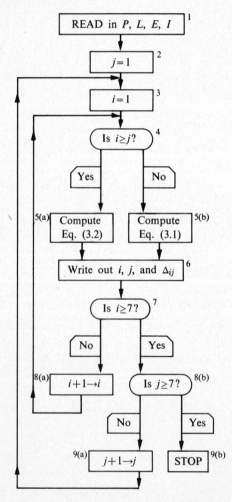

Fig. 3.7. Flow chart for beam deflection.

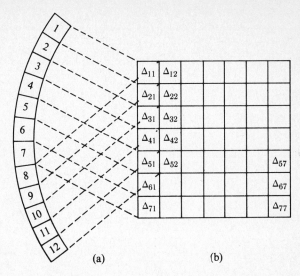

Fig. 3.8. Matrix Δ_{ij}. (a) Order of output. (b) Square matrix.

The equations for deflection can be found in standard books on strength of materials.† They are

$$\text{for } x \leq a: \quad \Delta = \frac{P}{EI}\frac{bx}{6L}(L^2 - b^2 - x^2), \tag{3.1}$$

$$\text{for } x \geq a: \quad \Delta = \frac{P}{EI}\frac{bx}{6L}(L^2 - b^2 - x^2) + \frac{P(x-a)^3}{6EI}, \tag{3.2}$$

where

$$a = \frac{jL}{8}, \qquad b = L - \frac{jL}{8}, \qquad x = \frac{iL}{8}.$$

Figure 3.7 shows a sample flow chart. We observe that it has a loop within a loop; the inner loop principally controls the i's while the outer one controls the j's.

We shall now discuss this flow chart in detail. Step 1 is concerned with the reading in of the four constants (transfer of data). Steps 2 and 3 are for initialization (transfer of data). Step 4 makes a logical decision. If the load is on the right side of the point for which the deflection is sought, $i > j$, then we can go to step 5(a) and compute the deflection based on Eq. (3.2). If the load is on the left side of the point being considered, $i < j$, then we follow step 5(b) and compute the deflection, Eq. (3.1). The third, and last, possibility is that $i = j$, or that the deflection at the loading point is required. The use of either Eq. (3.1) or Eq. (3.2) should yield the same answer. However, the path in a flow chart has to be definite and specific. Hence, we arbitrarily

---

† For example, see Timoshenko, S., *Elements of Strength of Materials*, 4th ed. Van Nostrand, Princeton, N.J., 1962, p. 201.

assign Eq. (3.2) for $i = j$. Steps 5(a) and (b) are arithmetic operations. Step 6 causes the answer Δ_{ij} to be written out together with the identifying numbers i and j. It is clear that the first answer is Δ_{11}.

Steps 4 through 6 are the core of the flow chart. Steps 7 through 9 are concerned with comparing and updating. To complete the discussion, it should be mentioned that the sequence of answers is in the order shown in Fig. 3.8.

An array of numbers such as that shown in Fig. 3.8(b) is called a matrix. For a unit load ($P = 1$), this matrix is called the "flexibility matrix." It plays an important role in analysis of structures.

3.5 ALGORITHM

In the previous two sections, we discussed several flowcharts, each consisting of a series of steps. A *general* method of supplying a series of steps leading to a solution of a given problem is known as an *algorithm*. Consider, for example, the following problem: "Given two positive integers i and j, find their greatest common divisor (GCD)." A possible algorithm, devised by Euclid, consists of the following finite number of steps:

1. Read the two positive integers, i and j. Then go to the next step.

2. Compare the two integers under consideration. Examine whether they are equal; if not, find out which integer is larger. Go to the next step.

3. If the integers are equal, each is the answer; then stop. If they are not equal, go to the next step.

4. Subtract the small integer from the larger one, and replace the integers under consideration by the subtrahend (the smaller integer) and the remainder. Go to Step 2.

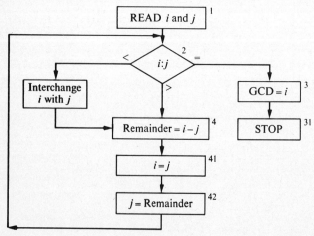

Fig. 3.9. A flowchart for finding the greatest common divisor of two positive integers i and j. ($i \neq 0$ or $j \neq 0$.)

In Step 4, the process of division is carried out by repeated subtraction. We have stated the Euclidean algorithm in ordinary language. By contrast, in Fig. 3.9 is shown a flowchart that provides a series of even more detailed steps by which the greatest common divisor of any two *positive* integers can be found. However, this flowchart does not take care of the following two important cases:

1. One of the given integers is zero.
2. One or both integers are given as negative quantities.

As an illustration of the first case, let us assume that $i = 14$ and $j = 0$. Since $i > j$, we can trace through the Steps 4, 41 and 42 in Fig. 3.9. At the exit of the Step 42, the new set of integers become: $i = 0, j = 14$. We now trace through the same Steps 2, 4, 41, and 42 for the second time. At the end of the Step 42, i and j become 14 and 0 respectively. Continuing in this manner, a permanent loop is formed if either i or j is equal to zero. To avoid this undesirable situation, the original integers must be checked. If one of them is zero, a message is printed and then a stop. This is shown in Fig. 3.10.

We turn now to the second case: One or both given integers are negative. If we assume that $i = -8$ and $j = 12$, the series of steps shown in Fig. 3.9 would fail to

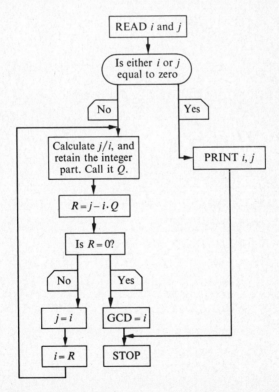

Fig. 3.10. A flowchart for finding the greatest common divisor of two integers i and j.

yield the correct GCD. A more general algorithm is stated in ordinary language as follows:

1. Read the two integers, i and j.

2. If either i or j is equal to zero, say so and then stop. Otherwise, go to the next step.

3. Calculate j/i; discard the portion of the quotient on the right of its decimal point. For example, if $j = 7$ and $i = 3$, the quotient is $7/3 = 2.333\ldots$, but $0.333\ldots$ is discarded. Call this result Q.

4. Calculate $j - i \cdot Q$. This is the remainder R.

5. If $R = 0$, the GCD is equal to i. Stop. Otherwise, go to the next step.

6. $j = i$.

7. $i = R$

8. Go to step 3.

A detailed flowchart following these steps is shown in Fig. 3.10, where the given integers i and j may be positive, negative, or equal to zero. The process of division is again rendered as repeated subtraction.

BIBLIOGRAPHY

Chapin, N., "Flowcharting with the ANSI standard: a tutorial," *Comput. Surv.*, **2**, no. 2, 119–146 (June 1970).

Farina, M., *Flowcharting*. Prentice-Hall, Englewood Cliffs, N.J., 1970.

Goldstine, H. H., and J. von Neumann, *Planning and Coding of Problems for an Electronic Computing Instrument* (mimeographed reports). Institute for Advanced Study, Princeton, N.J., 1946.

Hart, J. F., *et al.*, *Handbook of Computer Approximations*. Wiley, New York, 1968.

Hastings, C., *Approximations for Digital Computers*. Princeton University Press, Princeton, N.J., 1955.

Knuth, D. E., *The Art of Computer Programming, Vol. I*. Addison-Wesley, Reading, Mass., 1968.

Rice, H. G., "Recursion and iteration," *Commun. Assoc. Comput. Mach.*, **8**, 114–115 (1965).

Schriber, T. J., *Fundamentals of Flowcharting*. Wiley, New York, 1969.

Trakhtenbrot, B. A., *Algorithms and Automatic Computing Machines*. Translated by J. Kristian, J. D. Crawley, and S. A. Schmitt. D. C. Heath, Boston, 1963.

Flowchart Symbols and Their Usage in Information Processing. United States of America Standards Institute, New York, 1968.

PROBLEMS

1. In Fig. 3.1 the sum was obtained from $i = 0$ to $i = 8$. Draw a flow chart to obtain the sum $x = \sum_{i=0}^{8} c_i y^i$, working from $i = 8, 7, 6, \ldots, 0$.

2. The sum $x = \sum_{i=1}^{4} c_i y^i$ can be obtained by the following computer-oriented algorithm:

$$x = y\{c_1 + y[c_2 + y(c_3 + yc_4)]\}.$$

The computation is done from the inside out, or from yc_4. Draw a flow chart for this procedure.

3. Draw a flow chart to compute

$$c = \sum_{k=1}^{62} A_k B_k.$$

It is known that

$$\begin{aligned} A_k &= k \cos k, & k \le 30, \\ &= k^2 \sin k, & k > 30, \\ B_k &= e^k. \end{aligned}$$

4. Write a detailed flow chart to compute the product

$$\prod_{i=1}^{8} c_i x \cos ix,$$

where the values of x and the c_i's are read in as constants. The pi notation has the meaning

$$\prod_{k=1}^{n} x_k = x_1 \cdot x_2 \cdot x_3 \cdots x_n.$$

5. Draw a flow chart to obtain factorials 1 through K and their sum.

6. Draw a flow chart to compute the number e, using the formula

$$e = \sum_{n=0}^{\infty} (1/n!).$$

Terminate the process when $1/n! < 10^{-8}$.

7. Draw a flow chart to compute the hyperbolic tangent, using the following continued fraction approximation:

for $x < 0$, $\tan hx = -\tan h|x|$;

for $|x| \le 0.00034$, $\tan hx = x$;

for $0.00034 < |x| < 0.17$, $\tan hx = f\{A + f^2[B + C(D + f^2)^{-1}]\}^{-1}$,

where

$$f = 4x \log_2 e, \quad A = 5.7707801636, \quad B = 0.0173286795,$$
$$C = 14.1384114018, \quad D = 349.6699888;$$

for $0.17 \le |x| \le 11.14112$, $\tan hx = \dfrac{e^x - e^{-x}}{e^x + e^{-x}}$; for $|x| > 11.14112$, $\tan hx = 1$.

8. Draw a flow chart to read in many sets of three unequal positive numbers x, y, and z. If any one of the three numbers in a set is equal to or greater than 77, multiply x by $\tan 16°$ and print it. If all three numbers are less than 77, multiply y by $\cot 16°$ and print the new y. Stop the operation if x is a negative number.

9. Draw a flow chart to obtain the term grade average of K students, each taking N courses, with varying credits $c_1, c_2, \ldots, c_N$. Each course has a positive (nonzero) identification number $I_1, I_2, \ldots, I_m$. Compute also the percentage of students having a final-term grade average higher than 75.

10. Draw a flow chart to read in a set of x_i, y_i, z_k for $i = 1, 2, \ldots, 7$ and $k = 1, 2, \ldots, 6$, and compute

$$p = \left(\sum_{i=1}^{7} v_i \right) \left(\sum_{k=1}^{6} z_k \right),$$

where

$$\begin{aligned}
v_i &= x_i y_i \cos (\pi/8), &&\text{if} && |x_i| > |y_i|, \\
&= x_i y_i \sin (\pi/8), &&\text{if} && |x_i| < |y_i|, \\
&= 0, &&\text{if} && |x_i| = |y_i|.
\end{aligned}$$

Stop the operation when $z_k \geq 10^5$.

11. Write a flow chart to evaluate the expression

$$\sum_{j=0}^{5} \left(c_j x \prod_{i=0}^{7} x \cos ix \right).$$

Assume that the values of x and c_j are read in at the beginning of the program.

12. Write a flow chart to evaluate the following double summation:

$$\sum_{i=1}^{10} y_i \left(\sum_{j=1}^{15} c_j x_i^j \right),$$

where $y_i = 4i^3$, $c_j = (-3)^j + 2^j$, and $x_i = 2i$.

CHAPTER 4

Floating-Point Method

4.1 THE HANDLING OF DECIMALS

While it is not difficult to keep track of the decimal point in a number by using pencil and paper, the handling of decimal points by a computer does not appear to be a simple matter. In Section 2.6, we learned that a binary integer can be represented in a computer by a string of bits. We also discovered that decimal digits can be treated as EBCDIC digits. In either case, there was no explicit indication as to where a decimal point might be. This chapter deals with the techniques of locating the decimal point of a number in a computer.

Two techniques are available: the fixed-point method and the floating-point method. The latter method is a convenience and is widely used in scientific and engineering applications. The principal points vary in magnitude and number of significant places. Before we discuss the floating-point method, we should briefly describe the fixed-point method. In this method the programmer must be concerned with two things: (a) when adding or subtracting two numbers, he is responsible for aligning the decimal points, and (b) for multiplication or division, he is responsible for keeping track of the location of the decimal point in the answer.

4.2 FLOATING-POINT NUMBERS

Engineers and scientists often encounter numbers which vary widely in magnitude. Their literature reveals representative examples such as

(1) Young's modulus of steel $= E = 30 \times 10^6$ pounds per square inch, and
(2) The coefficient of thermal expansion of copper $= \alpha = 9.2 \times 10^{-6}$.
If we rewrite the values E and α in the forms

$$E = 0.3 \times 10^8 \text{ psi}$$

and
$$\alpha = 0.92 \times 10^{-5},$$

then they become floating-point numbers. A floating-point number has two parts: an exponent and a fraction. Although a misnomer, the fractional part is often referred to as "mantissa." In the above example for E, the exponent was 8 and the fractional part was 3. Similarly, for α the exponent was -5 and the fractional part was 92.

In general, a floating-point number N has the following form:

$$N = M \cdot \beta^k,$$

where

    $\beta = 16$ for IBM System/360 or System/370,
    $\beta = 2$ for many other binary computers,
    $\beta = 10$ for decimal computers,
    $k = $ the exponent, an integer, and
    M $= $ the mantissa, a value which must lie between $+1$ and -1.

Example

Consider the number: $N = -19.2 \times 10^{-8}$.

If this decimal number is rewritten as $N = -0.192 \times 10^{-6}$, then the exponent is equal to -6, an integer, the mantissa is equal to -0.192, and it lies between $+1$ and -1. Thus it meets the requirements of the general form of floating-point numbers.

It should be noted that in its original form, the value -19.2 lay outside the permissible range of M.

In addition to the above range limitation, $-1 < M < 1$, if the value M also satisfies one of the following two conditions:

$$\frac{1}{\beta} \le |M| < 1 \qquad \text{or} \qquad M = 0,$$

we can then say that $N = M \cdot \beta^k$ is a *normalized* floating-point number.

Consider, for example, the difference between the following two floating-point numbers:

$$
\begin{array}{r}
0.27143247 \times 10^7 \\
-0.27072236 \times 10^7 \\
\hline
0.00071011 \times 10^7
\end{array}
$$

Here, the difference itself is a floating-point number, but not a *normalized* floating-point number due to the presence of the three leading zeros. However, if one shifts the fractional part three places to the left, the resulting 0.71011×10^4 fulfills the requirements of a *normalized* floating-point number.

4.3 FLOATING-POINT DATA FORMATS

In the previous section, we learned the meanings of both a floating-point number and a normalized floating-point number, we ask an important question: What does a floating-point number look like in a memory location?

For the sake of simplicity, let us first take a *decimal* computer with a 10-digit word length and assume that the number in consideration is -0.000031415926. The floating-point notation of this number is $-0.31415926 \times 10^{-4}$. In order to avoid the negative exponent, we *arbitrarily* add 50 to the exponent and the number now becomes $-0.31415926 \times 10^{-4+50}$, or $-0.31415926 \times 10^{46}$. We often call this adjusted exponent, 46, the *characteristic*.

This number can be uniquely represented in a memory location in a form of 10 digits plus a sign, assuming that 8 digits are used for the fraction. (See Fig. 4.1).

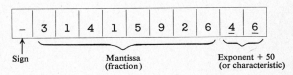

Fig. 4.1. Fraction and characteristic.

It should be noted that this method of representation fails for both very large and very small numbers. The characteristic e of a floating-point number is clearly limited by the following expressions: $-50 \le e \le 49$. The number in consideration has at best 8-digit accuracy.

The arbitrary number added to avoid a negative exponent may differ according to particular coding systems utilized. The characteristic, always a positive number, may also be placed in front of the mantissa. Coding systems will be discussed later in Chapter 5.

We now turn to the representation of a floating-point number in either an IBM System/360 or an IBM System/370. There are two different formats in these computers to represent a floating-point number—short format and long format. The short format (Fig. 4.2a) takes 32 bits: a sign bit (the 0 bit), 7 bits (the 1–7 bits) for the characteristic, and 24 bits (the 8–31 bits) for the fraction. On the other hand, the long format (Fig. 4.2b) contains 64 bits, as its fraction part is 56 bits (the 8–63 bits) in length.

A floating-point number using the short form may be called a *single-precision* floating-point number, or simply a *real* number. Likewise, a floating-point number specified by the long form is known as a *double-precision* floating-point number, which is often written as REAL*8 to indicate its 8-byte length.

It should be noted that in Fig. 4.2 no sign bit for the exponent is available. In order to take care of a negative exponent, we arbitrarily add $(64)_{10} = (40)_{16}$ to the exponent; this adjusted exponent is known as the *characteristic*.

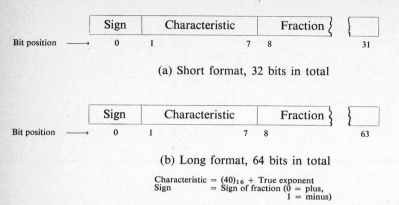

(a) Short format, 32 bits in total

(b) Long format, 64 bits in total

Characteristic = $(40)_{16}$ + True exponent
Sign   = Sign of fraction (0 = plus,
                                   1 = minus)

Fig. 4.2. Two floating-point formats.

In dealing with a floating-point number in either an IBM System/360 or an IBM System/370, each *4-bit* (a hexadecimal digit, see Appendix A) is considered as a basic unit. As a result, the magnitude of a floating-point number is the fraction times a power of 16.†

Example

Given: The short format representation as shown in Fig. 4.3. Find: (a) Characteristic and true exponent, (b) fraction in decimal, and (c) the total quantity in decimal.

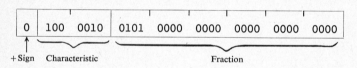

Fig. 4.3. Short format for a decimal number $+80$.

Solution

a) The characteristic is $(42)_{16}$ or $(66)_{10}$; the true exponent is

$$(42)_{16} - (40)_{16} = (66)_{10} - (64)_{10} = 2.$$

b) The fraction is

$$(0.0101)_2 = 2^{-2} + 2^{-4}, \quad \text{or} \quad (0.5)_{16} = 5 \times 16^{-1}.$$

c) The entire quantity is equal to:

$$16^2 \times (2^{-2} + 2^{-4}) = 80, \quad \text{or} \quad 16^2 \times (5 \times 16^{-1}) = 80.$$

---

† The remainder of this chapter may be skipped without loss of continuity. Students unfamiliar with binary and hexadecimal number systems will find a short account in Appendix A.

As shown in this example, the characteristic represents a power of 16, as each shift is done in increments of 4 bits. Thus, an IBM System/360 or a System/370 can be considered as a hexadecimal machine. As a result, a normalized floating-point number may contain a zero value in all first three high-order bits (bits 8, 9 and 10) in the fraction part.

Consider, for example, the representations of a given normalized floating-point number shown in Fig. 4.4 (a) and (b) respectively for a short and long form. In either representation, both bit positions 8 and 9 contain zeroes, yet this floating-point number is a *normalized* one.

Here, the true exponent is $(3D)_{16} - (40)_{16} = (-3)_{16}$; and the total quantity in decimal may be computed as follows:

$$16^{-3} \times (2 \times 16^{-1} + 15 \times 16^{-3} + 1 \times 16^{-4} + \cdots)$$
$$= 2 \times 16^{-4} + 15 \times 16^{-6} + 16^{-7} + \cdots$$
$$= 2 \times 2^{-16} + 15 \times 2^{-24} + 2^{-28} + \cdots$$
$$= 0.000031416$$

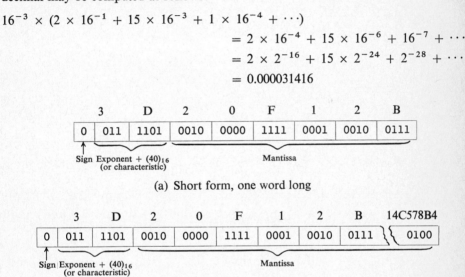

(a) Short form, one word long

(b) Long form, doubleword

Fig. 4.4. Floating-point notation for 0.000031416.

4.4 FLOATING-POINT ADDITION

In the previous section, we learned that a floating-point number is in the form of a fraction and an exponent. We shall now discuss the floating-point operations with particular emphasis on hexadecimal numbers.

When adding or subtracting two floating-point numbers, the exponents are first compared. If they are the same, the fractions are simply added or subtracted and the exponent is retained. For example:

$$
\begin{array}{r}
0.563 \times 16^2 \\
+)\,0.389 \times 16^2 \\
\hline
0.8EC \times 16^2
\end{array}
$$

When there is an overflow (a carry from the leftmost digit), both fractions are moved one place to the right, and the exponent is increased by 1, as shown in the following example:

$$0.C48 \times 16^2 \qquad\qquad 0.0C4 \times 16^3$$
$$+)\ 0.573 \times 16^2 \qquad\qquad +)\ 0.057 \times 16^3$$
$$\overline{\hphantom{xxxxxxxxxxx}} \qquad\qquad \overline{\hphantom{xxxxxxxxxxx}}$$
$$(1)\ 0.1BB \times 16^2 \qquad\qquad 0.11B \times 16^3$$

When the exponents are not equal, the larger exponent is retained and the smaller one is converted. For example:

$$0.587 \times 16^4 \qquad\qquad 0.587 \times 16^4$$
$$+)\ 0.743 \times 16^3 \qquad\qquad +)\ 0.074 \times 16^4$$
$$\overline{\hphantom{xxxxxxxxxxx}} \qquad\qquad \overline{\hphantom{xxxxxxxxxxx}}$$

If a leading zero appears in the answer, the number is normalized. For example:

$$0.538 \times 16^3$$
$$-)\ 0.526 \times 16^3$$
$$\overline{\hphantom{xxxxxxxxxxx}}$$
$$0.012 \times 16^3$$
$$\text{Normalized result} = \quad 0.120 \times 16^2$$

We see from the above discussions that floating-point arithmetic, as automatically performed by computers, greatly simplifies the scaling problem. It is interesting to note that many computers have built-in components capable of carrying out floating-point arithmetic.

In an IBM System/360 or System/370 there are four floating-point registers (Fig. 4.5) used only as accumulators. The registers are doublewords in length and are numbered 0, 2, 4, and 6. They are not a part of the 16 general registers.

All floating-point operations are performed in the four floating-point registers. The operation code in a given instruction dictates whether only the data in the left half of the register is to be used (short-form operation), or the data in the entire

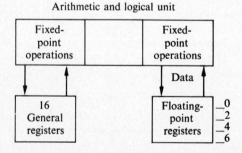

Fig. 4.5. Floating-point registers.

register is used (long-form operation).   Practically all scientific and engineering problems are solved on computers by using the floating-point system.

BIBLIOGRAPHY

Goldberg, I. B., "27 bits are not enough for 8 digit accuracy," *Communs. Assoc. Comput. Mach.*, **10**, 105–106 (1967).

Linz, P., "Accurate floating-point summation," *Communs. Assoc. Comput. Mach.*, **13**, no. 6 (June 1970).

PROBLEMS

1. Rewrite the following decimal numbers in the general form of floating-point numbers such that the mantissa part lies between 1 and -1.
   a) 2.7135.   b) -37.214.   c) 12×10^6.   d) 3.1627×10^{-2}.

2. What is the value of the exponent in each of the floating-point numbers in Problem 1? What is the adjusted exponent in each case if we arbitrarily add 49 to the original exponent?

3. Normalize the following decimal floating-point numbers:
   a) 0.0214×10^{-6}.   b) 12.01×10^4.

4. Figure 4.6 shows a short floating-point number, in binary, as used in an IBM System/360 or 370.

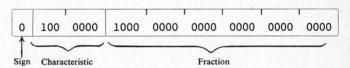

Fig. 4.6. A short floating-point number.

   a) Is this number a positive or a negative number?
   b) Express the fraction part in decimal.
   c) Express the characteristic in decimal.
   d) What is its true exponent?
   e) What is the value of the number in decimal?
   f) Is this floating-point number normalized?

5. Same as Problem 4, except that the short floating-point number is shown in Fig. 4.7.

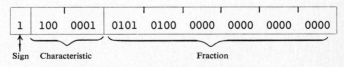

Fig. 4.7. A short floating-point number.

CHAPTER 5

Coding Systems
for Digital Computers

5.1 INTRODUCTION

Digital computers do not accept or operate from the flow charts discussed in Chapter 3. To convert flow charts into a language which a particular computer will recognize and accept is commonly called "coding."

Basically, each type of computer accepts only one particular language for computation—its own machine language. The general trend is to code in a language (source language) akin to our written language, and then utilize the computer to translate this language into the machine's language (target or object language). The computer then follows the commands in its own machine language and dutifully performs the necessary computations. It should be noted in this connection that computers are used for dual purposes: translation and computation. The human labor previously spent in coding the rather involved machine languages or machinelike languages is actually being saved at the expense of computer processing time.

There are a number of reasons why programming in machine languages or machinelike languages can be both time consuming and expensive: (1) human language has little in common with machine language, yet a programmer must somehow create a perfect set of machine instructions, completely without error; (2) it is often difficult to find and correct errors in a complex program; (3) it may take as much as a year to train a programmer to be really effective; (4) once trained, he will still have to be taught the intimate details of a complex problem; and (5) small changes in the definition of a problem can require extensive changes in the program, and changes in problem definition are quite common, especially in business data processing.

In an attempt to overcome these difficulties, manufacturers and users of computers have devoted much effort in a search for ways to shift the burden of programming from humans to computers. Their work has led to the development of "compilers"; a compiler is essentially a set of computer programs that directs the computer to translate a problem-oriented language (high-level language) into machine-oriented language. Thus, under the guidance of a compiler, a computer will translate a problem described in a simplified version of human language into a set of instructions that can later be executed by the machine. A brief statement in problem-oriented language can usually be translated into a lengthy set of computer instructions.

The problem-oriented languages, or high-level languages, in wide use include FORTRAN, COBOL, ALGOL, and PL/I. FORTRAN, an abbreviation for *For*mula *Tran*slation, is probably one of the most widely accepted high-level languages in engineering and science; a typical statement in FORTRAN might look like this:

$$BETA=1.+COS(2.*PI*A/ROOT)$$

COBOL is probably the most popular business-oriented high-level language; a typical COBOL statement is shown below:

```
IF MONTHLY FICA LESS THAN 16.00 GO TO SPECIAL FICA:
ELSE ADD MONTHLY FICA TO ANNUAL FICA.
```

PL/I was first proposed in 1964 and has only recently become available for use on some computers. It is a versatile language; a typical statement in PL/I is:

```
IF N=1 | N=0 THEN RETURN(1)
```

In the remainder of this chapter, we shall examine those facets of the FORTRAN IV language which will be required to implement the algorithms developed in Parts II and III. This subset of the FORTRAN IV is compatible with most computer systems for which some FORTRAN IV compilers are available. For the sake of convenience, we shall refer to this subset (IBM/360 basic FORTRAN IV language) simply as FORTRAN IV, or FORTRAN.

5.2 FORMULA TRANSLATION LANGUAGE—FORTRAN SYSTEM

The FORTRAN language is most easily described by presenting some concrete examples.

Example 5.1

Let X_1 and X_2 be the two roots† of the quadratic equation $aX^2 + bX + c = 0$, where $X_1 \geq X_2$.

It is desired to compute the unknowns D and E, where $D = X_1 \cos 15°$ and $E = X_2 \sin 15°$ for the four cases listed in the table below.

---

† The discussion is limited to the real roots, assuming that $b^2 - 4ac$ is nonnegative.

	Case I	Case II	Case III	Case IV
a	1.0	-22.41	23.12	-3.1415926
b	-3.0	-60.22	-3.14159	11.3232323
c	1.0	1.15	-22.14	0.2134567

We need to draw a detailed flow chart for the machine computation and write a FORTRAN program that will compute D and E, with each answer showing the values of D, E, a, b, and c.

Solution: One possible flow chart is shown in Fig. 5.1; this chart needs little explanation other than some comments on Step 4, which is used to determine whether

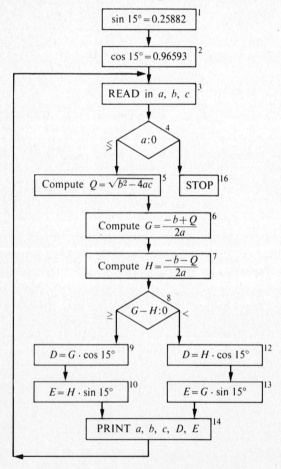

Fig. 5.1. Flow chart for Example 5.1.

the required computations have been finished by feeding in a fifth set of data *a*, *b*, *c*. In this last set, the value *a* is deliberately taken as zero. When these data are tested for zero, either Step 5 or Step 16 will be followed.

Questions may arise as to how the trigonometric functions in Steps 1 and 2 are obtained. In this case, where we only have two function values, it is simpler to obtain them from a mathematical table. However, should a great number of the trigonometric functional values be required in a single program, the computer must be instructed to compute them by means of the available subroutines. This will be discussed more fully later.

The commands as written in the FORTRAN languages are commonly known as "statements." The FORTRAN statements for our example are shown in Fig. 5.2. We shall now give detailed explanations of all statements in our program except statements 35 and 145, which will be treated more fully later in the chapter. Each statement will be referred to by its statement number. Single-precision floating-point numbers (real numbers) are used in all computations; these numbers were discussed in Section 4.3.

```
  1  SIN15=0.25882
  2  COS15=0.96593
  3  READ(5,35) A,B,C
 35  FORMAT(F10.4,F10.4,F10.4)
  4  IF(A) 5,16,5
  5  Q=SQRT(B*B-4.*A*C)
  6  G=(-B+Q)/(2.*A)
  7  H=(-B-Q)/(2.*A)
  8  IF(G-H) 12,9,9
  9  D=G*COS15
 10  E=H*SIN15
 11  GO TO 14
 12  D=H*COS15
 13  E=G*SIN15
 14  WRITE(6,145) D,E,A,B,C
145  FORMAT(5F10.4)
 15  GO TO 3
 16  STOP
     END
```

(a)

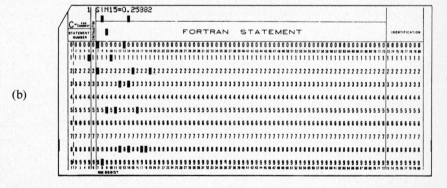

(b)

Fig. 5.2. (a) A FORTRAN program for Example 5.1. (b) FORTRAN statement card showing statement 1.

Statement 1. This statement serves to store the single-precision floating-point number (real number) in a specific storage location arbitrarily named SIN15. The automatic conversion of data to a real number is triggered by the presence of the decimal point in the datum 0.25882. In the FORTRAN system, any number with a decimal point is by necessity a floating-point number. The naming of a real variable is very flexible, but it must conform with the following two conditions:

(1) The name must have 6 or less letters or numbers (for example, SIN15, A3, . . .).

(2) The name must begin with alphabetic letters other than I, J, K, L, M, or N.

These six letters are reserved for identifying fixed-point variables.† These will be discussed later since there are no fixed-point variables involved in this particular program.

Statement 2. This statement transfers a real number into a specific storage location arbitrarily named COS15.

Statement 3. This is a "read-in" statement asking the computer to read in three real numbers A, B, and C from the data card. These three numbers will be stored in three reserved locations for subsequent use. The stored real numbers will be replaced only after the second data card is read in (refer to Statement 15 and the related loop). In this program the letters A, B, and C refer to the coefficients a, b, and c in the quadratic equation $aX^2 + bX + c = 0$.

There are two integers specified in the parentheses, 5 and 35. The integer 5 refers to a logical device number for a card reader. It should be noted that this number varies from installation to installation. Throughout the book we will use the following logical device numbers:

5 stands for a card reader

6 stands for an on-line printer

The second unsigned integer 35 refers to a format statement number. Any number not already used in the program could have replaced the number 35, so long as the statement number in the format statement would correspond to it. It should be noted that there should be no comma between the right parenthesis and A in this READ statement. Any number not already used in the program could have replaced the number 35, so long as the statement number in front of the format statement would correspond to it.

Statement 4. The computer is asked to make a decision as to whether or not the value of a is zero. This operation is accomplished by use of an IF statement. In this statement, if the value of a is less than zero, then the computer is instructed to go to the first statement number following the parenthesis (in this case, statement 5). If a is zero, the computer is directed to proceed to the second number after the parenthesis,

---

† Known as the predefined or implicit specification. In all examples that follow in this book, it is presumed that this specification applies unless otherwise noted.

or statement 16. If *a* is larger than zero, however, the computer is instructed to proceed to the third number or statement 5. Note that there must be *no* comma after the word IF in this statement.

Statement 5. This statement commands the computer to perform the operation of computing the term $\sqrt{b^2 - 4ac}$ and store the quantity in the word Q. The asterisk * indicates that a multiplication operation is to be performed, and the decimal point following the number 4 automatically converts it into a real number, as already discussed in connection with Statement 1.

It should be noted that A, B, C and 4. are in real mode. If the decimal point after 4 is dropped in this statement, the real constant 4. becomes a fixed-point constant, mixed with real terms in this statement. This is permissible as it is automatically converted to the real mode before the evaluation. A square-root subroutine designated by SQRT is called for in this statement. There are many subroutines available in this system, such as SIN for sine function, COS for cosine function, EXP for exponential function, etc. A list of many common subroutines for the FORTRAN system is shown in Table 5.1.

Table 5.1 Partial list of library subroutines for the FORTRAN system

Function name	Usual mathematical notation
COS(A)	$\cos A$
EXP(A)	e^A
ALOG(A)	$\ln A$
SIN(A)	$\sin A$
ATAN(A)	$\tan^{-1} A$
SQRT(A)	$\sqrt{A}$

We note that the term or terms to which the subroutine pertains—the argument—must be enclosed in parentheses. Finally, the designation Q stands for a location in which only a floating-point number is present, since the term does not begin with the letters I, J, K, L, M, or N. Should one unintentionally write statement 5 as

$$MQ = SQRT(B * B - 4. * A * C)$$

an incorrect answer will result. The computer will first perform the operation $\sqrt{b^2 - 4ac}$, in which, for example, the result might turn out as 16.2749. It then drops the digits after the decimal point and the result becomes 16. This would then be stored in a location called MQ. Obviously this is not a correct answer.

Statements 6 and 7. Note the decimal point after 2. It may be omitted.

Statement 11. This statement informs the computer to skip the next two statements and proceed directly to Statement 14.

Statement 14. This is a WRITE instruction which tells the machine what values or answers should be written out and in what order this should be done. In this case, the product of the larger root and cos 15° is the first answer in the output, the product of the smaller root and sin 15°, the second answer, and the coefficients *a*, *b*, and *c*, the third, fourth, and fifth answers, respectively. Thus a complete record of both the variables read in and the corresponding answers obtained is made available. These values are all in real form. The first integer in the parentheses, or 6, indicates that the results are to be written out on an on-line printer. The second integer, or 145, is a format statement number, where the printing format will be given in detail.

Statement 15. After the printing of the answers has been completed for this particular set of *a*, *b*, and *c*, the computer proceeds to Statement 3 and reads in a new card with a completely new set of variable coefficients. Calculations are then repeated using the new data.

Statement 16. The STOP statement commands the computer to stop all computations.

Statement 17. The END statement is required at the end of every FORTRAN program.

Additional explanation. We shall now discuss the key punching of statement cards for the source program and of data cards.

The statements are punched one to a card (Fig. 5.2b) using the following format: The statement itself is punched in a column field occupying columns 7 to 72. Columns 73 through 80 are either left blank or may contain any information which the programmer may want to use for identification purposes, and columns 1 to 5 are reserved for statement numbers. Column 1 is also used for punching C to indicate a comment card. This card will be ignored by the computer during the subsequent translation. Column 6 of the first card of a statement must be left blank. If the statement is longer than 66 characters including blanks, a second, third, or up to 20 cards may be used

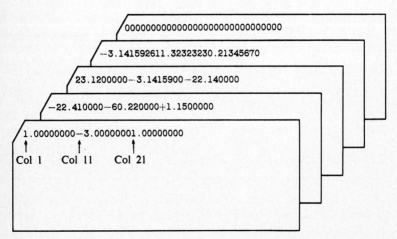

Fig. 5.3. Format for data cards for example of quadratic equation.

for the statement. These 19 continuation cards are usually designated by a non-zero or a non-blank character punched in column 6.

We now turn to the preparation of data cards. In the above example, the three variables a, b, and c must be punched on the *data card* in the order they will be read in by the READ statement (see Fig. 5.3). In this particular case, Statement 35, "FORMAT (F10.4, F10.4, F10.4)," requires that each variable occupy a total of 10 digits. Moreover, 4 digits are expected to be after the decimal point; this is not a rigid requirement, however, since where the decimal point is placed will not affect the input data so long as the number is punched within the first ten columns. Statement 35 can also be written as

35 FORMAT (3F10.4)

in FORTRAN language. It is important to note that all 80 columns of cards can be used for data input or punched output.

5.3 EXAMPLE IN FORTRAN—DART PROBLEM

The purpose of this section is to illustrate the usage of IF statements with a specific example.

A circle in the first quadrant of a coordinate system with axes x and y has its center at (a, b) and a radius of $r : a = 8.569$, $b = 12.000$, $r = 3.000$. Nine sets of (x, y) are given:

x	y	x	y
2.347	18.176	9.164	13.978
3.291	14.129	11.569	12.000
6.987	11.253	13.875	15.192
7.867	5.967	99999.000	0.000
8.569	9.000		

Use $x \geq 99998.000$ to terminate the problem.

We wish to write a program in FORTRAN to determine which sets of (x, y) will fall within the given circle. If the point falls on the circumference, we assume that it has fallen within the circle.

A possible procedure to solve this problem is (see Fig. 5.4):

a) Find the left-hand (smallest x) and the right-hand (largest x) limits of the circle. (Note that these are the same in every case.)

b) Check to see whether the given x falls between or on these two limits, p and q (unshaded region). If not, then it is obvious that the point is not in the circle.

c) If it falls between or on the two limits p and q, find the two y-values (upper and lower) of the boundary of the circle at the given x. (Note that the two y's are the same if x is equal to either the left-hand or right-hand limits of the circle. Would this cause difficulties?)

Check to see whether the given y falls between or on the two boundary values s and t.

The two y-values, s and t, of the boundary of the circle may be calculated as follows:

The equation of the circle is $(x - a)^2 + (y - b)^2 = r^2$, but we know the value of x. Therefore,

$$(y - b)^2 = r^2 - (x - a)^2, \qquad y = b \pm \sqrt{r^2 - (x - a)^2}$$

or,

$$\overline{os} = b + \sqrt{r^2 - (x - a)^2} \qquad \overline{ot} = b - \sqrt{r^2 - (x - a)^2}.$$

The required printing format is as follows (a ▯ indicates that a blank space is to be left):

▯X▯=▯000002.347▯▯▯▯▯▯▯▯▯▯▯Y▯=▯000018.176

IT IS NOT IN THE CIRCLE

or

IT IS IN THE CIRCLE

The required input format is F10.3; it is shown in Fig. 5.5. A possible flow chart and a FORTRAN program are presented in Figs. 5.6 and 5.7, respectively.

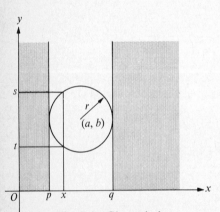

Fig. 5.4. Given circle.

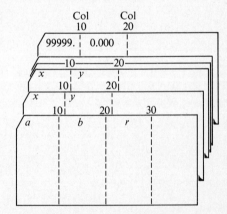

Fig. 5.5. Input card format.

In the FORTRAN program, we used five IF statements; each branches to one of two statement numbers. Consider, for example, the following statement:

8   IF(X–XR) 9,9,6

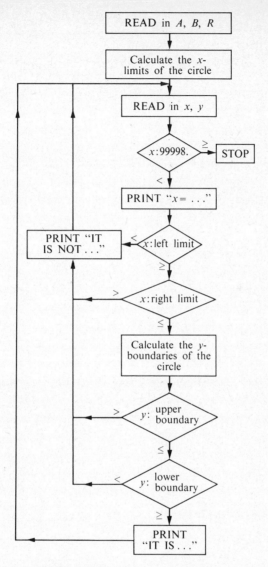

Fig. 5.6. Flow chart for dart problem.

which provides a means of branching to the next statement (statement number 9), if X–XR is less than or equal to zero. It will go to statement number 6, if X is greater than XR. Another way to express this branch is as follows:

```
8  IF (X.GT.XR) GO TO 6
```

where .GT. stands for "greater than." (This is known as a relational operator and must be preceded and followed by a period.) If X is not greater than XR, it will branch

```
C       DART PROGRAM
        READ(5,1) A,B,R
      1 FORMAT(3F10.3)
        XL=A-R
        XR=A+R
     31 READ(5,3) X,Y
      3 FORMAT(2F10.3)
        IF(X-99998.0) 4,100,100
      4 WRITE(6,5) X,Y
      5 FORMAT(' X = ',F10.3,10X,' Y =  ',F10.3/)
        IF(X-XL) 6,8,8
      6 WRITE(6,7)
      7 FORMAT(3X,' IT IS NOT IN THE CIRCLE'//)
        GO TO 31
      8 IF(X-XR) 9,9,6
      9 YT=B+SQRT(R*R-(X-A)*(X-A))
        YB=B-SQRT(R*R-(X-A)*(X-A))
     22 IF(Y-YT) 10,10,6
     10 IF(Y-YB) 6,11,11
     11 WRITE(6,12)
     12 FORMAT(3X,' IT IS IN THE CIRCLE'//)
        GO TO 31
    100 STOP
        END
```

```
8.569       12.000      3.000
2.347       18.176
3.291       14.129
6.987       11.253
7.867        5.967
8.569        9.000
9.164       13.978
11.569      12.000
13.875      15.192
99999.000    0.000
```

Fig. 5.7. FORTRAN program and data for dart problem.

to the next statement (number 9). The other four IF statements can also be written each with a relational operator. They are shown below:

IF(X-99998.0)4,100,100	IF(X.GE.99998.0) GO TO 100
IF(X-XL)6,8,8	IF(X.GE.XL) GO TO 8
IF(Y-YT) 10,10,6	IF(Y.GT.YT) GO TO 6
IF(Y-YB) 6,11,11	IF(Y.LT.YB) GO TO 6

where .GE. stands for "greater than or equal to," and .LT. stands for "less than."

5.4 EXAMPLE IN FORTRAN—SERIES SUMMATION

In the example of Section 5.2, a loop was accomplished by using the statement GO TO 14 (Fig. 5.2a). We shall present a DO statement and a DIMENSION statement in the following example for a series summation and explain their use in connection with loops.

We desire to obtain the ζ vs. t curve from the expression

$$\zeta = \sum_{N=1}^{M} \frac{(-1)^N}{N} \sin(N\pi X) \left[1 + \frac{A_N}{1 + N^2\psi^2} \cos\left(\frac{2N\pi}{\sqrt{1 + N^2\psi^2}} t\right) \right],$$

where $M \leq 33$.

Write a FORTRAN program using $t = T_1$ to $t = T_2$ with interval $\Delta t = T_3$. The values of T_1, T_2, T_3, M, X, and ψ should read in as the first data card. Then read the available values for A_N. The following data should be used for the first production run:

$$T_1 = 0.$$
$$T_2 = 2.$$
$$T_3 = 0.01$$
$$M = 30$$
$$X = 0.500$$
$$\psi = 1.2051494 \times 10^{-5}$$
$$A_1 \text{ through } A_{33} = 1.0.$$

Solution: The flow chart and the corresponding FORTRAN program are shown in Figs. 5.8 and 5.9, respectively. The following statements in this program need some explanation:

```
C  EXAMPLE XI – T CURVE
```

This is a comment card, indicated by the "C" punched in the first column. The computer will ignore the card throughout the program.

```
DIMENSION A (33)
```

This statement instructs the computer to reserve 33 locations in which to store A_1 through A_{33}. This statement is not necessary if there is just a single A-value to be stored. A DIMENSION statement is also required if, in the program, a block of numbers is to be computed and stored.

The DIMENSION statements are also used to reserve a block of locations for an *array* of numbers. For example, if a block of 42 locations is used to store Δ_{ij} (where $i = 1, 2, \ldots, 7$ and $j = 1, 2, \ldots, 6$), then the reservation of these 42 locations is indicated by the statement DIMENSION DELTA (7,6).

Three-dimensional reservation, for example, DELTA (7, 6, 9), is also permitted.

```
DO 9 N = 1, M
```

This DO statement is a very powerful one to use for setting up loops. It instructs the computer to follow all the statements down to and including statement 9 for a definite number of times. In this case, it starts with $N = 1$, then $N = 2, 3, \ldots,$

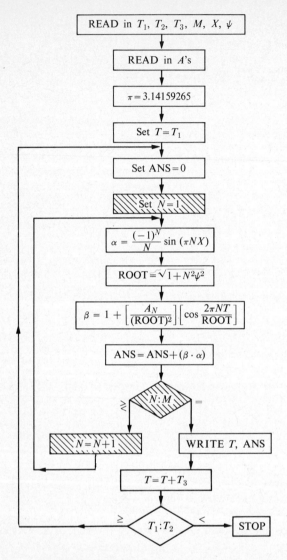

Fig. 5.8. Flow chart for series summation.

and finally, when N = M is completed, the computer will get out of the loop auto-matically and will execute the next statement, i.e., statement 81. From the flow chart (Fig. 5.8), we see that the three shaded boxes correspond to this single statement. When written in full, this statement is DO 9 N = 1, M, 1. The last number, 1, serves to indicate the increment by which N is increased for each loop. However, if the increment is 1, as in this case, the last number may be left off.

```
C      EXAMPLE  XI- T CURVE
       DIMENSION A(33)
       READ(5,10) T1,T2,T3,M,X,PSI
10 FORMAT(3F10.5,I10,F10.5,E14.9)
       DO 12 I=1,M
12 A(I)=1.
       PI=3.141593
       T=T1
17 ANS=0.
       DO 9 N=1,M
       AN=N
       ALPHA=(((-1.)**N)/AN)*(SIN(AN*PI*X))
       ROOT=SQRT(1.+AN*AN*PSI*PSI)
66 BETA=1.+(A(N)/(ROOT*ROOT))*COS(2.*PI*AN*T/ROOT)
 9 ANS=ANS+(BETA*ALPHA)
81 WRITE(6,1) T,ANS
 1 FORMAT(' T= ',F10.5,10X,'ANS= ',F10.5/)
       T=T+T3
       IF(T2.GE.T) GO TO 17
       STOP
       END
```

.00000   2.00000   0.01000   30   0.50000   .12051494E-4

Fig. 5.9. FORTRAN program and data for series summation.

The statement AN = N serves to convert the fixed-point variable N to a real variable AN. Note that the symbol AN is required in all of the arithmetic statements except as the power of an exponent.

It is important to note the difference between AN and A(N) in statement 66. AN is numerically equal to N, while A(N) is the symbol for n-th A, or A_n.

5.5 OUTPUT STATEMENTS FOR A MATRIX PROBLEM

Perhaps one of the most important, yet difficult, topics in FORTRAN programming is handling the class of statements for input and output: READ, WRITE, and FORMAT. This problem becomes more complex when an array of numbers or a matrix is involved. In this section some basic applications of output statements shall be reviewed in order to clarify any difficulties which may have been encountered in Sections 5.2 through 5.4. Then, using the flow chart presented in Fig. 3.7, we shall explain a FORTRAN program to compute the elements in a flexibility matrix. In addition, an improved version of this flow chart will be presented and its related FORTRAN program explained, with emphasis on the matrix output.

Let us review the output statements for single variables by means of the following example. Suppose that a fixed-point variable I is equal to 14 and a real variable A equals 3.141593. We can write the following two statements to print out I and A:

```
 5  WRITE (6,16) I,A
16  FORMAT(I3, F10.4)
```

The printout is

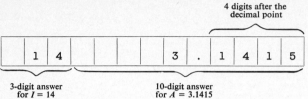

4 digits after the
decimal point

3-digit answer
for $I = 14$

10-digit answer
for $A = 3.1415$

In the above two statements, I in statement 5 stands for a fixed-point variable, while I3 in Statement 16 indicates that the answer must be printed out as a three-digit *integer* without a decimal point. The two are completely different in concept.

The FORMAT (Statement 16) calls for a total of thirteen spaces for output: three spaces are required by the I3 specification, and the next ten spaces are used for the F10.4 specification. The last digit of I, that is, 4, coincides with the rightmost position in these three spaces. If the value of I were, say, 7, it would be printed out in the rightmost space, preceded by two blank spaces as follows:

The specification F10.4 demands that the next ten spaces will be allotted to the value of A, including the four spaces for the decimal portion of A. Any decimal portion exceeding the FORMAT specification will be discarded and not printed out. Thus, the fifth and sixth decimal digits of A in this example are truncated, since the FORMAT specification allows only the first four digits after the decimal point to be printed.

We now proceed to present a FORTRAN program based on the flow chart shown in Fig. 3.7 to compute the 49 elements in the flexibility matrix. This program (see Fig. 5.10) will read in the given data, namely, P, ZL, E, AI, and N; choose Eqs. (3.1) or (3.2); and finally, print out the values of the 49 elements of the matrix, one value per line. The basic loop in this program is to (a) compute and (b) print out a value of each of the 49 elements in the matrix (see the output in Table 5.2).

The first statement in Fig. 5.10, DATA IREAD,IWRITE/5,6/ is equivalent to the following two statements:

$$IREAD=5$$
$$IWRITE=6$$

In FORMAT Statement 10, '1' is used to demand skipping to the top of the next page before printing. 9X means that 9 spaces in the horizontal printing line are to be skipped.

We note that a line is left between two consecutive answers. This is accomplished by simply writing a slash in the FORMAT statement (Statement 6). In general, k slashes in a FORMAT statement will cause k spaces in the printout if the slashes appear before the first or after the last specification. $N + 1$ consecutive slashes cause N blank lines if the slashes are between two specifications. An example illustrating the use of slashes between the specifications is given below.

Table 5.2 Output from the program of Fig. 5.10

I	J	DELTA(I,J)
1	1	.0039
1	2	.0065
1	3	.0077
1	4	.0076
.	.	.
.	.	.
.	.	.
7	4	.0076
7	5	.0077
7	6	.0065
7	7	.0039

THE END

```
C       THIS IS THE FORTRAN PROGRAM FOR THE PROBLEM INTRODUCED IN SECTION 3.4
C       FLEXIBILITY MATRIX PROBLEM
C       N=MAXIMUM NUMBER FOR I OR J
        DATA IREAD,IWRITE/5,6/
        WRITE(IWRITE,10)
    10 FORMAT('1',9X,'I',9X,'J',3X,'DELTA(I,J)'//)
        READ(IREAD,1) P,ZL,E,AI,N
     1 FORMAT(4F10.7,I10)
        ZNP1=N+1
        DIMENSION DELTA(10,10)
        DO 20 I=1,N
        ZI=I
        DO 20 J=1,N
        ZJ=J
        IF(ZI.GE.ZJ) GO TO 51
        DELTA(I,J)=((P/(E*AI))*(ZL**3)*(1.-ZJ/ZNP1)*ZI/(6.*ZNP1))*
     1   (1.-(1.-ZJ/ZNP1)**2-(ZI/ZNP1)**2)
        GO TO 16
    51 DELTA(I,J)=((P/(E*AI))*(ZL**3)*(1.-ZJ/ZNP1)*ZI/(6.*ZNP1))*
     1   (1.-(1.-ZJ/ZNP1)**2-(ZI/ZNP1)**2)+(P/(E*AI))*ZL*
     2   ZL*ZL*(ZI-ZJ)**3/((ZNP1**3)*6.)
    16 WRITE(IWRITE,6) I,J,DELTA(I,J)
     6 FORMAT(2I10,F10.4/)
    20 CONTINUE
        WRITE(IWRITE,25)
    25 FORMAT(' THE END')
        STOP
        END
1.0        1.0        1.0        1.0                7
```

Fig. 5.10. Flexibility matrix problem.

Example 1

```
10  FORMAT(F10.4, F12.4,/F5.2, F12.6)
```

This FORMAT specification causes the first two variables to be printed on the first line and the second two variables to be printed on the second line.

Example 2

Statements 28 through 30 cause three lines to be skipped:

```
28  IWRITE = 6
29  WRITE (IWRITE, 30)
30  FORMAT(///)
```

Example 3

The following two statements cause the word GOOD to be printed and then three lines to be skipped:

```
129  WRITE (6, 130)
130  FORMAT (5H□GOOD///)
```

The FORMAT Statement 130 can be replaced by the following format which does not need careful counting:

```
1130  FORMAT ('□GOOD'///)
```

The message is written between the two apostrophes.

Example 4

The following two statements have the effect that the values of I, J, and the element Δ_{ij} are printed on a single line, and that one line is then skipped:

```
16  WRITE (6,6) I,J,DELTA(I,J)
6  FORMAT(2I10, F10.4/)
```

Frequently, the printout of a matrix problem is required to be in matrix form. In other words, the answers are printed out row by row as shown in Fig. 5.11. The following two changes in the original program (Fig. 5.10) are therefore necessary:

a) computation of all elements and printing them out after all of them are computed;

b) printing the matrix row by row.

In order to compute all 49 elements without immediate printout, we simply move CONTINUE (Statement 20) so that it immediately follows Statement 51.

$$
\begin{array}{ccccccc}
\Delta_{11} & \Delta_{12} & \Delta_{13} & \Delta_{14} & \Delta_{15} & \Delta_{16} & \Delta_{17} \\
\Delta_{21} & \Delta_{22} & \Delta_{23} & \Delta_{24} & \Delta_{25} & \Delta_{26} & \Delta_{27} \\
\Delta_{31} & \Delta_{32} & \Delta_{33} & \Delta_{34} & \Delta_{35} & \Delta_{36} & \Delta_{37} \\
\Delta_{41} & \Delta_{42} & \Delta_{43} & \Delta_{44} & \Delta_{45} & \Delta_{46} & \Delta_{47} \\
\Delta_{51} & \Delta_{52} & \Delta_{53} & \Delta_{54} & \Delta_{55} & \Delta_{56} & \Delta_{57} \\
\Delta_{61} & \Delta_{62} & \Delta_{63} & \Delta_{64} & \Delta_{65} & \Delta_{66} & \Delta_{67} \\
\Delta_{71} & \Delta_{72} & \Delta_{73} & \Delta_{74} & \Delta_{75} & \Delta_{76} & \Delta_{77}
\end{array}
$$

Fig. 5.11. Printout in matrix form.

Printing out the matrix row by row requires the following four statements:†

```
      DO 120 I=1,N
432   WRITE (6,116) (DELTA(I,J), J=1,N)
116   FORMAT(7F10.4/)
120   CONTINUE
```

Perhaps Statement 432 needs some explanation. It serves to print out

$$\Delta_{i1}, \Delta_{i2}, \ldots, \Delta_{in}$$

on the ith row. In other words, this statement can be replaced by the following lengthy statement, if we take, for example, $N = 3$:

```
9432  WRITE (6,116) DELTA(I,1), DELTA(I,2), DELTA(I,3)
```

By means of a DO loop, the answers will be printed out row by row: row 1 ($i = 1$), row 2($i = 2$), etc.

---

† Or simply use the two statements

```
111   WRITE (6,116) ((DELTA(I,J),J = 1,N),I = 1,N)
116   FORMAT(7F10.4)
```

to obtain the identical printout for $N \leq 7$.

```
C        THIS IS A MODIFIED FORTRAN PROGRAM FOR A ROW BY ROW MATRIX PRINTOUT
C        N=MAXIMUM NUMBER FOR I OR J
         DATA IREAD,IWRITE/5,6/
         READ(IREAD,1) P,ZL,E,AI,N
       1 FORMAT(4F10.7,I10)
         ZNP1=N+1
         DIMENSION DELTA(10,10)
       7 DO 20 I=1,N
         ZI=I
       8 DO 20 J=1,N
         ZJ=J
         IF(ZI.GE.ZJ) GO TO 51
      50 DELTA(I,J)=((P/(E*AI))*(ZL**3)*(1.-ZJ/ZNP1)*ZI/(6.*ZNP1))*(1.-(1.-
      1    ZJ/ZNP1)**2-(ZI/ZNP1)**2)
         GO TO 20
      51 DELTA(I,J)=((P/(E*AI))*(ZL**3)*(1.-ZJ/ZNP1)*ZI/(6.*ZNP1))*(1.-(1.-
      1    ZJ/ZNP1)**2-(ZI/ZNP1)**2)+(P/(E*AI))*ZL*ZL*ZL*(ZI-ZJ)**3/
      2    ((ZNP1**3)*6.)
      20 CONTINUE
         DO 120 I=1,N
         WRITE(IWRITE,100) I
         WRITE(IWRITE,6) (DELTA(I,J),J=1,N)
       6 FORMAT(7F10.4/)
     100 FORMAT(23X,' FLEXIBILITY MATRIX ROW',I2/)
     120 CONTINUE
         WRITE(IWRITE,25)
      25 FORMAT(' THE END')
         STOP
         END


1.0         1.0         1.0         1.0                    7
```

Fig. 5.12. Program and data for row-by-row printing.

It is convenient to print out a heading for each row and use appropriate spacing as illustrated below:

FLEXIBILITY MATRIX ROW 1

Skip 1 line here $\longrightarrow$

$$\Delta_{11} \quad \Delta_{12} \quad \Delta_{13} \quad \Delta_{14} \quad \Delta_{15} \quad \Delta_{16} \quad \Delta_{17}$$

Skip 1 line here $\longrightarrow$

FLEXIBILITY MATRIX ROW 2

Skip 1 line here $\longrightarrow$

$$\Delta_{21} \quad \Delta_{22} \quad \Delta_{23} \quad \Delta_{24} \quad \Delta_{25} \quad \Delta_{26} \quad \Delta_{27}$$
$$\vdots \qquad\qquad\qquad\qquad\qquad\qquad \vdots$$

FLEXIBILITY MATRIX ROW 7

Skip 1 line here $\longrightarrow$

$$\Delta_{71} \quad \Delta_{72} \quad \Delta_{73} \quad \Delta_{74} \quad \Delta_{75} \quad \Delta_{76} \quad \Delta_{77}$$

This elegant printout can readily be accomplished by using the following five statements:

```
        DO 432 I=1,N
        WRITE (6,108) I
108     FORMAT(23X,23H FLEXIBILITY MATRIX ROW, I2/)
432     WRITE (6,116) (DELTA(I,J),J=1,N)
116     FORMAT(7F10.4/)
```

The Statement 116 means "print out a maximum of seven numbers per line, each number having ten digits of which four digits are behind the decimal point; and after printing this complete row, skip one line."

Table 5.3 is the printout for the improved version of the FORTRAN program, which is shown in Fig. 5.12.

We note that in Table 5.3 the computed values of all elements are symmetrical about the main diagonal, for example,

$$DELTA(1,4) = DELTA(4,1),$$

$$DELTA(2,3) = DELTA(3,2), \text{ etc.}$$

Table 5.3 Row-by-row output of a matrix

FLEXIBILITY MATRIX ROW 1

| .0039 | .0065 | .0077 | .0076 | .0065 | .0048 | .0025 |

FLEXIBILITY MATRIX ROW 2

| .0065 | .0117 | .0142 | .0143 | .0124 | .0091 | .0048 |

FLEXIBILITY MATRIX ROW 3

| .0077 | .0142 | .0183 | .0190 | .0168 | .0124 | .0065 |

FLEXIBILITY MATRIX ROW 4

| .0076 | .0143 | .0190 | .0208 | .0190 | .0143 | .0076 |

FLEXIBILITY MATRIX ROW 5

| .0065 | .0124 | .0168 | .0190 | .0183 | .0142 | .0077 |

FLEXIBILITY MATRIX ROW 6

| .0048 | .0091 | .0124 | .0143 | .0142 | .0117 | .0065 |

FLEXIBILITY MATRIX ROW 7

| .0025 | .0048 | .0065 | .0076 | .0077 | .0065 | .0039 |

THE END

Therefore it is possible to print out either the lower or upper triangular matrix for the output. A lower triangular matrix has the following appearance:

$$\Delta_{11}$$
$$\Delta_{21} \quad \Delta_{22}$$
$$\Delta_{31} \quad \Delta_{32} \quad \Delta_{33}$$
$$\Delta_{41} \quad \Delta_{42} \quad \Delta_{43} \quad \Delta_{44}$$
$$\Delta_{51} \quad \Delta_{52} \quad \Delta_{53} \quad \Delta_{54} \quad \Delta_{55}$$
$$\Delta_{61} \quad \Delta_{62} \quad \Delta_{63} \quad \Delta_{64} \quad \Delta_{65} \quad \Delta_{66}$$
$$\Delta_{71} \quad \Delta_{72} \quad \Delta_{73} \quad \Delta_{74} \quad \Delta_{75} \quad \Delta_{76} \quad \Delta_{77}$$

This matrix excludes any element to the right of the main diagonal line. To print out the elements for a lower triangular matrix, we can use the following three statements:

```
C       PROGRAM ADDITIONS FOR A LOWER TRIANGULAR MATRIX OUTPUT
        DO 220 I=1,N
  220   WRITE (6,221) (DELTA(I,J),J=1,I)
  221   FORMAT(7F10.4/)
```

We note that the limiting value for J in Statement 220 is I *rather than* N. Thus, when I = 1, J will be incremented to a limiting value of 1. Hence DELTA(1,1) is printed on the first line. Similarly, when I = 2, J will be equal to 1 and 2, respectively. This will cause the printout of DELTA(2,1) and DELTA(2,2) on the second line, etc.

Similarly, an upper triangular matrix may be printed out by using the following three statements:

```
C       PROGRAM ADDITIONS FOR AN UPPER TRIANGULAR MATRIX OUTPUT
        DO 222 I=1,N
  222   WRITE (6,223) (DELTA(I,J),J=I,N)
  223   FORMAT(7F10.4/)
```

Before turning to the next section, we remark on a new format for output, the E format. This format is most suitable when the order of magnitude of the answers to be printed is not clearly known ahead of time. For example, the following two statements may be used advantageously to print, on the same line, both the value of I, for example 17, and the variable ANS, say, having a value of -0.03141592:

```
        WRITE (6,12) I, ANS
  12    FORMAT(I3, E14.7)
```

The output will look like

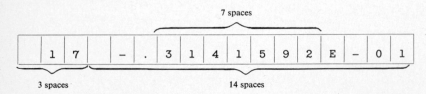

The format E14.7 needs some explanation. Fourteen spaces are required for this answer in which 7 places, or 3141592, are used for the mantissa (see Section 4.2), and the four rightmost places are used for, respectively, the letter E, the sign of the exponent, and the two-digit true exponent XX.

5.6 INPUT STATEMENTS FOR A MATRIX PROBLEM

We shall proceed, by means of specific examples, to discuss how to read in a matrix. Essentially, the line of reasoning for matrix input is the same as that for matrix output. Suppose, for example, that an array of numbers in six rows and eight columns is to be read in. This can be accomplished by the following four statements:

```
        C       READ IN A 6 X 8 MATRIX
                DIMENSION A(6,8)
                DO 128 I=1,6
        128     READ (5, 129) (A(I,J),J=1,8)
        129     FORMAT(7F10.4)
```

In this case, we need two data cards for each row. The first card has seven elements, namely, A(1,1), A(1,2),..., A(1,7); and the second card is punched only A(1,8) in the first 10 columns. The remaining 70 columns are left blank.

Frequently, we wish to read in a lower triangular matrix, say, of five rows and five columns. We can use the following four statements:

```
        C       READ IN A LOWER TRIANGULAR MATRIX. IREAD = 5.
                DIMENSION A(5,5)
                DO 226 I=1,5
        226     READ (IREAD,227) (A(I,J),J=1,I)
        227     FORMAT(7F10.4)
```

It is important to note that we need five data cards in this example. The first card has one element only, namely A(1,1), in the first 10 columns. The second has two elements, A(2,1) and A(2,2), in the first 20 columns. The third card has three elements, A(3,1), A(3,2), and A(3,3), in the first 30 columns. The fourth data card has four elements. Finally, the fifth card has five elements of the last row in the matrix. The last card is punched only in the first 50 columns as per FORMAT Statement 227.

Alternatively, we can simply use the following three statements to read in the same lower triangular matrix:

```
                DIMENSION A(5,5)
                READ (IREAD,227) ((A(I,J),J=1,I),I=1,5)
        227     FORMAT(7F10.4)
```

Here only three data cards are needed. The first card has seven elements, namely, A(1,1), A(2,1), A(2,2), A(3,1), A(3,2), A(3,3), and A(4,1), each having the F10.4

format. The second card has seven elements. They are A(4,2), A(4,3), A(4,4), A(5,1), A(5,2), A(5,3), and A(5,4). The remaining element, A(5,5), is punched in the first 10 columns of the third card.

One can readily see the similarity between the preceding group of statements and the output statements for a lower triangular matrix discussed in the previous section.

There are occasions when data cards are punched prior to the writing of a computer program. Thus, if we wish the computer not to read the even-numbered cards, i.e., the second, the fourth, etc., then we use the *slash* introduced in the previous section.

```
C       READ IN A 6 X 6 MATRIX
        DIMENSION A(6,6)
        DO 128 I=1,6
128     READ (IREAD,129) (A(I,J),J=1,6)
129     FORMAT(7F10.4/)
```

A total of twelve data cards are required here. The six odd-numbered cards are read in, and the other six cards are skipped.

Before concluding this section, we digress to remark on an elegant echo print statement. We utilize an echo print when we wish to read in a card and print out the same information. This can be accomplished by the following five statements and one data card, which together form a *complete* program.

```
        READ (5,20)
20      FORMAT(49H□□□□□□□□□□□□□□□□□□□□□□□□□□□□□□□□□□□□□□□□□□□□□□□□□□□)
        WRITE (6,20)
        STOP
        END
```

It is important to note the flexibility of this program in that the information on the data cards may be changed at ease without affecting the program. In this particular example, the format specification is a Hollerith field for which a maximum of 49 spaces may be used per FORMAT statement.

5.7 FORTRAN SUBROUTINE SUBPROGRAMS

In writing a computer program, it is often necessary to repeat the same group of statements many times within the same program. For example, suppose that we wish to write a FORTRAN program to perform a 90-degree clockwise rotation of the elements in the following array of numbers:

2	7	6
9	5	1
4	3	8

This operation may be carried out by interchanging the first row with the third column, the second row with the second column, etc. Thus we have the new array of numbers

4	9	2
3	5	7
8	1	6

A FORTRAN program related to this operation and also to the printout is shown in Fig. 5.13, where the array SWAP indicates the temporary storages, and A, the given array of numbers.

```
C       PROGRAM FOR MATRIX ROTATION
C       N=NUMBER OF ROWS AND COLUMNS OF THE GIVEN MATRIX
        DATA IREAD,IWRITE/5,6/
        READ(IREAD,2) N
      2 FORMAT(I2)
        DIMENSION A(10,10),SWAP(10,10)
        DO 5 I=1,N
      5 READ(IREAD,6) (A(I,J),J=1,N)
      6 FORMAT(10F5.2)
      7 DO 9 I=1,N
        DO 9 J=1,N
      9 SWAP(I,J)=A(I,J)
        DO 13 I=1,N
        DO 13 K=1,N
        J=N-K+1
     13 A(I,K)=SWAP(J,I)
        DO 15 I=1,N
     15 WRITE(IWRITE,16) (A(I,J),J=1,N)
     16 FORMAT(26X,10F5.2)
     17 STOP
        END


     03
      2.0    7.0    6.0
      9.0    5.0    1.0
      4.0    3.0    8.0
```

Fig. 5.13. FORTRAN program and data for matrix rotation.

If we now wish to make two or more successive 90-degree clockwise rotations of the given square matrix with a printout after each rotation, we would have to repeat Statements 7 through 16 two or more times. Figure 5.14 shows a long, repetitious, and therefore undesirable FORTRAN program for successive rotations. Imagine how much worse the repetition can become when the group of statements to be repeated is very long to begin with!

To avoid this undesirable repetition we often use a SUBROUTINE subprogram, sometimes called a *subroutine*. A subroutine is a collection of statements, or "packaged statements," for the purpose of achieving some specific objectives, such as rotating a given matrix and writing out the results. Figure 5.15 is such a subroutine.

```
C          PROGRAM FOR SEVERAL ROTATIONS
C          N=NUMBER OF ROWS AND OF COLUMNS OF THE GIVEN MATRIX
C
           DATA NREAD,NWRITE/5,6/
           READ(NREAD,3) N
     3 FORMAT(I2)
           DIMENSION A(10,10),SWAP(10,10)
           DO 5 I=1,N
     5 READ(NREAD,8) (A(I,J),J=1,N)
     8 FORMAT(10F5.2)
C
C          FIRST ROTATION AND PRINTING OUT
C
    12 DO 15 I=1,N
           DO 15 J=1,N
    15 SWAP(I,J)=A(I,J)
           DO 16 I=1,N
           DO 16 K=1,N
           J=N-K+1
    16 A(I,K)=SWAP(J,I)
           DO 19 I=1,N
    19 WRITE(NWRITE,20) (A(I,J),J=1,N)
    20 FORMAT(26X,10F5.2)
C
C          SECOND ROTATION AND PRINTING OUT
C
           DO 115 I=1,N
           DO 115 J=1,N
   115 SWAP(I,J)=A(I,J)
           DO 116 I=1,N
           DO 116 K=1,N
           J=N-K+1
   116 A(I,K)=SWAP(J,I)
           DO 119 I=1,N
   119 WRITE(NWRITE,20) (A(I,J),J=1,N)
C
C          THIRD ROTATION AND PRINTING OUT
C
           DO 215 I=1,N
           DO 215 J=1,N
   215 SWAP(I,J)=A(I,J)
           DO 216 I=1,N
           DO 216 K=1,N
           J=N-K+1
   216 A(I,K)=SWAP(J,I)
           DO 219 I=1,N
   219 WRITE(NWRITE,20) (A(I,J),J=1,N)
           STOP
           END


03
   2.0   7.0   6.0
   9.0   5.0   1.0
   4.0   3.0   8.0
```

Fig. 5.14. FORTRAN program and data for three matrix rotations.

```
      SUBROUTINE ROTATE(A,N)
      DATA IWRITE/6/
      DIMENSION A(10,10),SWAP(10,10)
    7 DO 9 I=1,N
      DO 9 J=1,N
    9 SWAP(I,J)=A(I,J)
      DO 13 I=1,N
      DO 13 K=1,N
      J=N-K+1
   13 A(I,K)=SWAP(J,I)
      DO 15 I=1,N
   15 WRITE(IWRITE,16) (A(I,J),J=1,N)
   16 FORMAT(26X,10F5.2)
      RETURN
      END
```

Fig. 5.15. Subroutine rotate

Called SUBROUTINE ROTATE, its Statements 7 through 16 are taken directly as a "package" from Fig. 5.13 (see the shaded portion). To complete this subroutine it is necessary to add (1) the NAME statement, (2) the DIMENSION statement, (3) the RETURN statement, and (4) the END statement to the above-mentioned package. We shall now discuss these statements in detail.

The first statement in a subroutine, the NAME statement, serves two purposes, namely, to identify and to present the arguments involved. The name required for identification has several limitations:

a) It must not be more than six characters in length.

Example

Correct: SUBROUTINE ROTATE
Incorrect: SUBROUTINE REVOLVE

b) It must begin with an alphabetic character.

Example

Correct: SUBROUTINE SEA7
Incorrect: SUBROUTINE 7SEA

The parentheses following the name of the subroutine enclose the arguments which specify the variables in the subprogram. The arguments may be (1) real variables, for example ANS; (2) fixed-point variables, for example N; or (3) subscripted variables, for example A, but not A(I,J), which represents the entire array.

It may be well to note that in the first statement of Fig. 5.15 there are only two arguments, A and N. An inspection of the complete listing shows, however, that five more variables, SWAP, I, J, K, and JJ, are also used in this subroutine. These variables are omitted from the parentheses because they are not independent variables. The array SWAP, for example, is defined in Statement 9 of the subroutine by the current values of the array A. In other words, the array SWAP is completely defined once the array A is known. Hence, using SWAP as an argument is a superfluous step; however, it will not cause any difficulties during compilation. Consider now I, J, and K in the subroutine. These are used exclusively as indices of DO loops, and hence are not counted as arguments. The variable JJ is defined by the two variables N and K and therefore is not required as an argument either. We now see that the only arguments required for the NAME statement in SUBROUTINE ROTATE are A and N.

The next statement in this subroutine is a DATA statement. In this subroutine, it serves to define the logical unit number of the on-line printer. As mentioned before, we choose to use the number 6 for the printer. There is no need to write a DATA statement in this subroutine if there are no READ or WRITE statements involved in it.

The next statement in this subroutine is a DIMENSION statement. This statement is required in a subroutine in two situations:

1. When an argument representing an array already appears in the NAME statement of the subroutine; for example, A(10,10) represents a two-dimensional array.

2. When *new* subscripted variables are introduced in the subroutine; for example, The array SWAP, a new subscripted variable in the rotation subroutine, is dimensioned SWAP(10,10).

The arrays mentioned here may be one-dimensional, for example X(20), Y(16); two-dimensional, for example, A(10,10), SWAP(10,10), Z(10,10); or three-dimensional, such as B(2,4,4). However, since SUBROUTINE ROTATE involves only two-dimensional arrays, its dimension statement is

DIMENSION A(10,10), SWAP(10,10)

or

DIMENSION SWAP(10,10), A(10,10)

The RETURN statement signifies an exit from the subroutine. In our example, SUBROUTINE ROTATE, it serves to pass the control to the first statement immediately following the shaded portion of Fig. 5.13. This is statement 17, which is STOP. Every subroutine requires at least one† RETURN statement. The last statement of any subroutine must be the END statement. It may not be omitted.

5.8 MAIN PROGRAM

We now proceed to introduce the concept of a main program, using the specific example of matrix rotation. Consider the program in Fig. 5.13. We recall that the shaded portion was removed to form the subroutine ROTATE. If we insert a single CALL STATEMENT in place of this missing portion,

CALL ROTATE (A, N)

then the resulting program (Fig. 5.16) is designated as the main program. Digressing momentarily, we observe that SWAP has been omitted from the DIMENSION statement

---

† See, for example, Fig. 12.15, where three RETURN statements are used in one subroutine.

```
C
C      N=NUMBER OF ROWS AND COLUMNS OF GIVEN MATRIX
       DATA IREAD,IWRITE/5,6/
       READ(IREAD,2) N
   2   FORMAT(I2)
       DIMENSION A(10,10)
       DO 5 I=1,N
   5   READ(IREAD,6) (A(I,J),J=1,N)
   6   FORMAT(10F5.2)
       CALL ROTATE(A,N)
       STOP
       END
       SUBROUTINE ROTATE(A,N)
       DATA IWRITE/6/
       DIMENSION A(10,10),SWAP(10,10)
   7   DO 9 I=1,N
       DO 9 J=1,N
   9   SWAP(I,J)=A(I,J)
       DO 13 I=1,N
       DO 13 K=1,N
       J=N-K+1
  13   A(I,K)=SWAP(J,I)
       DO 15 I=1,N
  15   WRITE(IWRITE,16) (A(I,J),J=1,N)
  16   FORMAT(26X,10F5.2)
       RETURN
       END
```

```
03
2.0   7.0   6.0
9.0   5.0   1.0
4.0   3.0   8.0
```

Fig. 5.16. A complete program consisting of the main program, one subroutine, and the required data.

in the main program since it does not appear explicitly in any of the nine statements of the main program.

We now see that the original program (Fig. 5.13) has been replaced by a main program and its accompanying subprogram, SUBROUTINE ROTATE. The statement CALL ROTATE in the main program transfers the control of execution to SUBROUTINE ROTATE; that is, the computer proceeds to SUBROUTINE ROTATE and performs whatever operations are indicated by the package of statements. The RETURN statement then turns the control of execution back to the statement succeeding the CALL statement, i.e., Statement 17 in the main program.

At first glance, the concept of partitioning a main program and its subroutine seems unwarranted; the partitioned program appears more complex than the original (Fig. 5.13). The usefulness of this concept can be illustrated, however, by making, say, three successive rotations of a given matrix with a printout after each rotation. We remember that the FORTRAN program previously prepared for this problem (Fig. 5.14) was long, repetitious, and undesirable. However, as shown in Fig. 5.16, our task is simplified when we employ subroutines; we just place three CALL ROTATE statements in the main program. If we now compare the total number of statements in the two schemes mentioned above, that is,

1) the single program shown in Fig. 5.13, and

2) the main program plus SUBROUTINE ROTATE (Fig. 5.16),

then we can readily conclude that scheme 2 is far shorter than scheme 1 principally because it avoids repeating a large group of statements.

```
C      M=THE SIZE OF THE GIVEN (M*M) MATRIX B
C      MAIN PROGRAM FOR THREE ROTATIONS USING SUBTOUTINE
C
       DATA IREAD,IWRITE/5,6/
       READ(IREAD,2) M
     2 FORMAT(I2)
       DIMENSION B(10,10)
       DO 5 ICOL=1,M
       READ(IREAD,9) (B(ICOL,JROW),JROW=1,M)
     5 WRITE(IWRITE,9) (B(ICOL,JROW),JROW=1,M)
     9 FORMAT(26X,10F5.2)
       CALL ROTATE(B,M)
       CALL ROTATE(B,M)
       CALL ROTATE(B,M)
       STOP
       END
```

Fig. 5.17. Main program for three rotations using subroutine.

In the main program shown in Fig. 5.17, each variable (both single and subscripted) has been deliberately renamed (for example, B,M) and therefore differs from its counterpart (for example, A,N) in the accompanying subroutine ROTATE.

In order to link the main program with the required subroutine, we note the following rule:

The arguments in the calling statement of a main program must agree in number, order, and mode with the arguments in the first statement (NAME) of a subroutine.

Examples

Correct:	CALL ROTATE(B,M)	in main program
	SUBROUTINE ROTATE(A,N)	in subroutine
Correct:	CALL ROTATE(M,B)	in main program
	SUBROUTINE(N,A)	in subroutine
Correct:	CALL ROTATE(3,B)	in main program (if M = 3 in the main program)
	SUBROUTINE ROTATE(N,A)	in subroutine
Incorrect:	CALL ROTATE (B,M)	in main program
	SUBROUTINE ROTATE(N,A)	in subroutine
Incorrect:	CALL ROTATE(B,M)	in main program
	SUBROUTINE ROTATE(A,AN)	in subroutine
Incorrect:	CALL ROTATE(B,M)	in main program
	SUBROUTINE ROTATE(A,N,SWAP)	in subroutine

Before concluding this section, let us emphasize a few of the advantages of using subroutines.

First, a subroutine can be summoned any number of times by the main program (or by another subroutine). This procedure allows the repeated use of a package of statements without actually writing them out in the main program or the calling program each time they are needed. Second, many useful subroutines are available through various computer users' organizations or at computer centers. Finally, main programs and subroutines can be compiled individually, even at different times; thus it is possible to significantly reduce the debugging time spent by programmers.

5.9 FUNCTION SUBPROGRAM AND FUNCTION STATEMENT

A FUNCTION subprogram may be considered as a smaller version of the SUBROUTINE subprogram discussed in Section 5.7. A SUBROUTINE may have more than one result, while a FUNCTION is limited to computing only one value. Just as in a SUBROUTINE, in each FUNCTION subprogram many arguments may be enclosed in the parentheses. The essential features will be illustrated with an example. Let us compute

$$\text{DOT}(A,B,K) = \sum_{i=1}^{K} A_i B_i, \tag{5.1}$$

where both A and B are one-dimensional arrays of 25 elements each ($K \leq 25$). It is assumed that Eq. (5.1) is used frequently in a given program.

```
   FUNCTION DOT(A,B,K)
   DIMENSION A(25),B(25)
   DOT=0.
   DO 90 I=1,K
   SAK=A(I)*B(I)
90 DOT=DOT+SAK
   RETURN
   END
```

Fig. 5.18. FUNCTION subprogram.

A possible FUNCTION DOT program is shown in Fig. 5.18. It may be noted that, like a SUBROUTINE, a FUNCTION subprogram invariably contains the following four types of statements:

1) FUNCTION name statement

2) DIMENSION, if one or more arrays are involved

3) RETURN

4) END

To obtain the numerical value of the expression

$$\alpha = \sqrt{\left(\sum_{i=1}^{17} C_i D_i \right) + 1} + \sum_{i=1}^{5} C_i D_i,$$

where C and D are one-dimensional arrays, we may write

$$\text{ALPHA = SQRT (1. + DOT(C,D,17)) + DOT(C,D,5)}$$

The above statement will bring the FUNCTION subprogram into operation without a CALL statement.

The essential differences between a SUBROUTINE and a FUNCTION subprogram are tabulated below:

SUBROUTINE subprogram	FUNCTION subprogram
Name: One to six letters or digits. The first character must be a letter.	*Name:* One to six letters or digits, the first of which must be I, J, K, L, M, or N if and only if the value of the function is a fixed point.
Calling into operation: By a CALL statement.	*Calling into operation:* By writing the name of the function in an expression where numerical value is to be determined.
Number of arguments: From *none* to some maximum.	*Number of arguments: From* 1 to 15.
Number of values returned: From *none* to some maximum.	*Number of values returned:* 1

Both SUBROUTINE and FUNCTION subprogram can be compiled independently from their main programs and both can consist of many statements.

We now proceed to describe a third type of function, the function statement, which has only one arithmetic statement and therefore *cannot* be compiled independently.

Example

It is known that
$$\phi(x) = x^2 - 2x + 7, \qquad \theta(y) = 3y^2 - 14,$$
$$Q(x, y, z) = 3\phi(x)\sqrt{\theta(y)} - 2.4z^2.$$

We wish to compute T = Q(A, B, C), where A = 1.46, B = 28.3, and C = 31.5. A program to compute T consists of the following statements:

```
PHI(X) = X * X - 2. * X + 7.
THETA(Y) = 3. * Y * Y - 14.
Q(X,Y,Z) = 3. * PHI(X) * SQRT(THETA(Y)) - 2.4 * Z * Z
T = Q(1.46, 28.3, 31.5)
```

The use of a function statement is subject to the following four restrictions:

1) The name may have one to six characters. The first character must be alphabetic. Those functions beginning with I, J, K, L, M, or N will be computed as fixed-point numbers; all others will be computed as floating-point numbers.

2) The function statement must be placed at the beginning of the program before the first executable statement.

3) Each function is defined as a *single arithmetic* statement.

4) The arguments in parentheses may not be subscripted.

In summary, we see that there are four types of subprograms in FORTRAN language, namely, SUBROUTINE subprograms (Sections 5.7 and 5.8), FUNCTION subprograms (Section 5.9), arithmetic function statements (Section 5.9), and library functions (Table 5.1).

FUNCTION subprograms are, in a sense, smaller versions of SUBROUTINE subprograms. The output from a FUNCTION subprogram is limited to a single variable, whereas that from a SUBROUTINE subprogram may yield many variables. Thus it is impossible for a FUNCTION subprogram to handle a matrix output.

An arithmetic function statement is quite similar to a library function. The latter, often referred to as the built-in routine or library subroutine, is supplied by the FORTRAN system; the arithmetic function statement is written by the user.

5.10 SUMMARY OF THE FORTRAN COMPILER LANGUAGE

As mentioned before, FORTRAN compiler language comprises a number of types of statements that a programmer may write to code a particular problem. This compiler language was initially prepared for the IBM 704 computer and has subsequently been extended to many other types. As it exists today, FORTRAN language is a product of evolution. There are many versions. The essential portion of the one known as IBM System/360 Basic FORTRAN IV language is summarized in this section. Programs written in this version of FORTRAN are compatible with any of the IBM System/370 or System/360. They can also be run on many other computers which accept programs compatible with the American Standards Association (USAS) Basic FORTRAN†.

A. Types of Numbers and Variables

1. *Fixed-point constants.* The largest positive integer allowable is $2^{31} - 1$, or 2,147,483,647, while the largest negative one is -2^{31}, or $-2,147,483,648$. These numbers are used primarily for counting and similar processes. Integral (fixed-point) constants are written in decimal notation without a decimal point (for example, -256, 95473, -1, 1, and -98675). The plus sign may not be required. If a fixed-point constant does not lie within the interval, its binary equivalent will be truncated; only its 32 rightmost bits are retained. (A practical application is discussed in Section 14.4).

---

† The IBM Basic FORTRAN IV is compatible with and encompasses USAS Basic FORTRAN. The extensions permit the mixed mode expressions, three-dimensional array, double-precision constants, skipping the printer by carriage control.

2. *Fixed-point variables.* Names of variables which take on fixed-point values have from one to six alphabetic or numerical characters, starting with I, J, K, L, M, or N (for example, I, KAT, L90, and NIM). In other words, a variable is declared as fixed point if it is represented by I, J, K, L, M or N, or by a name starting with one of those six letters. This is known as implicit specification, which can be annulled by an explicit declaration.

3. *Single precision floating-point constants.* Floating-point numbers are primarily used for data involved in arithmetic operations. Single-precision floating-point constants, also known as *real* constants, are written in one of the following two ways:
 a) In decimal form with a sign and a decimal point, for example, 22.976, -0.0003, or -7654321. In this form, it can have from 1 to 7 decimal digits.
 b) In decimal exponent form (a one- or two-digit fixed-point constant) preceded by an E, for example,

5.1E3	means	5.1×10^3,
5.1E+3	means	5.1×10^3,
5.12E–71	means	5.12×10^{-71}.

The valid range is 16^{-63} through 16^{63}, or approximately from 10^{-75} to 10^{75}.

4. *Double-precision floating-point constants.* A floating-point number written with from 8 to 16 digits is interpreted as a double-precision floating-point number. For example, 3.14159265358979 or .0000000037 are valid double-precision floating-point numbers. As an alternative, a double-precision constant can be written as 1 to 16 digits followed by a D exponent, for example,

Double-Precision Constant in FORTRAN	Its Value
.314159265358979D1	3.14159265358979
314159265.358979D–6	314.159265358979
9.D2	
9.0D2	900.
9.0D+2	
9.D–2	.009

5. *Single-precision floating-point (real) variables.* Names of real variables have from one to six alphabetic and numerical characters beginning with a character *not* I, J, K, L, M, or N. For example, EQ76A, A, BATE, TIME, DELT, DXDT, or $7SEA are valid names. A programmer may declare a variable as real by using this implicit or predefined specification. Or, he may use an explicit specification (see E5(c) below) which overrides the predefined specification.

6. *Double-precision floating-point variables.* A programmer must declare a double-precision variable with an explicit type specification.

B. Arrays

Arrays may be one-, two-, or three-dimensional and may contain as elements either fixed- or floating-point numbers. (The same name rules mentioned above for naming variables apply also to fixed- and floating-point arrays.) A particular element in an array is designated by subscripts, which must be in one or more of the following forms:

$$c, \quad v, \quad v \pm c, \quad c' * v \pm c,$$

where c and c' are fixed-point constants, v is a fixed-point variable, and the sign $*$, $+$, and $-$ are interpreted as multiplication, addition, and subtraction, respectively. For notational purposes, subscripts are enclosed in parentheses and separated by commas, for example,

A(2,I-5), B(5*KMAX-2,9*J), IC(I,J), BATE(10),

or

X(I,J+4,K*2)

C. Expressions

Arithmetic expressions may be formed by using parentheses and five connectives $+$, $-$, $*$, $**$, and $/$ for addition, subtraction, multiplication, exponentiation, and division, respectively. The allowable constituents of an expression are constants, variables, elements of arrays, and functions. The following examples show the correct usage of notation.

FORTRAN IV Expression	Mathematical Notation
A*B	AB
A*(-B)	$-AB$
-A*B	$-AB$
A+2. or A + 2	$A + 2$
A**(B+2.)*C or A**(B+2)*C	$A^{B+2}(C)$
A**(I-2)	A^{I-2}
A*B/(C*D)	AB/CD
A*(X-B*(X+C))	$A(X - B(X + C))$
((A-B)/C)**2.5	$(A - B)^{2.5}/(C)^{2.5}$

Two operation symbols may not appear together, but must be separated by parentheses such as was done in the second example above. The use of parentheses also helps to determine the hierarchy of operations in an expression, as is the case in conventional mathematical usage.

During the evaluation of an arithmetic expression in FORTRAN IV, the rank of the hierarchy of operators is assumed to be as follows:

Rank	Symbol	Operation
First	**	Exponentiation
Second	/ and *	Division and multiplication
Third	+ and −	Addition and subtraction

In the same expression, if two or all of the following three types of variables or constants are involved: integer, real and double-precision, then the expression is said to be of mixed mode. Such expressions are evaluated so that all variables or constants are converted to the highest in the hierarchy (Table 5.4) prior to evaluation. For example, if real and integer variables or constants are mixed in an expression, the integers are converted to the real mode before the expression is evaluated.

D. Library Functions

Library functions are prewritten subroutines. They are single valued, and have only one argument; that is, they use one value to compute a function of that value; for example, Y = SIN(X). The names have from one to six characters and arguments of functions may be arithmetic functions. The following internal functions are available:

FUNCTION name	Explanation	Type of argument and function
SIN(X)	Sine of X in radians	
COS(X)	Cosine of X in radians	
SQRT(X)	Square root of X	Single-precision
ATAN(X)	Arctangent of X (angle is in radians)	
LOG(X)	Natural logarithm of X	Floating-point
EXP(X)	Exponential of X (meaning e^X)	
ABS(X)	Absolute value of X	
DSIN(X)	Sine of X in radians	
DSQRT(X)	Square root of X	Double-precision
DLOG10(X)	Logarithm of X (base 10)	Floating point
MINI(X,Y,Z,...)	The smallest value of X, Y, Z,...	Arguments: Real; Function: Integer

Table 5.4 Hierarchy of types in mixed-mode expression

Highest	Double floating-point variable or constant
High	Real variable or constant
Low	Integer variable or constant

These functions are used in the same manner as constants and variables, for example,

```
Y = SQRT(B*FAT) - EXP(SIN(G(10)))
I = MINI(A,B,C)
```

E. Statements

A complete program consists of main program and possibly one or more subprograms, each of which is made up of a sequence of FORTRAN statements. These statements fall into five main types.

1. *Arithmetic statements.* These statements gave the form

$$a = b,$$

where the variable or array element a is replaced by the value of the expression b. The area elements a and b may be of the same mode (fixed-point, real, or double-precision floating-point), or they may be in different modes. It is thus possible to change from fixed- to floating-point, and vice versa.

FORTRAN	Mathematical expression
R=(A–B*X)/(C–D*X)	$R = (A - BX)/(C - DX)$
FY=X*(X**2–Y**2)/(X**2+Y**2)	$FY = X(X^2 - Y^2)/(X^2 + Y^2)$
PI=3.1415926	$\pi = 3.1415926$
M=2*M–10*J	$M_{new} = 2M_{old} - 10J$

2. *Control statements.* This type of statement is used to govern the flow of control in the program.

 a) The GO TO statement when executed causes the computer to transfer control to the statement numbered N. (In columns 1 to 5, statements are numbered arbitrarily from 1 to 99999. They do not have to be in sequence and not every card need be numbered.) For example,

GO TO 3

 b) The computed GO TO statement, GO TO $(n_1, n_2, \ldots, n_m)$, i is such that when executed, control is transferred to statement n_j, where j is the value of the fixed-point variable $i(1 \leq j \leq m)$. For example,

GO TO (12, 957, 34, 21, 16), L

when L = 1 applies, GO TO statement 12; when L = 2, then GO TO statement 957, ... ;
and finally, when L = 5, then GO TO statement 16.

c) IF $(A)n_1, n_2, n_3$ is a conditional statement causing transfer to statement n_1, n_2,
or n_3 depending on whether the value of the expression A is negative, zero, or positive.
Some examples are

```
IF (I) 5,5,2
IF (D**2-SIN(X**2)) 2,3,3
IF (I-5*J-6) 998, 678, 678
IF (C(J,K)-B) 10,4,30
```

The expression A may be a logical one† which may consist of a single element
(constant, variable, function reference).   In the first two examples above, if the
statement 2 is taken to be the next sequential statement for each of the two IF statements, they may be written as follows:

```
IF (I.LE.0) GO TO 5
IF ((D**2).GE.(SIN(X**2))) GO TO 3
```

where LE and GE are relational operators.   A table of frequently used relational
operators appears in Table 5.5.   The enclosing periods are part of the operator and
must be explicitly written.

Table 5.5 Relational operator

Operator	Relation
.GT.	Greater than
.GE.	Greater than or equal to
.LT.	Less than
.LE.	Less than or equal to
.EQ.	Equal to
.NE.	Not equal to

d) The DO statement may be in one of two forms:

$$DO\ n \quad i = m_1, m_2$$
$$DO\ n \quad i = m_1, m_2, m_3$$

where n is a statement number, i is a fixed point variable, and $m_1, m_2,$ and m_3 are
fixed-point variables or constants, but not expressions. When m_3 is not stated, as in
the first form above, it is assumed to be 1. All statements between the DO statement
and n are first executed with the variable i set equal to m_1. The quantity i is then
incremented by m_3 and the same procedure is repeated until i is greater than m_2.

---

† Not included in the ASA Basic FORTRAN.

When this occurs, control is transferred to the statement following n. For example,

```
        L(1) = 1
        DO 26 I = 1, 5          or      DO 17 LITE = K, J, 20
        K(I) = L(I) + 2                 DO 89 I = 4, 299
    26  L(I + 1) = L(I) + 1
```

The restrictions on a DO loop are:

1) The first statement after the DO statement must not be an END, CONTINUE, DIMENSION, or FORMAT.

2) Statement n must not be any of those shown below:

   CALL EXIT   IF
   COMMON   PAUSE
   DIMENSION   RETURN
   Another DO   STOP
   EQUIVALENCE   Explicit Specification Statement
   FORMAT   Subprogram Statement
   GO TO

3) Entrance into a DO loop must be through the DO statement.

4) If there is a DO within a DO, all statements in the inner DO must be within the outer DO loop. This is called a nest of DO's.

In a nest of DO's the innermost loop must be satisfied first. The following example of a nest of DO's might be helpful:

$$\sigma = \sum_{i=0}^{5} \sum_{j=0}^{N} \sum_{k=1}^{N} \left[kC + (B - i) + A^j \right],$$

where A, B, C are given constants. A FORTRAN program for this is shown in Fig. 5.19. In this example, the innermost loop is satisfied first and the outermost last. It is important to note that a nest of DO loops may have a common ending point,

```
        READ(5,2) A,B,C,N
      2 FORMAT(3F10.4,I2)
        SIGMA=0.
        NP1=N+1
        DO 10 I=1,6
        ZI=I-1
        BMI=B-ZI
        DO 10 JP1=1,NP1
        AJ=A**(JP1-1)
        DO 10 K=1,N
        ZK=K
     10 SIGMA = SIGMA + ZK*C+BMI+AJ
     20 WRITE(6,4) A,B,C,SIGMA,N
      4 FORMAT(4(F10.4,2X),I2)
        STOP
        END
```

Fig. 5.19. A nest of three DO loops.

but that the inner loop cannot end after the outer loops. The following example shows loops incorrectly ended:

Incorrect loops: ⌐Outer loop
                   ⌐Inner loop

5) The loop variable should not be redefined within the range of a DO loop. Here is an example of the error showing that the loop variable J is forced to change:

$$
\begin{array}{l}
\text{DO 22 J = 1,N}\\
\qquad \vdots\\
\text{J = J + 3}\\
\text{A = P(J) - P(J + 1)}\\
\qquad \vdots\\
\text{22 CONTINUE}
\end{array}
$$

e) The CONTINUE statement is a dummy statement made to ensure that the last statement in a DO loop is not a transfer of control.

f) The STOP statement is used to terminate the execution of the program. It is equivalent to a CALL EXIT statement on some computer systems.

g) END is the last card, physically, of all programs, both main and subprogram. It tells the compiler that this is the last card in the program. The object program will not be compiled if an END statement does not appear as the *last* statement in the program.

3. *Functions and Subprogram Statements.*

SUBPROGRAMS

a) CALL NAME (argument list) is a main program statement which generates a calling sequence to the named subprogram with the listed arguments as input. These arguments may be expressions; for example, CALL LULU (.5,X+Y,Z**5) creates a calling sequence to the subprogram LULU with the values of the three expressions as input arguments. Other examples include

$$
\begin{array}{l}
\text{CALL MAT (X, A, I, .000982)}\\
\text{CALL MULT (I, 16.3)}\\
\text{CALL ADDER (X, Y, C)}
\end{array}
$$

b) SUBROUTINE NAME $(a_1, a_2, \ldots, a_n)$, where NAME is the symbolic name of the subroutine and the arguments $a_1, \ldots, a_n$ are nonsubscripted variables. The name of the subroutine contains from one to six alphabetic and numerical characters, the first of which must be alphabetic. The name must not include special characters and it must not be used as the name of a variable.

The SUBROUTINE statement always comes first in a subprogram and defines it as a subroutine subprogram. A subroutine is brought into use by a CALL statement. At the time of execution of the subroutine, the variables in the definition are replaced by the constants, variables, and expressions in the CALL statement. (The arguments may be arrays.) Some examples are:

```
SUBROUTINE LULU (A, B, C)
SUBROUTINE MAT (A, D, G, K)
SUBROUTINE ANS77
```

FUNCTIONS

c) The FUNCTION statement, always first in the subprogram, defines it as a FUNCTION subprogram. The basic form is

$$\text{FUNCTION NAME } (a_1, a_2, \ldots, a_n)$$

where NAME is the symbolic name of the subprogram and the arguments are non-subscripted variables. The name consists of one to six alphabetic and numerical characters, the first being alphabetic. The name must appear either in the input statement list or at least once on the left-hand side of an arithmetic statement.

The function subprograms are used in the same manner as the internal functions described previously. Their names are used as though the internal functions were variables. The desired arguments are enclosed in parentheses. Some examples are:

```
FUNCTION ARCSN (RADS)
FUNCTION ROOT (B, A, C)
FUNCTION INTRT (RATE,YEARS)
FUNCTION H2Q (K,A)
```

The name of the function must start with I, J, K, L, M, or N if the value of the function is fixed point. This is predefined by the first letter of the function subprogram name. This implicit declaration may be annulled by using the following general form:

$$\text{Type FUNCTION NAME } (a_1, a_2, \ldots, a_n)$$

where *Type* represents the word INTEGER, REAL, or DOUBLE PRECISION. For example,

```
REAL FUNCTION INTRT (RATE,YEARS)
INTEGER FUNCTION ROOT (B, A, C)
```

d) The RETURN statement returns control from the subprogram to the main program. It stops further computation by the subprogram and returns the values calculated for use in the main program. It is the last executed statement in the SUBROUTINE and FUNCTION subprograms. The statement appears in the program as follows:

```
RETURN
```

4. *Input-Output Statements.*

a) FORMAT (specification) is a statement which tells the computer how information is to be read in, printed, or punched out. The data may be either numerical or alphabetical.

The Hollerith FORMAT is used for reading, printing, or punching out alphabetic information such as column headings, etc. An example is

FORMAT (5HA□DOG)

The H signifies that a Hollerith specification follows. Every character following the H (including blanks) is printed or punched out exactly as it is written, and the total number of spaces to be used is placed before the H.

FORMAT (30H□JOHN□JONES□□DECEMBER□31,□1978)
FORMAT (24H□THIS□IS□ALPHAMERIC□DATA//)

The symbol □ in the above examples indicates a blank.

To avoid possible counting errors, literal data are often used. In this method, the characters to be printed, or punched, are written between apostrophes (' ') within the parentheses of FORMAT statements. The above three examples now become:

FORMAT('A□DOG')
FORMAT('JOHN□JONES□□DECEMBER□31,□1978')
FORMAT('THIS□IS□ALPHAMERIC□DATA'//)

To write an apostrophe, two successive apostrophes must be used. For example, the following two statements

WRITE (5,14)
14 FORMAT('DON''T')

will cause DON'T to be printed.

For numerical quantities the mode (fixed, real or double-precision) and the number of spaces to be allotted to each number must be specified by the FORMAT statement, for example,

FORMAT (I5,F9.2,E14.7,D12.5)

This statement tells the computer how to read, print, or punch four numbers. The first would be in the fixed-point mode and would be allowed five spaces. The second is without a power of 10, for example, −27561.09, for which there are, at most, nine spaces allowed, and two places are to the right of the decimal point. The third has a power of 10, for example, −.9562937E−21. There are 14 spaces allowed for this number, and seven places are to the right of the decimal point. The fourth and the final number is a double-precision floating-point number with an exponent, for example, −.40826D−10. There are 12 spaces reserved for this number: four spaces for the exponent, 5 spaces for the digits representing the value, one for the decimal point, and one space for the sign.

A comma separates each FORMAT specification:

FORMAT (I2,10H□DEGREE□=□,F10.7)
FORMAT (F18.9,I10,E22.9)

However, a comma may not be needed after a Hollerith or X-specification. A variation of the numerical FORMAT is

FORMAT (5I3,2E9.2,3F7.2)

This means that there are five numbers read (printed) according to I3, two according to E9.2, and three according to F7.2. Another example is

FORMAT(7H□RESULT,3F9.6)

Another type of FORMAT specification is the X-notation. For example:

FORMAT(I5,22X,F12.7)

The 22X means that 22 spaces will be skipped between the number read (printed) as I5 and the number read (printed) as F12.7.

 The final type of FORMAT is the tab format specification, which may be followed in a FORMAT statement by any other type of format specification. The tab format specification is used to define the column numbers with which reading or writing begins. For example,

WRITE(2,99),I,J
99 FORMAT(I3,T32,I4)

would cause the second answer J, which is a four-digit integer, to be punched in columns 32, 33, 34 and 35. Here, the logical device number 2 stands for a card punch.

 b) The READ (i, n), LIST statement will instruct the computer to read, according to FORMAT statement n, the list of fixed-and/or floating-point variables. The unit designation i, an unsigned integer or integer variable, refers to a logical input/output device number, which may vary from one installation to another. The device numbers used in this book will be 5 for the card reader, 6 for the on-line printer, and 2 for punch. For example, the following READ statement causes one data card to be read.

READ(5,33), A, B, K, L(1)

 c) The WRITE (i, n) LIST statement causes the list variables to be printed (punched) according to FORMAT statement n.

5. *Specification Statements.*

 a) Comment. Any card with a C in column 1 is called a *comment card* and is ignored by the compiler. These cards may contain arbitrary comments giving the programmer's name, the name of the program, or any other useful information, for example,

C JOHN JONES

 b) DIMENSION A(20,40), B(15,2), C(109), D(4,4,5) is a sample of a dimension statement giving the amount of storage space to be reserved for the two-dimensional array A, the two-dimensional array B, the one-dimensional array C, and the three-dimensional array D. An array must be dimensioned before it is first used in the program. When a main program and subprograms are used jointly, corresponding arrays must have the same dimensions.

c) Primarily used to override the implied types of the names, the INTEGER statement, the REAL statement, and the DOUBLE PRECISION† statement are known collectively as the *explicit* specification statements. They are used to specify the type of name of the following: a variable, an array, or the expression in a function. In the case of an array, they are also used to specify its dimension; for example,

```
INTEGER A,ROOT
DOUBLE PRECISION B(20,15), J(4,5,5)
```

The first statement specifies that the names A and ROOT are integer names to override their original, implied types as two single-precision floating-point variables. The second example illustrates that both B and J are arrays; all elements are in double-precision floating-point. Thus, J(1,3,5) is not an integer.

d) COMMON and EQUIVALENCE statements will be discussed in Section 8.9.

e) The DATA statement‡ is used to define the initial values for variables. These values are compiled into the object program. The following two statements, for example,

```
INTEGER A(2), B(2,2)
DATA I,A,B,C/3,7,1,6,9,7,4,-3.14E2/
```

would replace the following sequence of 8 statements:

```
        I = 3
     A(1) = 7
     A(2) = 1
   B(1,1) = 6
   B(2,1) = 9
   B(1,2) = 7
   B(2.2) = 4
        C = -3.14E2
```

5.11   COMMON ERRORS IN FORTRAN

Despite the fact that the FORTRAN compiler language uses mathematical notation and the English language, there are many pitfalls which trouble both new and experienced programmers. In this section, some common errors will be presented along with examples of incorrect programming taken from actual runs. It must be emphasized that logical errors are not discussed here. These consist principally of incorrect original equations or an incorrect sequence of statements. Furthermore, errors in key punching are rather commonplace. In particular, the letter O is often mistaken for the number 0; or an I for a 1.

---

† DOUBLE PRECISION is also written as REAL*8 in some systems.
‡ NOT included in the ASA Basic FORTRAN.

In this section, we shall consider only the most frequent errors occurring in the following three general areas:

A. Errors associated with arithmetic and control statements
B. Errors associated with input and output statements
C. Errors associated with subprograms

A. Rules to Follow with Arithmetic and Control Statements

1. The number of left parentheses must equal the number of right parentheses.

Examples

EULER = ((SIN(X)*H/EULER) + EULER	Incorrect
EULER = (SIN(X)*H/EULER) + EULER	Correct

2. Once a statement number is used, it may not be used again.

Examples

30 IF(Y − Z) 36, 36, 34	
30 X(I + 101) = X(I + 100) + 0.2	Incorrect

30 IF(Y − Z) 36, 36, 34	
29 X(I + 101) = X(I + 100) + 0.2	Correct

3. Subscripts must be in fixed-point mode.

Examples

X(I + 100) = X(I + 99.) + .2	Incorrect
X(I + 100) = X(I + 99) + .2	Correct

4. If a statement number is mentioned, there must be a corresponding statement in the program.

Example (Incorrect)

$$28\ IF(Y - YSTOP)\ 36,\ 36,\ 34$$
$$\vdots$$

No Statement 36.

5. When a variable is raised to a *negative* power, using the ** notation, the exponent preceded by a minus sign must be enclosed in parentheses.

Examples

Y = A + B** −2	Incorrect
Y = A + B**(−2)	Correct

6. No two operation symbols should appear in sequence.

Examples

AREA = RH/−B	Incorrect
AREA = RH/(−B)	Correct

7. The indices of a DO loop must never be reset by one of the statements within the DO.

Example

```
          DO 12 I = 1, 5
          DO 11 I = 1, 15
             ⋮
    11  CONTINUE                              Incorrect
             ⋮
    12  CONTINUE
```

8. If a computed GO TO statement is used, the value of the index (K in the example below) must be no greater than the total number of statements referred to.

Examples

```
          K = 5                              Incorrect
          GO TO (15, 16), K
```

9. The * cannot be omitted in a multiplication.

Examples

```
          Y = 2.PI*X                         Incorrect
          Y = 2.*PI*X                        Correct
```

10. If a variable is to be used, it must be defined previously either in a READ statement or on the left-hand side of an equal sign.

Examples

```
      C   START                             Incorrect
          Z = 2.*PI
      C   START
          PI = 3.141519                      Correct
          Z = 2.*PI
```

B. Rules to Follow with Input and Output Statements

1. A comma must not be present before the list in a READ or WRITE statement.

Examples

```
          READ (5,48), X,Y,Z                 Incorrect
          READ (5,48)   X,Y,Z                Correct
```

2. Dimensions must be large enough for the intended computation.

Example

```
          DIMENSION A(3, 3)
          DO 15 I = 1, 5                     Incorrect
    15  READ (5,6) (A(I,J), J = 1, 3)
```

3. *Data* in a data card should begin in column 1, not 7.

4. The required data cards must agree *exactly* with the supplied data cards in total number and in sequence.

5. Integers must be punched in the rightmost position of the field.

Example

If we wish to read N = 3, we write

$$\text{READ (5,4) N}$$
$$\text{4   FORMAT (I4)}$$

If the number 3 is punched in column 4, N will be read as   3
If the number 3 is punched in column 3, N will be read as   30
If the number 3 is punched in column 2, N will be read as   300
If the number 3 is punched in column 1, N will be read as 3000

6. No comment cards are allowed among the data cards.

Examples

```
C       DATA
.0471   .591
.4198   .724                        Incorrect
C       NEXT TRIAL
.0125   .781
.6324   .992                        Incorrect
C       END OF DATA
```

7. Floating-point variables should not be read in with an I-format.
8. Hollerith characters in a format statement should be counted carefully.

Example (Correct)

```
FORMAT(6X,3HA□=,F9.5,5HFAB□=,F9.5)
```

(The symbol □ denotes a blank.)

9. When using the X-specification for blank fields, place the number denoting the number of blanks before the X. Also, 1X must be written (not just X) to obtain one blank.

Example

```
FORMAT(X6,F10.5, X, F3.2)           Incorrect
```

10. Column 1 of the output on *certain* off-line printers is used for carriage controls and not for general output.

Example

```
WRITE (6,6)
6  FORMAT (7HANSWER=,F10.5)          Incorrect
```

Note: The corrected Statement 6 should read:

```
6  FORMAT (8H□ANSWER=,F10.5)
```

or

```
6  FORMAT (5X,7HANSWER=,F10.5)
```

or

```
6  FORMAT (5X,'ANSWER=',F10.5)
```

C. Rules to Follow with Subprograms

1. The required subprogram must be provided by the programmer or stored in the library.
2. There must be an END card for each subprogram.
3. When separately compiled main program and subprograms are used jointly, they must have the same numerical dimensions for corresponding arrays.

Example (Incorrect)

```
C  MAIN PROGRAM              SUBROUTINE ROTATE(A,N)
   DIMENSION A(12, 12)       DIMENSION A(24, 24)
```

4. The arguments in CALL statements must agree exactly in number, order, and mode with those in the subroutines.

Example (Incorrect)

```
C  MAIN PROGRAM              SUBROUTINE ROTATE(A,N)
   :
   CALL ROTATE(A, .6,B)
```

5.12 COMMUNICATION WITH A COMPUTATION CENTER

Due to the recent rapidly expanding role of computation in education, research, and development, the number of computation centers is ever increasing.

As computer operating procedures have become increasingly sophisticated, the knowledge required to use a *computation center*, rather than *computers*, varies from one installation to another, depending on the set up in the individual center. In addition, this knowledge is in many cases either gained by personal trial-and-error procedure or passed on by word of mouth. The purpose of this section is to provide the reader with a detailed check list (Table 5.6) for the hardware, software, and operating procedures of his own computation center.

Table 5.6

I. Hardware

A. Digital Computer

1. Type of digital computer to be used
   decimal computer ☐
   binary computer ☐
   Model _____

2. Core memory size_____
   in words ☐
   in bytes ☐

3. Internal memory cycle
   Less than 2 microseconds (μsec) ☐
   2–10 μsec ☐
   Above 10 μsec ☐

4. Time to add two numbers
   Less than 25 μsec ☐
   25–100 μsec ☐
   Above 100 μsec ☐

B. Peripheral Equipment

1. Punched card equipment
   a) Key punch   Type _____
   b) Reproducer   Type _____
   c) Sorter   Type _____
   d) Verifier   Type _____

2. High-speed printer
   _____ lines per minute

3. Additional memory units
   a) Tape transport   yes ☐   no ☐
      Recommended tape density is _____ bits per inch.
      Seven-track ☐, or nine-track ☐.
   b) Magnetic drum   yes ☐,   no ☐
      additional words _____
                or bits _____
   c) Disk drives   yes ☐   no ☐   Model _____
      additional words _____, or _____ cylinders, or _____
      tracks will be available to a user.
   d) Data cell   yes ☐   no ☐   Model _____
      additional words _____, or bytes _____.

4. Readers
   Optical characters   yes ☐   no ☐
   Magnetic characters   yes ☐   no ☐

5. Plotter   yes ☐
          no ☐   Type _____

6. Cathode ray tube plotting device   yes ☐   no ☐   Model _____
   Output: paper ☐. 35 mm. film ☐

7. Oscilloscope display unit   yes ☐
                            no ☐   Type _____

8. Time-sharing terminal   yes ☐
                       no ☐   Type _____. No. of terminals
                                                       available to user _____

9. Converters
   a) Card to magnetic tape   yes ☐   no ☐
   b) Card to paper tape   yes ☐   no ☐
   c) Magnetic tape to card   yes ☐   no ☐
   d) Paper tape to card   yes ☐   no ☐

(cont.)

Table 5.6

II. Software

A. Compilers available

FORTRAN IV
USAS Basic ☐   IBM Basic FORTRAN IV ☐   FORTRAN IV G ☐
FORTRAN IV H ☐
COBOL ☐   PL/I ☐   ALGOL ☐
WATFOR ☐   WATFIV ☐   Assembler Language ☐
BASIC ☐   MAD ☐   others ☐

B. The logical device number used

	In this book	In our installation
Card reader input	5	_____
On-line printer output	6	_____
Punch output	2	_____

C. The following FORTRAN statements are not permitted

```
PRINT _____
STOP _____
PAUSE _____
CALL EXIT _____
Mixed Mode such as A = I * Y _____
Explicit specification such as INTEGER A, ROOT _____
DOUBLE PRECISION B(20, 15) _____
REAL*8 B(20,15) _____
DATA IREAD,IWRITE/5,6/
```

D. Program Libraries available

	yes	no
1. FORTRAN built-in functions, such as SIN, COS, SQRT, etc.	☐	☐
2. Routines used to generate plotter tapes	☐	☐
3. Scientific Subroutine Package (SSP)	☐	☐
4. General Purpose System Simulator (GPSS)	☐	☐
5. BioMeDical Statistical Routines (BMD)	☐	☐
6. PDUMP subroutine	☐	☐

E. Operating system used

DOS ☐
OS ☐
    Primary Control Program (PCP) ☐
    Multiprogramming with a Fixed Number of Tasks (MFT) ☐
    Multiprogramming with a Variable Number of Tasks (MVT) ☐
    Shared Main Storage Multiprocessing (M65MP) ☐

Others _____
    Release Version _____

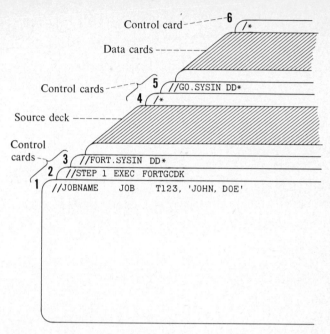

Fig. 5.20. A typical run deck. The source program is in FORTRAN language. Compile, link-edit, and execute. The object deck is required.

III. Routine Operation of Computation Center

A. *Type of Operation*

   Open shop ☐   Closed shop ☐   Partially closed and partially open ☐

B. *Consulting supervisor at the center* _____

C. *My identification number, if required, for using the computer* _____

D. *My problem number, if required* _____.

E. *Maximum allowable running time for each program* _____ *min.*

F. *Normal turn-around time* _____ *hours*

G. *User manual or new user's packet available*   *yes* ☐   *no* ☐

IV. Job Control Cards

In order to run a FORTRAN program on a computer, say, IBM System/360, a user must submit the following three items:

1. Job control cards
2. Source deck(s) or object deck(s)
3. Data, if any.

In Fig. 5.20 is shown a sample set-up at a particular 360 installation to obtain final answers for a FORTRAN program. The detailed column description for each job control

(cont.)

Table 5.6 *(concl.)*

card is shown in Table 5.7. So that it may be easier for a beginner to follow, the sample column description has been deliberately given far less flexibility.

Obtain a copy of the standard job "set-up" designed for your installation from your instructor or the programming consultant.

Table 5.7 Column description on job control cards (as shown in Fig. 5.20) to compile, link-edit, and execute a FORTRAN source deck, and to punch object deck

Control Cards	Columns		Description
	For 4-character problem number	For 7-character problem number	
1	1–2	1–2	//
	3–8	3–8	User must supply Job Name (no blanks allowed)
	9	9	blank
	10–12	10–12	JOB
	13–16	13–16	blank
	17–20	17–23	User must supply his problem number
	21	24	,(comma)
	22	25	Single quote (') or multi 8,5 punch
	23–42	26–42	User must supply Programmer's Name
	43	43	Single quote (') or multi 8,5 punch
2	1–7		//STEPI
	8–9		blank
	10–13		EXEC
	14–15		blank
	16–23		FORTGCDK
3	1–12		//FORT.SYSIN
	13–15		blank
	16–19		DD *
4	1–2		/*
5	1–10		//GO.SYSIN
	11–15		blank
	16–19		DD *
6	1–2		/*

BIBLIOGRAPHY

General

Carr, J. W., III, "Digital computer programming," *Handbook of Automation, Computation and Control*, E. M. Grabb, S. Ramo, and D. E. Wooldridge, eds., **2**, Ch. 2. Wiley, New York, 1959.

Gill, S., "Current theory and practice of automatic programming," *Comput. J.*, **2**, 110–114 (1959).

Lord Bowden, "The language of computers," *Am. Scient.*, **58**, no. 1, 43–53 (Jan.–Feb. 1970).

Rosen, S., *Programming Systems and Languages.* McGraw-Hill, New York, 1967.

Sammet, J. E., *Programming Languages: History and Fundamentals.* Prentice-Hall, Englewood Cliffs, N.J., 1969.

Wegner, P., *Programming Languages, Information Structures and Machine Organization.* McGraw-Hill, New York, 1968.

FORTRAN Programming

American Basic FORTRAN (X3.10–1966). United States of America Standards Association, New York, 1966.

American Standard FORTRAN (X3.9–1966). United States of America Standards Association, New York, 1966.

Golden, J. T., *FORTRAN IV Programming and Coding.* Prentice-Hall, Englewood Cliffs, N.J., 1965.

Haag, J. N., *Comprehensive FORTRAN Programming.* Hayden, New York, 1965.

Kennedy, M., and M. B. Solomon, *Ten Statement FORTRAN plus FORTRAN IV.* Prentice-Hall, Englewood Cliffs, N.J., 1970.

Lee, R. M., *A Short Course in FORTRAN IV Programming: Based on the IBM System/360.* McGraw-Hill, New York, 1967.

McCracken, D. D., *A Guide to FORTRAN IV Programming.* Wiley, New York, 1965.

Organick, E. I., *A FORTRAN IV Primer.* Addison-Wesley, Reading, Mass., 1966.

Stein, M. L., and W. D. Munro, *A FORTRAN Introduction to Programming and Computers.* Academic Press, New York, 1966.

COBOL Programming

Cobol—Proposed USA Standard. United States of America Standards Institute, New York, 1967.

McCracken, D. D., and U. Garbassi, *A Guide to COBOL Programming*, 2nd ed. Wiley, New York, 1970.

Spitzbarth, L. M., *Basic COBOL Programming.* Addison-Wesley, Reading, Mass., 1970.

PL/I Programming

PL/I: Language Specifications. Form C28-6571, IBM Corp. New York, 1965.

Pollack, S. V., and T. D. Sterling, *A Guide to PL/I.* Holt, Rinehart, and Winston, New York, 1969.

Scott, R. C., and N. E. Sondak, *PL/I for Programmers.* Addison-Wesley, Reading, Mass., 1970.

Weinberg, G. M., *PL/I Programming Primer.* McGraw-Hill, New York, 1966.

ALGOL Programming

Andersen, C., *Introduction to ALGOL 60.* Addison-Wesley, Reading, Mass., 1964.

Backus, J. W., *et al.,* "Report on the algorithmic language ALGOL 60," *Num. Math.,* **2,** 106–136 (1960), or *Commun. Assoc. Comput. Mach.,* **3,** 299–314 (1960).

Bauer, H., S. Becker, and S. Graham, *ALGOL W Implementation*. CS 98, Computer Science Department, Stanford University, Palo Alto, 1968.

Baumann, R., and M. Feliciano, *Introduction to ALGOL*. Prentice-Hall, Englewood Cliffs, N.J., 1964.

Dijkstra, E. W., *A Primer of ALGOL 60 Programming*. Academic Press, New York, 1962.

McCracken, D. D., *A Guide to ALGOL Programming*. Wiley, New York, 1962.

"Revised report on the algorithmic language ALGOL 60," *Commun. Assoc. Comput. Mach.*, **6,** 1–17 (1963).

Other High-Level Languages

Iverson, K. E., *A Programming Language*. Wiley, New York, 1962.

Kemeny, J. G., and T. E. Kurtz, *BASIC Programming*. 2nd. ed. Wiley, New York, 1971.

Assembler Language Programming

Chapin, N., *360 Programming in Assembly Language*. McGraw-Hill, New York, 1968.

Kuo, S. S., *Assembler Language for FORTRAN, COBOL and PL/I Programmers*. Addison-Wesley, Reading, Mass. (in press).

Leeds, H. D., and G. M. Weinberg, *Computer Programming Fundamentals*. McGraw-Hill, New York, 1961.

McCracken, D. D., *Digital Computer Programming*. Wiley, New York, 1957.

Struble, G., *Assembler Language Programming: The IBM System/360*. Addison-Wesley, Reading, Mass., 1969.

Compiler Construction

Gries, D., *Compiler Construction*. Wiley, New York, 1971.

Hopgood, F. R. A., *Compiler Techniques*. American Elsevier, Inc., New York, 1969.

Lee, J. A. N., *The Anatomy of a Compiler*. Reinhold, New York, 1967.

McKeeman, W. M., J. J. Horning, and D. B. Wortman, *A Compiler Generator Implemented for the IBM System/360*. Prentice-Hall, Englewood Cliffs, N.J., 1970.

PROBLEMS

1. Translate the following mathematical expressions into FORTRAN expressions

(a) $X + (Y - C)$.

(b) $\dfrac{\pi}{e - 7.24}$.

(c) x^{b-2}.

(d) $\sqrt{b^2 - 4ac}$.

(e) $\dfrac{\pi}{6} h(3r^2 + h^2)$.

2. Debug each of the following unrelated FORTRAN statements (write the corrected statement under each given statement).

```
Col 6
  ↓
     A = B + (ZM - 3)C
     DO 17, JACK = 0, 100, 5
  A  SACKSON(5) = 6./(2 + ZM)
  2  IF (3.14159 - X) 3, 0.75, 6
42345  PI = 3.1416F3.5
     CONTINUE
     PI = (3.14 * (2/ZA) = 2.
     2 = 8. * ZN
     ROOT + SQRTF(1. + AN * AN * PSI * PSI).
     GO TO END
```

3. Convert Fig. 3.1 into a complete FORTRAN program.

4. The rectangular coordinates x, y of a set of points are:

x	1	2	3	4...	25
y	1	4	9	16...	625

Write a FORTRAN program to calculate the corresponding polar coordinates for each point. The following equations are given:

$$\rho = \sqrt{x^2 + y^2}, \qquad \theta = \tan^{-1}(y/x).$$

5. Write a program to calculate

$$y = 1/\sqrt{1 + x^2}$$

for $x = 0(0.1)10$. (*Note:* This means that x starts at zero and ends at 10 with interval 0.1.)

6. An array A of N numbers is given. Sort the numbers in descending order so that the largest number is in A(1), the next smaller number in A(2), ..., the smallest number in A(N). Write a FORTRAN program for this task.

7. Write a FORTRAN program to compute 100 elements in a single-dimensioned fixed-point array I in the following manner:
a) Place 1 in each element of the array,
b) Add 2 to the elements 2, 4, 6, ..., 100,
c) Add 3 to the elements 3, 6, 9, ..., 99,
   ⋮
k) Add k to the elements k, $2k$, $3k$, ...
Print the 100 answers in 10 rows of 10 numbers each.

8. Matrix A:

$$a_{ij} = m \sin(i^2 j^2 \pi/p) + n \cos(q i^3 j^3), \qquad i \le j,$$

$$a_{ij} = \sqrt{m} \cos(i^2 j^2 \pi/p), \qquad i > j,$$

where

$$i = 1, 2, 3, \qquad m = 8.2913, \qquad p = 6.21,$$

$$j = 1, 2, 3, \qquad n = 0.528, \qquad q = 3.142.$$

Symmetric matrix B:

$$b_{11} = 1.342, \qquad b_{22} = 7.293,$$
$$b_{12} = 6.87, \qquad b_{23} = 1,$$
$$b_{13} = 0, \qquad b_{33} = 3,$$
$$c_{ij} = a_{ij} + b_{ij}, \qquad i = 1, 2, 3; \quad j = 1, 2, 3.$$

Write a program in FORTRAN which, if necessary, will handle a 10×10 matrix and will do the following:

a) Compute and print all elements of matrix A row by row (see Table 5.3).
b) Print the *complete* matrix B row by row.
c) Compute and print matrix C row by row.
d) Print the lower triangular matrix B.

Matrix B information: The input format will be 7F10.4. The first data card will contain three numbers, namely b_{11}, b_{12}, and b_{13}. The second data card will contain two numbers, b_{22} and b_{23}, and the last card will contain only one bumber, b_{33}.

9. Given an $N \times N$ square matrix A. Find those off-diagonal elements which are symmetric, or

$$|a_{ij} - a_{ji}| \le \varepsilon,$$

where ε is an assigned small number. Print them out in the following format:

```
I = xx     J = xx      A(I,J) = xx.xxx
  ⋮
```

10. Write a subroutine to compute N-factorial by using the following calling statement:

```
CALL FAC(N, IANS)
```

where N is the number and IANS is the result. Assume that $N > 0$ and $N_{max} = 25$.

11. Hats in a check room are hopelessly scrambled. The check girl hands back the total k hats at random. The probability p that no man gets his own hat is

$$p = \frac{1}{2!} - \frac{1}{3!} + \frac{1}{4!} - \cdots \pm \frac{1}{k!}$$

where the $-$ sign is chosen if k is odd and $+$ sign if k is even. Write a complete FORTRAN program to compute p for $k = 2, 3, \ldots, 8$. Assume that a subroutine

```
FAC(N, IANS)
```

is already available to compute N-factorial where IANS stores the result of N! (see Problem 10).

12. Write a subroutine FLIP to perform a "flipping" of the following array of numbers and to print the results row by row.

2	7	6
9	5	1
4	3	8

This operation can be readily carried out by interchanging the first and third columns.

13. Write a main program to flip the array of Problem 12 consecutively five times.

14. Write a FORTRAN program to print, VOTE FOR ME, twice on one page. One message is on the top of a page, and the other is in the middle of the same page. Both messages should be centered.

Three consecutive pages should be printed with two messages on each page.

15. It is known that the complete program shown in Fig. 5.21 has logical errors. Debug the program.

```
C        PROGRAM BUGS (INPUT,OUTPUT)
         DIMENSION N(20)
    1    DO 10 K=1,20
   10    N(K)=10-K
   11    DO 20 I=1,10
         IP1=I+1
         DO 20 J=IP1,10
         IF(N(I).LE.N(J)) GO TO 20
         ITEMP=N(I)
         N(I)=N(J)
         N(I)=ITEMP
   20    CONTINUE
   30    WRITE(6,40) (N(I),I=1,10)
   40    FORMAT(1X,10I2)
         STOP
         END
```

Fig. 5.21. Bugs in a program.

COMPUTER-ORIENTED NUMERICAL METHODS

Part II deals principally with selected mathematical methods which are either extensively used on, or particularly suitable for, digital computers. Many problems in science and engineering frequently reduce to one or two standard mathematical problems. Subroutines for these standard problems are usually available in program libraries or programming assistance offices commonly affiliated with computer centers. However, it is dangerous, and therefore not recommended, to use a subroutine without understanding the related numerical method. It would obviously be impossible to encompass the whole field of numerical methods in a book of moderate size. Only selected methods of known popularity are set forth in detail.

The bibliographic references to many other "best" methods are listed at the end of each chapter. These references may be of interest to students who wish to study some special methods in more detail.

CHAPTER 6

Computer Solution of Polynomial and Transcendental Equations

6.1 INTRODUCTION

In scientific and engineering analysis it is often necessary to solve transcendental or higher-degree polynomial equations. Two typical examples are:

$$e^{-x} - \sin(\pi x/2) = 0 \tag{6.1}$$

and

$$x^4 - x^3 - 10x^2 - x + 1 = 0. \tag{6.2}$$

Equation (6.1) represents a transcendental equation since sine and exponential functions are involved. Equation (6.2) is a fourth-order polynomial equation.

The problem may be stated in two ways: first, to obtain a more accurate value for a real root whose approximate location is already known by other means, including graphical determination; and second, to find *all* roots, both real and complex, of the given equation without any prior information about their approximate locations. The second type of problem is obviously more difficult.

This chapter is concerned with a discussion of the possible methods of numerical solution of transcendental or higher-degree polynomial equations.

6.2 HALF-INTERVAL SEARCH

As an example, let us solve the transcendental equation

$$e^{-x} - \sin(\pi x/2) = 0. \tag{6.1}$$

This equation has a single solution between 0 and 1, as may be seen from the intersection of the two curves shown in Fig. 6.1. The basic problem is to find the root according to a specified tolerance. We shall employ the notation $f(x)$ to represent the

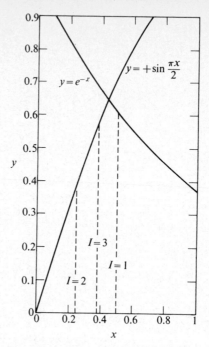

Fig. 6.1. Half-interval search for the root of $e^{-x} - \sin(\pi x/2) = 0$.

function appearing on the left-hand side of the equation. Thus in Eq. (6.1), $f(x)$ is equal to $e^{-x} - \sin(\pi x/2)$.

Using this method, one must compute the left-hand side of Eq. (6.1), or $f(x)$, at $x = 0.5$ and determine whether the solution is to the right or to the left of $x = \frac{1}{2}$. The location of the root is then known to be in an interval of length $\frac{1}{2}$ instead of 1. By computing the given expression $f(x)$ at the midpoint of this interval, the solution can further be localized to an interval of length $\frac{1}{4}$. This procedure can then be repeated to

Table 6.1 Example of half-interval search

(1)	(2)	(3)	(4)	(5)	
Step number i	x_i	e^{-x}	$\sin(\pi x/2)$	(3) − (4)	
				(+)	(−)
1	0.50	0.6065	0.7071		✓
2	0.25	0.7788	0.3827	✓	
3	0.375	0.6873	0.5556	✓	
4	0.4375	0.6456	0.6344	✓	
5	0.46875	0.6258	0.6716		✓
⋮					

locate the root to an accuracy of any reasonable number of significant digits. What we are trying to determine in each step is whether the required root lies in the right or left half of the previous interval. Some representative computations are shown in Table 6.1.

Figure 6.2 shows a possible flow chart for this half-interval search procedure, also known as binary search. Data A and B are the initial left- and right-hand limits, respectively, in which the root lies. Data ϵ represents the tolerance of $f(x)$. In other words, the search is continued until the condition $|f(x_i)| \leq \epsilon$ is satisfied, where ϵ is some arbitrarily chosen small number.

As a practical programming detail, it is worth while to note that the product of $f(A)$ and $f(x)$ is used to compare the sign of two quantities. This is shown in block 3 in

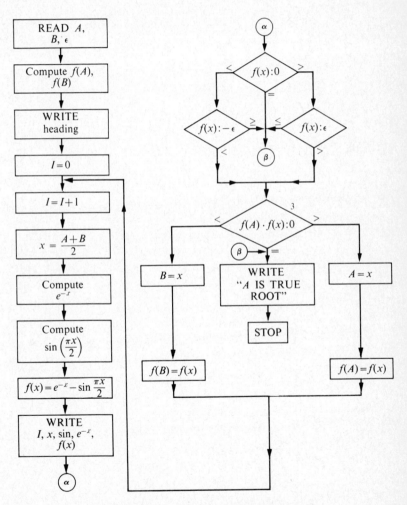

Fig. 6.2. Flow chart for half-interval search.

Fig. 6.2. Here $f(A)$ stands for the value of $f(x)$ when $x = A$. Similarly $f(B)$ represents the value of $f(x)$ for $x = B$. When the product has a positive value, $f(A)$ and $f(x)$ must have the same sign. Conversely, if the product is negative, $f(A)$ and $f(x)$ are of different signs. If the product happens to be equal to zero, $f(A)$ must be equal to zero. Therefore, A is the true root.

```
C
C       HALF INTERVAL SEARCH
C       A AND B ARE INITIAL LEFT- AND RIGHT-HAND LIMITS, RESPECTIVELY
C       EPS REPRESENTS THE TOLERANCE OF F(X), IT IS A SMALL NUMBER
C
        DATA IREAD,IWRITE/5,6/
        READ(IREAD,22) A,B,EPS
     22 FORMAT(3F10.4)
        PI=3.14159265
C
C       EVALUATE F(A) AND F(B)
C
        FA=EXP(-A)-SIN(PI*A/2.)
        FB=EXP(-B)-SIN(PI*B/2.)
        WRITE(IWRITE,81)
     81 FORMAT('1',5X,'I',9X,'X(I)',11X,'SIN',12X,'EXP',17X,'F(X)'//)
        I=0
      7 I=I+1
C
C       COMPUTE THE VALUE OF F(X) AT THE MIDPOINT OF THE INTERVAL
C
        X=(A+B)/2.
        EX=EXP(-X)
        SI=SIN(PI*X/2.)
        F=EX-SI
        WRITE(IWRITE,99) I,X,SI,EX,F
     99 FORMAT(I6,3F15.8,F20.8//)
C
C       TEST TO SEE WHETHER THE VALUE OF F(X) IS SATISFIED THE SPECIFIED
C       TOLERANCE. IF IT IS, STOP THE COMPUTATION
C
        IF(ABS(F).LE.EPS) GO TO 10
C
C       IF THE PRODUCT OF F TIMES FA IS NEGATIVE, THE TRUE ROOT WOULD LIE
C       IN THE INTERVAL BETWEEN A AND THE MIDPOINT. IF NOT, IT IS IN THE
C       OTHER INTERVAL. IF IT IS ZERO, A IS THE TRUE ROOT.
C
        IF(F*FA) 5,8,6
      5 B=X
        FB=F
        GO TO 7
      6 A=X
        FA=F
        GO TO 7
      8 WRITE(IWRITE,9)
      9 FORMAT(' A IS TRUE ROOT')
     10 STOP
        END


        .0          1.0          .0001
```

Fig. 6.3. FORTRAN program for a half-interval search.

In Fig. 6.2, the graphical devices $\textcircled{\alpha}$ and $\textcircled{\beta}$ should be noted. They are the connector symbols, often used when a flowchart is too long to finish in the same column, or when actual connection with lines between two boxes will make a flowchart difficult to read. Thus, between two connector symbols $\textcircled{\alpha}$, there should be a line connected. However, it is not actually drawn for the sake of clarity.

It is required that the results be printed out in this format:

I	X(I)	SIN	EXP	F(X)
(16)	(F15.8)	(F15.8)	(F15.8)	(F20.8)

Here "I" represents the number of iterations and the figures in parentheses represent the format to be used in the FORTRAN program. Figure 6.3 shows the corresponding FORTRAN program, together with the required data. From the answer sheet (Table 6.2), we see that the value of the required root is approximately equal to 0.4436. In this computation we used $\epsilon = 0.0001$.

Table 6.2 Answers for half-interval search

I	X(I)	SIN	EXP	F(X)
1	.50000000	.70710681	.60653069	−.10057612
2	.25000000	.38268344	.77880079	.39611735
3	.37500000	.55557026	.68728929	.13171903
4	.43750000	.63439329	.64564857	.01125528
5	.46875000	.67155898	.62578403	−.04577495
6	.45312500	.65317286	.63563873	−.01753413
7	.44531250	.64383156	.64062406	−.00320750
8	.44140625	.63912448	.64313140	.00400692
9	.44335938	.64148104	.64187650	.00039546
10	.44433594	.64265703	.64125000	−.00140703
11	.44384766	.64206924	.64156320	−.00050604
12	.44360352	.64177518	.64171981	−.00005537

6.3 METHOD OF FALSE POSITION (*REGULA FALSI*)

The half-interval search procedure discussed in the last section is one of several computer methods available for obtaining an approximate solution to an equation $f(x) = 0$. All the methods are iterative, since an initial approximation, $x = x_0$, to the root is usually made and a sequence of approximations, $x_1, x_2, \ldots, x_n$, is generated

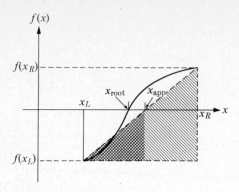

Fig. 6.4. Method of false position.

to get closer to the true root. A second iterative procedure is the method of *false position*. With this method we seek to find an interval $x_L < x < x_R$ in which $f(x)$ changes its sign (Fig. 6.4). The linear interpolation is then used to find the next iterated value x_{app} of the true root x_{root}, where x_{app}, the approximate value, lies between x_L and x_R. First, by the similar triangles shaded in Fig. 6.4, we have

$$\frac{f(x_L)}{x_{app} - x_L} = \frac{f(x_L) - f(x_R)}{x_R - x_L}$$

or

$$x_{app} = x_L + \frac{f(x_L)(x_R - x_L)}{f(x_L) - f(x_R)}. \tag{6.3}$$

We now seek to determine which of the two subintervals, $x_L < x < x_{app}$ or $x_{app} < x < x_R$, contains the true root x_{root}. This can be done by checking the sign of $f(x_{app})$. If $f(x_R) \cdot f(x_{app}) < 0$, then the right-hand interval contains the root, and x_{app} becomes the new x_L for the next iteration. On the other hand, if $f(x_L) \cdot f(x_{app}) < 0$, then $x_L < x_{root} < x_{app}$, and x_{app} becomes the new x_R for the next iteration.

$$|f(x_{app})| < \epsilon,$$

where ϵ is a preassigned, small positive member, say 10^{-5}. The iteration may also be terminated by narrowing down the interval, that is,

$$(x_R - x_L) < \epsilon',$$

where ϵ' is some arbitrarily chosen small number.

Example

Solve $e^{-x/c} - \sin(\pi x/2) = 0$, using the method of false position. Since there is a root between 0 and 1, we take

$$x_L = 0,$$
$$x_R = 1,$$

and

$$\epsilon = 0.0001.$$

A sample flow chart is presented in Fig. 6.5, and the related FORTRAN program and its results are shown in Fig. 6.6 and Table 6.3, respectively.

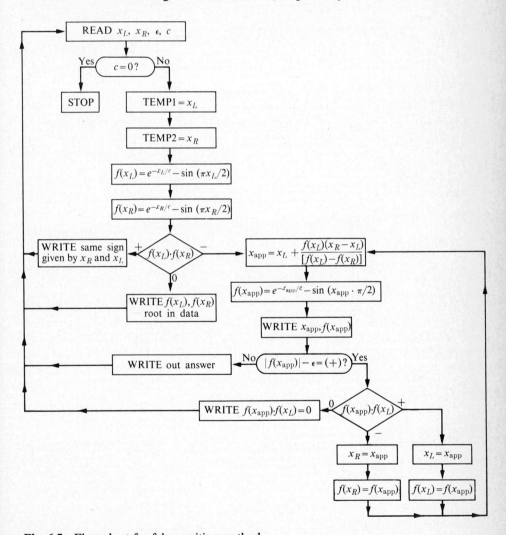

Fig. 6.5. Flow chart for false-position method.

```
C
C         METHOD OF FALSE POSITION
          DATA IREAD,IWRITE/5,6/
          PI=3.14159265
C
C         READ IN THE INITIAL INTERVAL,TOLERANCE, AND C
C
       25 READ(IREAD,10) XL,XR,EPS,C
       10 FORMAT(4F10.4)
          IF(C.EQ.0.) GO TO 21
       20 TEMP1=XL
          TEMP2=XR
C
C         COMPUTE THE VALUES OF F(X) AT X = XL, AND X = XR
C
          FXL=EXP(-XL/C)-SIN(PI*XL/2.)
          FXR=EXP(-XR/C)-SIN(PI*XR/2.)
C
C         IF THE PRODUCT OF FXL AND FXR IS NEGATIVE, THE TRUE ROOT WOULD LIE
C         IN THE INTERVAL BETWEEN XL AND XR.  IF IT IS EQUAL TO ZERO, PRINT
C         THE DATA.
C
          IF(FXL*FXR) 22,23,24
       24 WRITE(IWRITE,11)
       11 FORMAT('1',5X,'SAME SIGN GIVEN BY XR AND XL')
          GO TO 25
       23 WRITE(IWRITE,12) FXL,FXR
       12 FORMAT(2E14.8,' ROOT IN DATA')
          GO TO 25
       22 WRITE(IWRITE,13)
       13 FORMAT(7X,' XAPP',9X,' FXAPP')
C
C         FIGURE OUT THE APPROXIMATE ROOT AND F(XAPP)
C
       31 XAPP=XL+FXL*(XR-XL)/(FXL-FXR)
          FXAPP=EXP(-XAPP/C)-SIN(XAPP*PI/2.)
          WRITE(IWRITE,14) XAPP,FXAPP
       14 FORMAT(2F15.8)
C
C         IF THE ABSOLUTE VALUE OF FXAPP IS LESS THAN THE SPECIFIED
C          TOLERANCE, STOP THE COMPUTATION
C
          IF(ABS(FXAPP).GT.EPS) GO TO 27
       26 WRITE(IWRITE,18)
       18 FORMAT(//)
          WRITE(IWRITE,15) C,TEMP2,TEMP1
       15 FORMAT(5X,'C=',F5.2,2X,'XR=',F6.2,2X,'XL=',F6.2/)
          WRITE(IWRITE,16) XAPP,EPS
       16 FORMAT(' ROOT IS',F8.4,2X,'BASED ON EPSILON = ',F7.4)
          GO TO 25
C
C         FOLLOWING 10 STATEMENTS TO SEARCH FOR A NEW INTERVAL
C
       27 IF(FXAPP*FXL) 28,29,30
       28 XR=XAPP
          FXR=FXAPP
          GO TO 31
       29 WRITE(IWRITE,17)
       17 FORMAT(' FXAPP*FXL = 0')
          GO TO 25
       30 XL=XAPP
          FXL=FXAPP
          GO TO 31
       21 STOP
          END


          .0          1.0          0.0001      1.0
          .0          0.0          0.0         0.0
```

Fig. 6.6. FORTRAN program.

Note that $c = 1$ is used in Fig. 6.6. (Question: how should the data be prepared so that $c = 1, 2, 3, \ldots, 9$, respectively, for nine different cases?)

Table 6.3 Answers by method of false position

XAPP	FXAPP
.61269986	-.27869499
.47916032	-.06428183
.45021941	-.01222181
.44478336	-.00223222
.44379273	-.00040465
.44361324	-.00007329

```
       C= 1.00   XR=   1.00   XL=   0.00
  ROOT IS      .4436   BASED ON EPSILON =      .0001
```

6.4 NEWTON-RAPHSON METHOD

The solution to an equation $f(x) = 0$ may often be found by a simple procedure known as the Newton-Raphson method. This method consists of drawing the tangent to the curve at the point A, as shown in Fig. 6.7. The x-intercept of the tangent, or x_2, is then used as the first approximation. Figure 6.7 shows the use of the Newton-Raphson method to solve $f(x) = 0$ to the root x_T. From the figure, we have

$$f'(x_1) = \tan \theta = \frac{f(x_1)}{x_1 - x_2}, \tag{6.4}$$

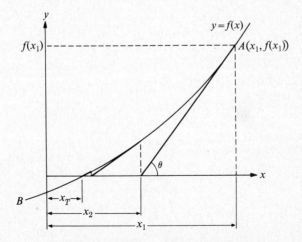

Fig. 6.7. Newton-Raphson method.

where $f'(x_1)$ denotes the derivative of $f(x)$ evaluated at $x = x_1$. Hence

$$x_2 = x_1 - \frac{f(x_1)}{f'(x_1)}. \tag{6.5}$$

An iterative sequence can now be set up as follows:

$$x_{n+1} = x_n - \frac{f(x_n)}{f'(x_n)}. \tag{6.6}$$

This formula can be repeatedly used to find improved approximations to the real root x_T. It should be noted that this process does not work well when the slope of the curve becomes very small; it will not work at all if the slope of the curve becomes zero along the arc AB. It is further assumed in this method that there is no inflection point along the arc AB.

The recurrence relation, as expressed in Eq. (6.6), may also be derived from the Taylor-series expansion for $f(x_T)$ about x:

$$f(x_T) = f(x) + (x_T - x)f'(x) + \cdots + \frac{f^{(k)}(x)}{k!}(x_T - x)^k + R. \tag{6.7}$$

Taking only the first two terms, we obtain

$$f(x_T) \doteq f(x) + (x_T - x)f'(x),$$
$$0 \doteq f(x) + (x_T - x)f'(x),$$

and from this, Eq. (6.6) follows.

Example

The iterative equation (6.6) can be readily used to approximate the square root of C by solving $x^2 - C = 0$. Here we take

$$f(x) = x^2 - C,$$
$$f'(x) = 2x.$$

The iterative sequence then becomes

$$x_{n+1} = x_n - \frac{x_n^2 - C}{2x_n},$$

or

$$x_{n+1} = \frac{1}{2}\left(x_n + \frac{C}{x_n}\right). \tag{6.8}$$

Let us carry this process through to compute $\sqrt{24}$. First we take

$$x_1 = 1 \text{ (arbitrarily)}.$$

Then

$$x_2 = \tfrac{1}{2}(1 + 24) = 12.5,$$
$$x_3 = \tfrac{1}{2}(12.5 + 24/12.5) = 7.21,$$
$$x_4 = \tfrac{1}{2}(7.21 + 24/7.21) = 5.2693,$$
$$x_5 = \tfrac{1}{2}(5.2693 + 24/5.2693) = 4.9119,$$
$$x_6 = \tfrac{1}{2}(4.9119 + 24/4.9119) = 4.8989,$$
$$\vdots$$

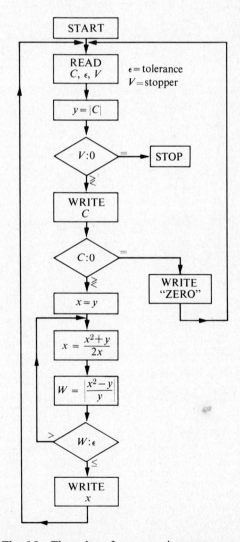

Fig. 6.8. Flow chart for computing square root.

Figure 6.8 shows the flow chart for this method of solving $x^2 - C = 0$. A FORTRAN program is presented in Fig. 6.9. It should be noted that the iteration continues until the following relation is established:

$$\left| \frac{x^2 - C}{C} \right| \leq \epsilon,$$

```
C       NEWTON-RAPHSON METHOD FOR FINDING SQUARE ROOT
        DATA IREAD,IWRITE/5,6/
   31 READ(IREAD,2) C,E,V
    2 FORMAT(3F10.5)
      Y=ABS(C)
      IF(V.EQ.0.) GO TO 40
   10 WRITE(IWRITE,7) C
    7 FORMAT(10X,'C= ',F12.5)
      IF(C.NE.0.) GO TO 11
   12 WRITE(IWRITE,17)
   17 FORMAT(' X=ZERO'//)
      GO TO 31
   11 X=Y
   14 X=(X*X+Y)/(2.*X)
      W=ABS((X*X-Y)/Y)
      IF(W.GT.E) GO TO 14
   30 WRITE(IWRITE,50) X
   50 FORMAT(10X,'X= ',F12.5//)
      GO TO 31
   40 STOP
      END


   24.0           0.00001        1.
    0.             0.             0.
```

Fig. 6.9. FORTRAN program for Newton-Raphson method.

where ϵ is a preassigned, small positive number. In this example, we take

$$\epsilon = 10^{-5}.$$

The Newton-Raphson formula can readily be extended to obtain the rth root of the number C. Making use of Eq. (6.6), we have

$$x_{n+1} = x_n - \frac{x_n^r - C}{r x_n^{r-1}},$$

or

$$x_{n+1} = \frac{1}{r}\left[(r-1)x_n + \frac{C}{x_n^{r-1}} \right]. \tag{6.9}$$

It can be seen that when $r = 2$, Eq. (6.9) becomes Eq. (6.8). Equation (6.9) is the basic iterative formula used to obtain the rth root of a given number C. It should be noted that only real roots will be obtained from Eq. (6.9).

As a second example, let us consider a circle of radius R ($R = 3.00$) and six different values of A, the shaded area as shown in Fig. 6.10.

$$
\begin{aligned}
A &= 28.274333 \\
&= 30.000000 \\
&= 14.137166 \\
&= 7.100000 \\
&= 4.999999 \\
&= 0.000000.
\end{aligned}
$$

Using the Newton-Raphson method, write a FORTRAN program to find θ in degrees for the corresponding value of A. As shown in the flow chart (Fig. 6.11), the program is terminated when A equals zero. Figure 6.12 shows the FORTRAN program for this example.

The computer carries out all mathematical operations with a finite number of significant digits. All excess digits are generally truncated (chopped off). Therefore, the equation

$$
\theta_n - \theta_{n+1} = 0
$$

may never be satisfied. We will introduce a tolerance ϵ and substitute the equation

$$
|\theta_n - \theta_{n+1}| - \epsilon \leq 0.
$$

For this problem we will use $\epsilon = 0.001$. Data cards for this example require the following format:

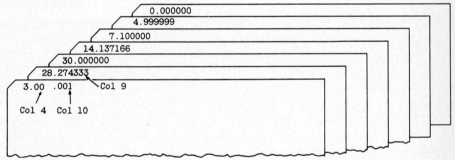

Table 6.4 shows a sample answer sheet.

Table 6.4 Answers for angles at center

A	ANGLE
28.274333	DEGREES = 359.46544
30.000000	AREA GREATER THAN CIRCLE
14.137166	DEGREES = 179.99999
7.100000	DEGREES = 132.58525
4.999999	DEGREES = 115.41348
0.000000	DEGREES = 0.00000

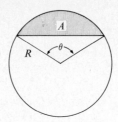

Fig. 6.10. Computation of angle at center.

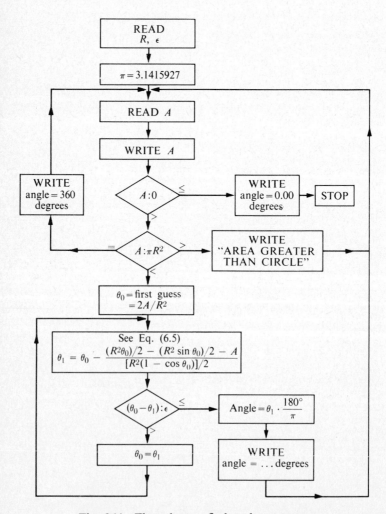

Fig. 6.11. Flow chart to find angle at center.

```
C
C       ANGLE AT CENTER FROM GIVEN AREA OF THE CIRCLE
        DATA IREAD,IWRITE/5,6/
        READ(IREAD,101) R,EPS
  101 FORMAT(F4.2,F6.3)
        WRITE(IWRITE,100)
  100 FORMAT(15X,'A',30X,'ANGLE'//)
        PI=3.14159265
C
C       READ IN THE AREA OF SEGMENT
C
    1 READ(IREAD,102) A
  102 FORMAT(F9.6)
        IF(A.LE.0.) GO TO 99
        ARCLE=PI*R*R
C
C       TEST TO SEE WHETHER THE AREA IS GREATER THAN CIRCLE
C
        IF(A-ARCLE) 7,5,6
    5 ANGLE=360.0
   21 WRITE(IWRITE,111) A,ANGLE
        GO TO 1
  111 FORMAT(10X,F10.6,10X,'DEGREES = ',F13.5//)
    6 WRITE(IWRITE,112) A
  112 FORMAT(10X,F10.6,10X,'AREA GREATER THAN CIRCLE'//)
        GO TO 1
    7 THETO=2.*A/(R*R)
C
C       COMPUTE THET1 BY NEWTON-RAPHSON METHOD
C
    3 THET1=THETO-(THETO-SIN(THETO)-2.*A/(R*R))/(1.-COS(THETO))
        Z=THETO-THET1
C
C       TEST TO SEE IF Z IS SATISFIED THE SPECIFIED TOLERANCE. IF NOT,
C       CONTINUE THE ITERATION
C
        IF(ABS(Z).GT.EPS) GO TO 11
   10 ANGLE=THET1*180./PI
        GO TO 21
   11 THETO=THET1
        GO TO 3
   99 ANGLE=0.
        WRITE(IWRITE,111) A,ANGLE
        STOP
        END


 3.0   .001
28.274333
30.000000
14.137166
 7.100000
 4.999999
  .000000
```

Fig. 6.12. FORTRAN program for angles at center.

6.5 BAIRSTOW'S METHOD FOR POLYNOMIAL EQUATIONS WITH REAL COEFFICIENTS

In the last section we considered some examples for the solutions of transcendental equations. We shall now discuss a method for the solution of a polynomial equation of any degree with real coefficients:

$$y^n + a_1 y^{n-1} + a_2 y^{n-2} + \cdots + a_n = 0. \tag{6.10}$$

This equation has the following properties:

a) It has n roots, single or repeated.

b) It always contains a single real root if n is a positive odd integer.

c) The complex roots present themselves in conjugate complex pairs, say, $D + Ej$ and $D - Ej$, where D and E are real and $j = \sqrt{-1}$.

d) Descartes' rule of signs applies to this equation.

The algorithms discussed so far in this chapter cannot compete with algorithms designed specifically for polynomials. For example, Newton-Raphson method is not competitive as the basis of a polynomial zero-finder. The method we shall discuss was suggested by Bairstow[†] in 1914 and modified by Hitchcock[‡] in 1944. The fundamental principle of the method may best be indicated by an example. Consider the polynomial equation

$$y^5 - 17y^4 + 124y^3 - 508y^2 + 1035y - 875 = 0. \tag{6.11}$$

Dividing the left-hand side of Eq. (6.11) by a quadratic factor $(y^2 + py + q)$ leads to the identity

$$\begin{aligned} y^5 - 17y^4 &+ 124y^3 - 508y^2 + 1035y - 875 \\ &= (y^2 + py + q)(y^3 + B_1 y^2 + B_2 y + B_3) + Ry + S. \end{aligned} \tag{6.12}$$

Equating the coefficients, we get

$$\begin{aligned} p + B_1 &= -17, \\ B_2 + B_1 p + q &= 124, \\ B_3 + B_2 p + B_1 q &= -508, \\ R + B_3 p + B_2 q &= 1035, \\ S + B_3 q &= -875. \end{aligned} \tag{6.13}$$

From the first three equations in (6.13) we find

$$\begin{aligned} B_1 &= -17 - p, \\ B_2 &= 124 + 17p + p^2 - q, \\ B_3 &= -508 - 124p - 17p^2 - p^3 + 2pq + 17q. \end{aligned}$$

† Bairstow, L., "Investigations Relating to the Stability of the Aeroplane," *Reports and Memoranda No. 154*, Advisory Committee of Aeronautics, 1914.

‡ Hitchcock, F. L., "An Improvement on the G.C.D. Method for Complex Roots," *J. Math. Phys.*, **23**, pp. 69–74 (1944).

Substituting into the last two equations of Eq. (6.13), we have

$$p^4 + 17p^3 + 124p^2 + 508p + 1035 - 3qp^2 - 3468 - 124q - q^2 = R, \qquad (6.14)$$

$$p^3q + 17p^2q + 124pq - 2pq^2 - 17q^2 + 508q - 875 = S. \qquad (6.15)$$

The fundamental problem is to find p and q such that both R and S in Eqs. (6.14) and (6.15) become zero. Once p and q are known, we can then obtain two roots of the original equation (6.11) by solving the quadratic equation. In this example, p and q are found, by an iterative procedure to be described below, to be -4.0 and 5.0, respectively. Thus the quadratic factor is $(y^2 - 4y + 5)$, and the two roots are $2 \pm j$.

We shall now describe the step-by-step procedure for obtaining p and q, and cite some numerical examples. The detailed derivations will be discussed in Appendix C. The steps in the computation of p and q are as follows:

1) Select the initial values for p and q. ($A_1, \ldots, A_n$ are n given coefficients.)

2) Compute all B_k's from $B_1, \ldots, B_n$ from

$$B_k = A_k - pB_{k-1} - qB_{k-2}, \qquad (6.16)$$

where $B_{-1} = 0$, and $B_0 = 1$.

3) Compute all C_k's from $C_1, \ldots, C_{n-1}$ from

$$C_k = B_k - pC_{k-1} - qC_{k-2}, \qquad (6.17)$$

where $C_{-1} = 0$, and $C_0 = 1$.

4) Compute $\bar{C}_{n-1} = C_{n-1} - B_{n-1}$.

5) Using the values of C_{n-3}, C_{n-2}, $\bar{C}_{n-1}$, B_n and B_{n-1} obtained in steps 2, 3, and 4, compute Δp and Δq:

$$\Delta p = \frac{B_{n-1}C_{n-2} - B_nC_{n-3}}{(C_{n-2}^2 - \bar{C}_{n-1}C_{n-3})}, \qquad (6.18)$$

$$\Delta q = \frac{B_nC_{n-2} - B_{n-1}\bar{C}_{n-1}}{(C_{n-2}^2 - \bar{C}_{n-1}C_{n-3})}. \qquad (6.19)$$

6) Increment p and q by the amounts Δp and Δq:

$$p_{i+1} = p_i + \Delta p, \qquad (6.20)$$

$$q_{i+1} = q_i + \Delta q, \qquad (6.21)$$

where i is the number of iterations.

7) Test for convergence:

$$M = |\Delta p| + |\Delta q| < \epsilon.$$

If $M > \epsilon$, return to Step 2 and repeat the process (an upper limit on i is imposed for possible nonconvergent cases). If $M \leq \epsilon$, then p and q are satisfactory values of the coefficients in the desired quadratic factor $(y^2 + py + q)$.

Table 6.5 Bairstow's method.

FIRST ITERATION				
k	A_k	B_k	C_k	$p = 0, \quad q = 0, \quad n = 5$ (fifth-order equation)
				A_k = given coefficients, $k = 0, 1, \ldots, 5$
0	1	1	1	B_k and C_k can be computed from Eqs. (6.16) and (6.17)
1	−17	−17	−17	$\overline{C}_4 = C_4 - B_4 = 0$
2	124	124	124	$\Delta p = \dfrac{(1035)(-508) - (-875)(124)}{(-508)^2 - (0)(124)} = -1.617$
3	−508	−508	−508	
4	1035	1035	1035	$\Delta q = \dfrac{(-875)(-508) - (1035)(0)}{(-508)^2 - (0)(124)} = 1.722$
$n = 5$	−875	−875		$p + \Delta p = -1.617, \quad q + \Delta q = +1.722$

SECOND ITERATION				
k	A_k	B_k	C_k	$p = -1.617, \quad q = 1.722$
				$\overline{C}_4 = -76.739 - 343.323 = -420.063$
0	1	1	1	
1	−17	−15.383	−13.766	$\Delta p = \dfrac{(343.323)(-181.573) - (238.221)(73.422)}{(-181.573)^2 - (-420.063)(73.422)} = -1.251$
2	124	97.404	73.422	
3	−508	−324.005	−181.573	$\Delta q = \dfrac{(238.221)(-181.573) - (343.323)(-420.063)}{(-181.573)^2 - (-420.063)(73.422)} = 1.582$
4	1035	343.323	−76.739	
5	−875	238.221		$p + \Delta p = -2.868, \quad q + \Delta q = 3.305$

THIRD ITERATION				
k	A_k	B_k	C_k	$p = -2.868, \quad q = 3.305$
				$\overline{C}_4 = -337.614$
0	1	1	1	
1	−17	−14.132	−11.264	$\Delta p = \dfrac{(106.5)(-66.4) - (195)(44.6)}{(-66.4)^2 - (-337.6)(44.6)} = -0.810$
2	124	80.165	44.555	
3	−508	−231.386	−66.379	$\Delta q = \dfrac{(195)(-66.4) - (106.5)(-337.6)}{(-66.4)^2 - (-337.6)(44.6)} = 1.183$
4	1035	106.468	−231.146	
5	−875	195.004		$p + \Delta p = -3.678, \quad q + \Delta q = 4.487$

FOURTH ITERATION				
k	A_k	B_k	C_k	$p = -3.678, \quad q = 4.487$
				$\overline{C}_4 = -235.323 - 23.919 = -259.242$
0	1	1	1	
1	−17	−13.322	−9.644	$\Delta p = \dfrac{(23.9)(-33.2) - (60.5)(30.6)}{(-33.2)^2 - (-259.2)(30.6)} = -0.293$
2	124	70.513	30.555	
3	−508	−188.865	−33.205	$\Delta q = \dfrac{(60.5)(-33.2) - (23.9)(-259.2)}{(-33.2)^2 - (-259.2)(30.6)} = 0.465$
4	1035	23.919	−235.323	
5	−875	60.473		$p + \Delta p = -3.971, \quad q + \Delta q = 4.952$

FIFTH ITERATION				
k	A_k	B_k	C_k	$p = -3.971, \quad q = 4.952$
				$\overline{C}_4 = -234.083 - 2.023 = -236.107$
0	1	1	1	
1	−17	−13.029	−9.058	$\Delta p = \dfrac{(2)(-26.6) - (5.6)(26.4)}{(-26.6)^2 - (-236.1)(26.4)} = -0.029$
2	124	67.311	26.390	
3	−508	−176.196	−26.550	$\Delta q = \dfrac{(5.6)(-26.6) - (-234.1)(-236.1)}{(-26.6)^2 - (-236.1)(26.4)} = 0.048$
4	1035	2.023	−234.083	
5	−875	5.550		$p + \Delta p = -4.000, \quad q + \Delta q = 5.000$

In order to compute the two roots $D + Ej$ and $D - Ej$, the following expressions are used:

$$D = -p/2, \tag{6.22}$$

$$E = \sqrt{q - p^2/4}. \tag{6.23}$$

Example

Consider the polynomial

$$y^5 - 17y^4 + 124y^3 - 508y^2 + 1035y - 875 = 0. \tag{6.24}$$

We assume that the initial values of both p and q are zero. The procedure to be followed, i.e., the five iterations, are illustrated in Table 6.5. We find that the quadratic factor is $(y^2 - 4y + 5)$, and the two roots are $2 + j$ and $2 - j$.

We next solve the polynomial

$$y^3 - 13.029y^2 + 67.311y - 176.196 = 0,$$

where the coefficients are B_k's $(k = 1, 2, \ldots, 3)$, obtained in the fifth iteration. Using the same procedure, one obtains the following quadratic factor after four iterations:

$$(y^2 - 6.0y + 25.2).$$

The two roots corresponding to this quadratic factor are $3 + 4j$ and $3 - 4j$.

It should be noted that the product of the factors

$$(y^2 - 4y + 5)(y^2 - 6y + 25)(y - 7)$$

yields the left-hand expression in Eq. (6.24). This checks well with our results.

Figures 6.13 and 6.14 show a flow chart and FORTRAN program for computing all the roots of a given polynomial with real coefficients. Table 6.6 shows the answer for the five roots of Eq. (6.11).

Table 6.6 Solution of $x^5 - 17x^4 + 124x^3 - 508x^2 + 1035x - 875 = 0$

	REAL	IMAGINARY
X(5) =	2.0000	1.0000
X(4) =	2.0000	-1.0000
X(3) =	3.0000	4.0000
X(2) =	3.0000	-4.0000
X(1) =	7.0000	0.0000

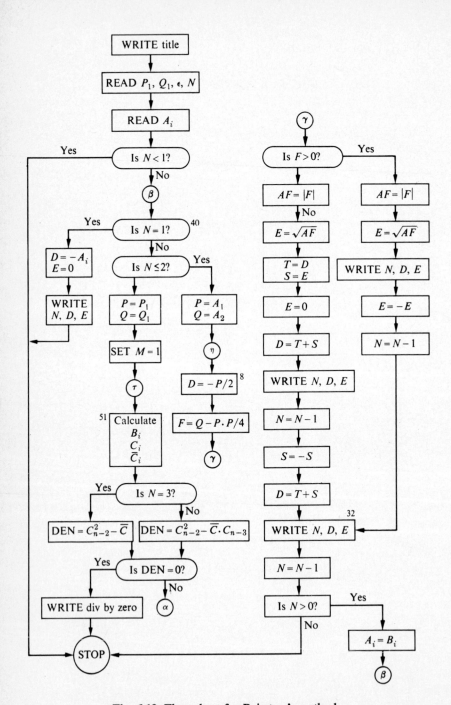

Fig. 6.13. Flow chart for Bairstow's method.

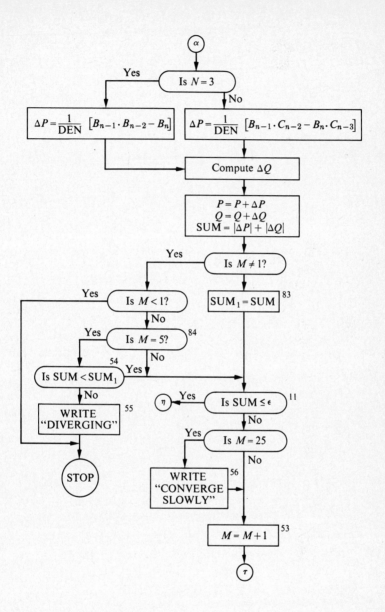

Fig. 6.13. *(concl.)*

```
C       BAIRSTOW'S METHOD FOR FINDING QUADRATIC FACTORS OF POLYNOMIALS.
C       SOLUTION OF POLYNOMIAL UP TO X**99 WITH REAL COEFFICIENTS
C       N DESIGNATES THE DEGREE OF THE POLYNOMIAL
C       P1 AND Q1 ARE INITIAL GUESSES.
C       EPSON REPRESENTS THE DESIRED ACCURACY.
        DATA IREAD,IWRITE/5,6/
        DIMENSION A(100),B(100),C(100)
        WRITE(IWRITE,12)
     12 FORMAT(20X,'REAL',20X,'IMAGINARY'//)
C
C       READ IN P1,Q1,EPSON,AND N
C
        READ(IREAD,10) P1,Q1,EPSON,N
     10 FORMAT(3F10.4,I2)
C
C       READ IN THE COEFFICIENTS OF THE POLYNOMIAL
C
        READ(IREAD,20) (A(I),I=1,N)
     20 FORMAT(9F8.4)
     40 IF(N-1)80,42,43
C
C       FOLLOWING 6 STATEMENTS TO SOLVE LINEAR EQUATION OF ONE VARIABLE
C
     42 D=(-1.)*A(1)
        E=0.
        WRITE(IWRITE,13) N,D,E
     13 FORMAT(7X,'X(',I2,') =',3X,2(F8.4,19X))
        GO TO 80
     43 IF(N.GT.2) GO TO 46
C
C       FOLLOWING 3 STATEMENTS FOR QUADRATIC EQUATION
C
     45 P=A(1)
        Q=A(2)
        GO TO 8
C
C       FOLLOWING 7 STATEMENTS TO COMPUTE B(K)'S
C
     46 P=P1
        Q=Q1
        M=1
     51 B(1)=A(1)-P
        B(2)=A(2)-P*B(1)-Q
        DO 6 K=3,N
      6 B(K)=A(K)-P*B(K-1)-Q*B(K-2)
C
C       FOLLOWING 5 STATEMENTS TO COMPUTE C(K)'S
C
        L=N-1
        C(1)=B(1)-P
        C(2)=B(2)-P*C(1)-Q
        DO 7 J=3,L
      7 C(J)=B(J)-P*C(J-1)-Q*C(J-2)
C
C       COMPUTE CBAR
C
        CBARL=C(L)-B(L)
        IF(N.NE.3) GO TO 70
     71 DEN=(C(N-2))**2-CBARL
        GO TO 72
     70 DEN=(C(N-3))**2-CBARL*C(N-3)
```

Fig. 6.14. FORTRAN program for Bairstow's method.

```
C
C        IF DEN IS ZERO, NEW VALUES OF P AND Q SHOULD BE TRIED
C
   72 IF(DEN.EQ.0) GO TO 48
C
C        FOLLOWING 5 STATEMENTS TO COMPUTE DELTP AND DELTQ
C
      IF(N.NE.3) GO TO 47
   73 DELTP=(B(N-1)*C(N-2)-B(N))/DEN
      GO TO 74
   47 DELTP=(B(N-1)*C(N-2)-B(N)*C(N-3))/DEN
   74 DELTQ=(B(N)*C(N-2)-B(N-1)*CBARL)/DEN
C
C        INCREASING P AND Q BY THE AMOUNTS DELTP AND DELTQ
C
      P=P+DELTP
      Q=Q+DELTQ
      ABSDP=ABS(DELTP)
      ABSDQ=ABS(DELTQ)
      SUM=ABSDP+ABSDQ
C
C        FOLLOWING 19 STATEMENTS TO TEST FOR CONVERGENCE.
C        TEST AT ITERATION=5. IF THE SUM OF DELTP AND DELTQ IS GREATER THAN
C        THAT OF BOTH INITIAL DELTP AND DELTQ, THE FUNCTIONS DIVERGE FOR
C        ASSUMED VALUES OF P AND Q. STOP THE COMPUTATION
C        TEST AT ITERATION=25. IF THE SUM OF DELTP AND DELTQ IS STILL
C        GREATER THAN THE DESIRED ACCURACY, THE FUNCTIONS ARE CONVERGING
C        SLOWLY. GO ON THE COMPUTATION
C
      IF(M-1)80,83,84
   83 SUM1=SUM
      GO TO 11
   84 IF((M.EQ.5).AND.(SUM.GE.SUM1)) GO TO 55
   11 IF(SUM.LE.EPSON) GO TO 8
      IF(M.EQ.25) GO TO 56
   53 M=M+1
      GO TO 51
   55 WRITE(IWRITE,57)
   57 FORMAT(9X,'FUNCTIONS DIVERGING FOR ASSUMED VALUES OF P AND Q'/)
      GO TO 80
   56 WRITE(IWRITE,59)
   59 FORMAT(20X,'THE FUNCTIONS ARE CONVERGING SLOWLY'/)
      WRITE(IWRITE,60) P,Q
   60 FORMAT(20X,'P=',E14.8,3X,'Q=',E14.8/)
      GO TO 53
    8 D=(-P)/2.
      F=Q-P*P/4.
      IF(F.GT.0.) GO TO 31
C
C        FOLLOWING 13 STATEMENTS TO COMPUTE TWO REAL ROOTS OF THE QUADRATIC
C        FACTOR AND THEN PRINT OUT.
C
   30 AF=ABS(F)
      E=SQRT(AF)
      T=D
      S=E
      E=0.0
      D=T+S
      WRITE(IWRITE,13) N,D,E
      N=N-1
      S=(-1.)*S
      D=T+S
```

Fig. 6.14 *(cont.)*

```
   32 WRITE(IWRITE,13) N,D,E
      N=N-1
      IF(N)80,80,81
C
C     FOLLOWING 6 STATEMENTS TO COMPUTE THE COMPLEX ROOTS OF THE
C     QUADRATIC FACTOR AND THEN PRINT OUT.
C     REAL NUMBER
C
   31 AF=ABS(F)
      E=SQRT(AF)
      WRITE(IWRITE,13) N,D,F
      E=(-1.)*E
      N=N-1
      GO TO 32
C
C     NEXT STEP IS TO SOLVE THE ANOTHER FACTOR
C     ( A NEW LOWER ORDER POLYNOMIAL )
   81 DO 82 I=1,N
   82 A(I)=B(I)
      GO TO 40
   48 WRITE(IWRITE,49)
   49 FORMAT(2X,'DIVIDED BY ZERO TRY NEW VALUES OF P AND Q'/)
   80 CALL EXIT
      END
```

```
0.          0.           .0001      5
     17.        124.       -508.     1035.     -875.
```

Fig. 6.14. *(concl.)*

BIBLIOGRAPHY

General

Frank, E., "On the calculation of the roots of equations," *J. Math. Phys.*, 187–197 (1955).

Hildebrand, F. B., *Introduction to Numerical Analysis.* McGraw-Hill, New York, 1956.

Householder, A. S., *Principles of Numerical Analysis.* McGraw-Hill, New York, 1953.

Kristiansen, G. K., "Zero of arbitrary function," *BIT*, **3**, 205–206 (1963).

Masaitis, C., "Numerical location of zeros," *Aberdeen Proving Ground Ordn. Comput. Res. Rep.*, **4**, 26–28 (1957).

Ostrowski, A. M., *Theory of the Solution of Equations and Systems of Equations.* Academic Press, New York, 1960.

Polynomials

Aitkin, A. C., "On Bernoulli's numerical solution of algebraic equations," *Proc. Roy. Soc. Edinburgh*, **46**, 289–305 (1926).

Bareiss, E. H., "Resultant procedure and the mechanization of the Graeffe process," *J. Assoc. Comput. Mach.*, **7**, 346–386 (1960).

Henrici, P., "Zeros of a polynomial by Q-D algorithm," *Commun. Assoc. Comput. Mach.*, **8**, 570–574 (1965).

Lehmer, D. H., "A machine method for solving polynomial equations," *J. Assoc. Comput. Mach.*, **8**, 15–162 (1961).

Lehmer, D. H., "Search procedures for polynomial equation solving," *Construction Aspects of the Fundamental Theorem of Algebra*, pp. 193–208, B. DeJon and P. Henrici, eds. Wiley, London, 1969.

Smallwood, J. L., "A Comparison of Five Numerical Methods for Solving Polynomial Equations with Real Coefficients." Clearinghouse, U.S. Dept. Commerce, Springfield, Va., AD 688947, 1969.

Traub, J. F., "The solution of transcendental equations," *Mathematical Methods for Digital Computers*, Vol. 2, pp. 171–184, A. Ralston and H. S. Wilf, eds. Wiley, New York, 1967.

Wilkinson, J. H., "The evaluation of the zeros of ill-conditioned polynomials," *Numer. Math.*, **1,** 150–180 (1959).

Transcendental Functions

Erdelyi, A., W. Magnus, F. Oberhettinger, and F. G. Tricomi, *Higher Transcendental Functions*, Vol. 2. McGraw-Hill, New York, 1953.

Lance, G. N., "Solutions of algebraic and transcendental equations on an automatic digital computer," *J. Assoc. Comput. Mach.*, **6,** 97–101 (1959).

Algebraic Equations

Brooker, R. A., "The solution of algebraic equations on the EDSAC," *Proc. Cambridge Phil. Soc.*, **48,** 255–270 (1952).

Frank, W. L., "Finding zeros of arbitrary functions," *J. Assoc. Comput. Mach.*, **5,** 154–165 (1958).

Heinrich, H., "Zur Vorbehandlung algebraischer Gleichungen," *Z. Angew. Math. Mech.*, **36,** 145–148 (1956).

Muller, D., "A method for solving algebraic equations using an automatic computer," *Math. Tables Aids Comput.*, **10,** 208–215 (1956).

Peltier, J., *Résolution numérique des équations algébriques*. Gauthier-Villars, Paris, 1957.

Porter, A., and C. Mack, "New methods for the numerical solution of algebraic equations," *Phil. Mag.*, **40,** 578–585 (1949).

Miscellaneous Methods

Caldwell, G., "A note on the downhill method," *J. Assoc. Comput. Mach.*, **6,** 223–225 (1959).

Everling, W., "Eine Verallgemeinerung des Hornerschen Schemas," *Z. Angew. Math. Mech.*, **37,** 74 (1957).

Gross, O., and S. M. Johnson, "Sequential minimax search for a zero or a convex function," *Math. Tables Aids Comput.*, **13,** 44–51 (1959).

Kulik, S., "A method of approximating the complex roots of equations," *Pacific J. Math.*, **8,** 277–281 (1958). "On the solution of algebraic equations," *Proc. Am. Math. Soc.*, **10,** 185–192 (1959).

Munro, W. D., "Some iterative methods for determining zeros of functions of a complex variable," *Pacific J. Math.*, **9,** 555–566 (1959).

Wynn, P., "Cubically convergent process for zeros," *Math. Tables Aids Comput.*, **10,** 164–169 (1956).

PROBLEMS

1. In the study of Fraunhofer diffraction, we encounter the equation

$$\left(\frac{\sin \alpha}{\alpha}\right)^2 = \frac{1}{2}.$$

Solve for α in radians.

2. The following equation describes the motion of a system of helical gears:

$$\tan \phi - \phi = c.$$

Find the angle ϕ for the two different cases: (a) $c = 0.01$ and (b) $c = 0.001$.

3. The following equation describes the motion of a planetary gear system used in an automatic transmission: $\sin \omega t - e^{-at} = 0$. Determine the smallest positive root for

Case 1: $\omega = 0.573$,   Case 2: $\omega = 0.01$,

$\quad\quad\quad \alpha = 0.01$. $\quad\quad\quad a = 0.1$.

4. Find the smallest positive root of the following equation used in the study of vibrations:

$$\tanh x + \tan x = 0.$$

5. Find the smallest positive value of kL which satisfies the following equation derived in a column-buckling problem: $\tan kL - kL = 0$.

6. In many circumstances tubular insulators are used for high-potential conducting through-walls (see Fig. 6.15). What must the ratio x of the external diameter $2R$ be to the bore width $2r$, to produce a minimum cross section Q? The cross section is defined as follows:

$$Q = \pi q^2 (x^2 - 1)/(\ln x)^2,$$

where q is the ratio of the line voltage to the maximum admissible field strength and is considered a constant in the problem.

7. The circuit shown in Fig. 6.16 has a current given by

$$i = 5e^{-5t} \sin (\pi/4) + 5 \sin (5t - \pi/4).$$

Find the time between 0.5 and 1.4 sec, at which the current is zero ($e = 2.71828$).

8. Calculate the first five roots of the following equation for a two-phase diffusion problem

$$\sin \left(\frac{\alpha}{0.2}\right) \cos \alpha + \cos \left(\frac{\alpha}{0.2}\right) \sin \alpha - \alpha \sin \alpha \sin \left(\frac{\alpha}{0.2}\right) = 0.$$

It is known that the first five roots are near 0.07, 0.14, 0.22, 0.31 and 0.41 respectively.

9. Find the real root of $xe^x = 1$ using Newton's method. Use $x = 1$ for the first approximation and repeat until $|(e^{-x} - x)/(x + 1)| < 10^{-6}$.

10. Find the root between 0 and -1 of the equation $e^{ct} + \sqrt{2/\pi t}/c^2 - t^2 = 0$, where $e = 2.71828$, for $c = 0.8$; $c = 0.9$.

11. Find the smallest positive root of the equation $2 - te^{2t} + 4t = 0$.

12. Find the root beteeen 0 and 1 of the equation $1 + \cos t - 4t = 0$.

13. Find the smallest root of the equation $e^{2t} \tan t = e^{-3\pi/2}$. [Hint: the smallest root is close to the value $e^{-3\pi/2}$.]

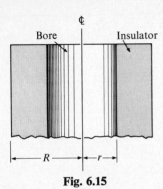

Fig. 6.15

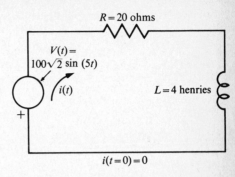

Fig. 6.16

Use Bairstow's method to find the roots of each of the following polynomial equations.

14. $x^4 - 6x^3 + 23x^2 - 34x + 26 = 0$.

15. $x^4 - x^3 - 2x^2 - 6x - 4 = 0$.

16. In pipe-flow problems one frequently encounters the equation

$$C_5 D^5 + C_1 D + C_0 F = 0.$$

Solve for D in feet, given that $C_5 = 8820$, $C_1 = -2.31$, $C_0 = -431$, and $F = 0.015$.

17. Two crossed ladders of unequal length lean against two walls as shown in Fig. 6.17. The lengths of the ladders are L_1 and L_2 respectively, and the height of their crossing point is h. In order to find the distance $u + v$ between two walls, one can use the following equation, which can be obtained by considering similar triangles:

$$y^4 - 2hy^3 + (L_1^2 - L_2^2)y^2 - 2h(L_1^2 - L_2^2)y + h^2(L_1^2 - L_2^2) = 0.$$

Solve y and $u + v$ for the following two cases:

1. $L_1 = 100$, $L_2 = 80$, $h = 10$.
2. $L_1 = 119$, $L_2 = 70$, $h = 30$.

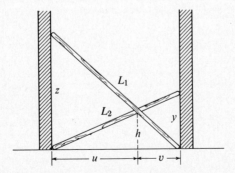

Fig. 6.17

CHAPTER 7

Ordinary
Differential Equations
with Initial Conditions

7.1 INTRODUCTION

The formulation of a large class of engineering and scientific problems in mathematical form leads to either ordinary or partial differential equations. In a great many cases, it is found that the resultant equations do not possess closed-form solutions; that is, specific expressions for the dependent variables cannot be found.

In this chapter, we shall discuss some useful numerical methods for solving ordinary differential equations on a digital computer. These methods are so well developed† that only a suitable program is necessary to obtain the required answers in tabulated form.

Depending upon the given conditions, ordinary differential equations derived from scientific and engineering problems may customarily be classified into two distinct types. The first type is called an initial-value problem and the second, a boundary-value problem.

An initial-value problem is characterized by the fact that the information given concerns all the conditions at a given point. Thus a differential equation of the nth order, together with the known values of $x, x', x'', \ldots, x^{n-1}$ at a specific point $t = t_0$, may be described as an initial-value problem. For example, the basic differential equation describing the motion of a spring-mass system (Fig. 7.1) is

$$m \frac{d^2x}{dt^2} + c \frac{dx}{dt} + R(x) = F(t), \tag{7.1}$$

---

† Except for some questions concerning the singular points of the solutions.

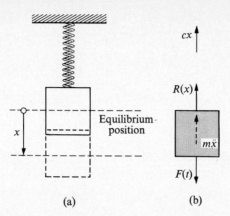

Fig. 7.1. Motion of a spring-mass system. (a) System. (b) Free body.

where

$$x = \text{displacement of mass } m,$$

$$t = \text{time},$$

$$c = \text{damping coefficient},$$

$$R(x) = \text{resistance of spring, function of } x,$$

$$F(t) = \text{applied forcing function},$$

and a given set of initial conditions is

$$t = 0, \qquad x = 1; \tag{7.2}$$

$$t = 0, \qquad \frac{dx}{dt} = 0. \tag{7.3}$$

Note that the two initial conditions in Eqs. (7.2) and (7.3) refer to $t = 0$. The fundamental problem is to find the x-t and $\dot{x}$-t relations as t increases with a given increment. Hence, the initial-value problem is sometimes referred to as a "marching problem." The solutions of Eq. (7.1) will be discussed in detail in Section 7.6.

In contrast to initial-value problems, a boundary-value problem is one in which the conditions are given at two or more distinct points. For example, a second-order differential equation

$$\frac{d^2y}{dx^2} - 2y = 7, \tag{7.4}$$

together with the boundary conditions

$$x = 0, \qquad y = 6; \tag{7.5}$$

$$x = 6, \qquad y = 17, \tag{7.6}$$

forms a boundary-value problem. Here we are seeking the y-x relation to satisfy Eqs. (7.4) through (7.6). This type of problem is often called a "jury problem."

Because a first-order differential equation can have only one given condition, it is always an initial-value, rather than a boundary-value, problem.

Since numerical solutions of the marching problems are generally much simpler than those of jury problems, we shall limit our discussion in this book to computer methods for solving initial-value problems.

7.2 RUNGE-KUTTA METHOD WITH RUNGE'S COEFFICIENTS

Among the vast number of numerical methods available for solving initial-value problems, the Runge-Kutta method is probably the one most frequently used on computers because it simplifies programming. Detailed derivations are given in the next section. Here we shall be primarily concerned with the procedures and some typical examples.

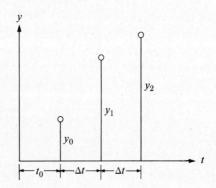

Fig. 7.2. Step-by-step solution of $y' = f(y, t)$.

Take a differential equation of the first order, $y' = f(t, y)$. The initial condition is given as (t_0, y_0). Our problem is to find the y_1 corresponding to a new $t = t_0 + \Delta t$ (Fig. 7.2). Once y_1 is found, a new y-value, y_2, can then be obtained that corresponds to a second t-value, say, $t_0 + 2\,\Delta t$. In this step-by-step manner, the required y-t relation can then be obtained and tabulated.

In the Runge-Kutta method, the basic formula with Runge's coefficients is

$$y_{n+1} = y_n + \Delta y_n, \tag{7.7}$$

where

$$\Delta y_n = \frac{\Delta t}{6}(k_0 + 2k_1 + 2k_2 + k_3)$$

and

$$k_0 = f(t_n, y_n),$$

$$k_1 = f\left(t_n + \frac{\Delta t}{2}, \; y_n + \frac{k_0}{2}\,\Delta t\right),$$

$$k_2 = f\left(t_n + \frac{\Delta t}{2}, \; y_n + \frac{k_1}{2}\,\Delta t\right),$$

$$k_3 = f(t_n + \Delta t, \; y_n + k_2\,\Delta t).$$

The geometrical interpretation of the values k_0 through k_3 is shown in Fig. 7.3(a). All four k-values represent the slopes at various points: the value k_0 is the slope at the starting point; k_3 is the slope at the right-hand point whose ordinate is $y_n + k_2\,\Delta t$; k_2 is one of the two slopes considered at the midpoint with the ordinate $y_n + \frac{1}{2}k_1\,\Delta t$; and finally, k_1 is the second slope at the midpoint whose ordinate is $y_n = \frac{1}{2}k_0\,\Delta t$.

It is interesting to compare the Runge-Kutta method with Euler's method, which is usually discussed in basic calculus courses. In the latter, the basic formula is simply

$$y_{n+1} = y_n + k_0\,\Delta t, \tag{7.8}$$

where $k_0 = y'_n = f(t_n, y_n)$. Thus in the Euler method the approximate solution y_{n+1} (Fig. 7.3b) depends on y_n and a slope k_0 at a point (t_n, y_n), whereas in the Runge-Kutta method, it depends not only on y_n and k_0, but also on three additional slopes at points other than (t_n, y_n). In this way, the Runge-Kutta procedure uses a weighted average of slopes, with those in the center receiving twice as much weight as those on the ends.

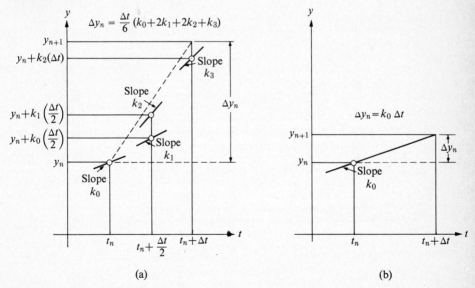

Fig. 7.3. Comparison of slopes used. (a) Four slopes used in Runge-Kutta method. (b) The slope k_0 used in Euler's method.

We shall illustrate the Runge-Kutta method with Runge's coefficients by the following example.

Example 7.1

It is known that the motion of a simple pendulum, as shown in Fig. 7.4, may be expressed by the following first-order differential equation:

$$\frac{d\omega}{d\phi} = \frac{-g}{L} \frac{\sin \phi}{\omega}, \tag{7.9}$$

where

$$\omega = \text{angular velocity in rad/sec,}$$
$$\phi = \text{angle (see Fig. 7.4),}$$
$$g = 32.2014 \text{ ft/sec}^2,$$
$$L = 2.5 \text{ ft.}$$

The initial condition is $\phi = 79°$, $\omega = 0.66488546$. Tabulate the solution at the interval of

$$\Delta\phi = -0.5°.$$

Solution by Runge-Kutta Method with Runge's Coefficients. With the given initial condition, we find that

$$k_0 = f(\phi_0, \omega_0) = \frac{-g}{L} \frac{\sin \phi}{\omega}$$

$$= \frac{-32.2014}{2.5} \frac{\sin 79°}{0.66488546} = -19.016673,$$

$$k_1 = f\left[\phi_0 + \frac{\Delta\phi}{2}, \omega_0 + \tfrac{1}{2}k_0 \Delta\phi\right]$$

$$= \frac{-32.2014}{2.5} \frac{\sin 78.75°}{0.66488546 + (-19.016673) \times (-0.25) \times (3.141593/180)}$$

$$= -16.892255,$$

$$k_2 = f\left[\phi_0 + \frac{\Delta\phi}{2}, \omega_0 + \tfrac{1}{2}k_1 \Delta\phi\right]$$

$$= \frac{-32.2014}{2.5} \frac{\sin 78.75°}{0.66488546 + (-16.892255) \times (-0.25) \times (3.1415926/180)}$$

$$= -17.104257,$$

$$k_3 = f(\phi_0 + \Delta\phi, \omega_0 + k_2 \Delta\phi)$$

$$= \frac{-32.2014}{2.5} \frac{\sin 78.5°}{0.66488546 + (-17.104257) \times (-0.5) \times (3.1415926/180)}$$

$$= -15.503294,$$

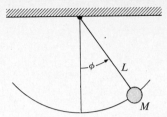

Fig. 7.4. Simple pendulum.

and

$$\Delta\omega = \frac{1}{6}\left(\frac{-0.5 \times 3.14159265}{180}\right) \times (k_0 + 2k_1 + 2k_2 + k_3)$$

$$= 0.14909917.$$

Therefore for $\phi = 78.5°$,

$$\omega = 0.66488546 + 0.14909917 = 0.81398463.$$

The computation may be conveniently arranged as follows:

ϕ	ω	k_0	k_1	k_2	k_3	$\Delta\omega$
79°	0.66488546	−19.016673	−16.892255	−17.104257	−15.503294	0.14909917
78.5°	0.81398463	−15.506411	−14.303566	−14.389225	−13.409645	0.12552070
78°	0.93950533					

Solution by Euler's Method. Let $\phi_i = 79° + i(\Delta\phi)$ and $\omega_i = \omega(\phi_i)$, where $i = 0, 1, 2, \ldots$ and approximate $d\omega/d\phi$ at ϕ_i by $(\omega_{i+1} - \omega_i)/\Delta\phi$. This yields the difference equation

$$\omega_{i+1} = \frac{-g}{L}\frac{\sin\phi_i}{\omega_i}\Delta\phi + \omega_i,$$

which is used to determine all values of ω_i. The initial value of ω is given as 0.66488546.

The Exact Solution. It should be noted that Eq. (7.9) does have an exact solution:

$$\omega = \sqrt{(2g/L)(\cos\phi - \cos 79°) + \omega_{79}^2}. \tag{7.10}$$

For the sake of comparison, the solutions based on both the Euler and the Runge-Kutta methods, as well as exact solutions are tabulated in Table 7.2. All these solutions were carried out on a high-speed digital computer. A sample flow chart and a FORTRAN program are given in Figs. 7.5 and 7.6, respectively. In the program it is required to read in g and L from the first data card and ϕ_{79}, ω_{79}, $\Delta\phi$, ϕ_{max} from the second card. It is also required to print the five results per line immediately after the related computations. This required printing format is as shown in Table 7.1 and the actual printout is as shown in Table 7.2.

Table 7.1

(1)	(2)	(3)	(4)	(5)	(6)
Angle	Closed	Euler	(2)-(3)	Runge	(2)-(5)
79.0	——	——	——	——	——
78.5	——	——	——	——	——
78.0	——	——	——	——	——
——	——	——	——		
——	——	——	——		
——	——	——	——		
ϕ_{max}					
(F6.2)	(F12.8)	(F12.8)	(F12.8)	(F12.8)	(F12.8)

Column 1 is the angle in degrees. Column 2 is the exact solution. Column 3 is Euler's solution. Column 5 is the Runge-Kutta solution (Runge's coefficients).

Table 7.2 Simple pendulum problem

(1) ANGLE	(2) CLOSED	(3) EULER	(4) (2)-(3)	(5) RUNGE	(6) (2)-(5)
79.00	.66488546	.66488546	0.00000000	.66488546	0.00000000
78.50	.81397756	.83083723	-.01685967	.81398457	-.00000701
78.00	.93949705	.96341139	-.02391434	.93950522	-.00000817
77.50	1.04992610	1.07753480	-.02760870	1.04993320	-.00000710
77.00	1.14961220	1.17937800	-.02976580	1.14961920	-.00000700
76.50	1.24114130	1.27224320	-.03110190	1.24114760	-.00000630
76.00	1.32619610	1.35815310	-.03195700	1.32620220	-.00000610
75.50	1.40594740	1.43845710	-.03250970	1.40595250	-.00000510
75.00	1.48124480	1.51411020	-.03286540	1.48125000	-.00000520
74.50	1.55273250	1.58581830	-.03308580	1.55273720	-.00000470
74.00	1.62090830	1.65412120	-.03321290	1.62091330	-.00000500
73.50	1.68617060	1.71944270	-.03327210	1.68617500	-.00000440
73.00	1.74883960	1.78212300	-.03328340	1.74884400	-.00000440
72.50	1.80918130	1.84244010	-.03325880	1.80918510	-.00000380
72.00	1.86741590	1.90062470	-.03320880	1.86741970	-.00000380
71.50	1.92373160	1.95687070	-.03313910	1.92373480	-.00000320
71.00	1.97828680	2.01134310	-.03305630	1.97829030	-.00000350
70.50	2.03122010	2.06418340	-.03296330	2.03122390	-.00000380
70.00	2.08265200	2.11551440	-.03286240	2.08265550	-.00000350
69.50	2.13268610	2.16544320	-.03275710	2.13268990	-.00000380
69.00	2.18141600	2.21406400	-.03264800	2.18141960	-.00000360
68.50	2.22892280	2.26146020	-.03253740	2.22892660	-.00000380
68.00	2.27528060	2.30770580	-.03242520	2.27528390	-.00000330
67.50	2.32055370	2.35286720	-.03231350	2.32055700	-.00000330
67.00	2.36480210	2.39700390	-.03220180	2.36480490	-.00000280
66.50	2.40807800	2.44016960	-.03209160	2.40808080	-.00000280
66.00	2.45043070	2.48241300	-.03198230	2.45043300	-.00000230
65.50	2.49190310	2.52377840	-.03187530	2.49190540	-.00000230
65.00	2.53253620	2.56430620	-.03177000	2.53253820	-.00000200

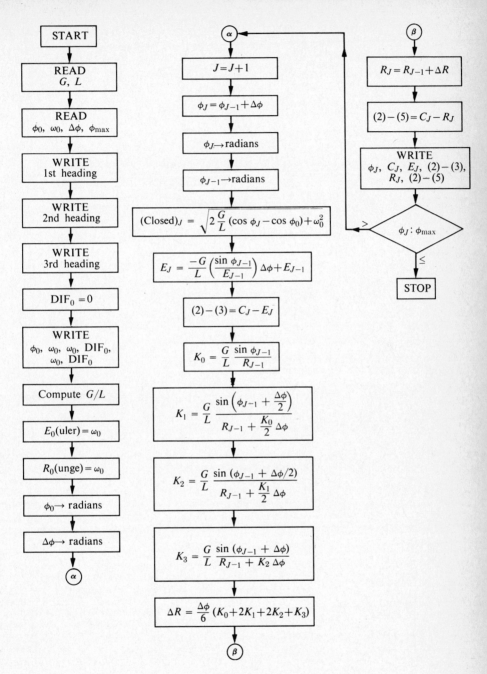

Fig. 7.5. Flow chart for simple pendulum problem.

```
C
C EXAMPLE 7.1
C USE EULER AND RUNGE-KUTTA METHODS FOR SIMPLE PENDULUM PROBLEM
C THE RESULTS ARE COMPARED WITH EXACT SOLUTION
C
C PHI=INITIAL ANGLE        PHMAX=FINAL ANGLE
C W=INITIAL OMEGA          H=INCREMENT OF ANGLE
C
C
      DATA IREAD,IWRITE/5,6/
C
C READ IN THE VALUE G AND THE LENGTH OF THE PENDULUM
C
      READ(IREAD,10) G,ZL
   10 FORMAT(F10.4,F6.2)
C
C READ IN THE ANGLE, ANGULAR VELOCITY,INCREMENT AND MAXIMUM ANGLE
C
      READ(IREAD,12) PHI,W,H,PHMAX
   12 FORMAT(F6.2,F10.8,F8.2,F6.2)
      WRITE(IWRITE,55)
   55 FORMAT('1',25X,'TABLE 7.2 SIMPLE PENDULUM PROBLEM'//)
      WRITE(IWRITE,66)
   66 FORMAT(2X,'(1)',8X,'(2)',10X,'(3)',10X,'(4)',10X,'(5)',10X,'(6)'/)
      WRITE(IWRITE,31)
   31 FORMAT(' ANGLE',5X,'CLOSED',8X,'EULER',21X,'RUNGE')
      WRITE(IWRITE,33)
   33 FORMAT(37X,'(2)-(3)',19X,'(2)-(5)'///)
C
C INITIALIZE THE DIFFERENCE
C
      DIF=0.
C
C PRINT OUT THE INITIAL VALUE OF THE ANGLE AND ANGULAR VELOCITY
C
      WRITE(IWRITE,51) PHI,W,W,DIF,W,DIF
   51 FORMAT(F6.2,F13.8,F13.8,F13.8,F13.8,F13.8)
C
C DEFINE THE RATIO OF ACCELERATION AND LENGTH OF THE PENDULUM
C
      GOL=G/ZL
C
C INITIALIZE THE EULER, RUNGE-KUTTA AND GILL SOLUTIONS
C
      EULER=W
      RUNGE=W
C
C THE NEXT THREE STATEMENTS DEAL WITH THE CONVERSION FACTOR
      CONV=3.14159265/180.0
      PCONV=PHI*CONV
      HCONV=H*CONV
C
C INCREMENT OF THE ANGLE
C
    9 PHI=PHI+H
      PP=PHI*CONV
```

Fig. 7.6. FORTRAN program for simple pendulum problem.

```
      PPP=(PHI-H)*CONV
C
C COMPUTE THE EXACT SOLUTION FOR EACH INCREMENT OF ANGLE
C
      CLOSE=SQRT((-2.0*GOL)*(COS(PP)-COS(PCONV))+W*W)
C
C COMPUTE EULER SOLUTION FOR EACH INCREMENT OF ANGLE
C
      EULER =(GOL*SIN(PPP)*HCONV/EULER)+EULER
C
C FIND THE DIFFERENCE BETWEEN EXACT AND EULER SOLUTIONS
C
      ANS4=CLOSE-EULER
C
C THE NEXT SIX STATEMENTS DEAL WITH RUNGE COEF. AND SOLUTION
C STEP BY STEP
C
      ZK0=GOL*SIN(PPP)/RUNGE
      ZK1=GOL*(SIN(PPP+0.5*HCONV)/(RUNGE+0.5*ZK0*HCONV))
      ZK2=GOL*(SIN(PPP+0.5*HCONV)/(RUNGE+0.5*ZK1*HCONV))
      ZK3=GOL*(SIN(PPP+HCONV)/(RUNGE+ZK2*HCONV))
      TEMP=(HCONV/6.0)*(ZK0+2.0*(ZK1+ZK2)+ZK3)
      RUNGE=RUNGE+TEMP
C
C FIND THE DIFFERENCE BETWEEN EXACT AND RUNGE-KUTTA SOLUTIONS
C
      ANS6=CLOSE-RUNGE
C
C PRINT OUT THE RESULTS
C
      WRITE(IWRITE,51) PHI,CLOSE,EULER,ANS4,RUNGE,ANS6
C
C TEST FOR COMPLETION
C
      IF(PHI.GT.PHMAX) GO TO 9
    8 STOP
      END


    32.2014    2.5
  79.000.66488546    -0.50 65.00
```

7.3 DERIVATION OF THE RUNGE-KUTTA FORMULA

In the last section, the procedure of the Runge-Kutta method was described. We turn now to the derivation of Eq. (7.7), the basic Runge-Kutta formula with Runge's coefficients. It should be noted that Eq. (7.7) is only one of many possible forms of the Runge-Kutta equations. All of them may be obtained from the following Taylor-series expansion:

$$y_{n+1} = y_n + y_n'(t_{n+1} - t_n) + \frac{y_n''}{2!}(t_{n+1} - t_n)^2$$

$$+ \frac{y_n'''}{3!}(t_{n+1} - t_n)^3 + \frac{y_n^{iv}}{4!}(t_{n+1} - t_n)^4 + \cdots \qquad (7.11)$$

Assuming that $t = t_{n+1} = t_n + \Delta t$, we find that Eq. (7.11) becomes

$$y_{n+1} = y_n + y_n'(\Delta t) + \frac{y_n''}{2!}(\Delta t)^2 + \frac{y_n'''}{3!}(\Delta t)^3 + \frac{y_n^{iv}}{4!}(\Delta t)^4 + \cdots \qquad (7.12)$$

If we call

$$\Delta y_n = y_{n+1} - y_n, \qquad (7.13)$$

then

$$\Delta y_n = y_n'(\Delta t) + \frac{y_n''}{2!}(\Delta t)^2 + \frac{y_n'''}{3!}(\Delta t)^3 + \frac{y_n^{iv}}{4!}(\Delta t)^4 + \cdots, \qquad (7.14)$$

and we note that

$$y' = f(t, y), \qquad (7.15)$$

$$y'' = f' = \frac{\partial f}{\partial t} + \frac{\partial f}{\partial y}\frac{dy}{dt} = f_t + f_y f, \qquad (7.16)$$

$$y''' = f'' = \frac{\partial f'}{\partial t} + \frac{\partial f'}{\partial y} f$$

$$= [f_{tt} + (f_{yt}f + f_y f_t)] + [f_{ty} + (f_{yy}f + f_y^2)]f, \qquad (7.17)$$

$$y^{iv} = f_{ttt} + \cdots, \qquad (7.18)$$

where the subscripts t and y designate the differentiation with respect to t and y.

Substituting Eqs. (7.15) through (7.18) into Eq. (7.14), we have

$$\begin{aligned}
\Delta y_n = {} & f_n(\Delta t) + (1/2!)(f_t + f_y f)_n(\Delta t)^2 \\
& + (1/3!)[f_{tt} + 2f_{ty}f + f_{yy}f^2 + (f_t + f_y f)f_y]_n(\Delta t)^3 \\
& + (1/4!)[f_{ttt} + \cdots]_n(\Delta t)^4,
\end{aligned} \qquad (7.19)$$

where the subscript n denotes that the functions are to be evaluated at the point (t_n, y_n).

Because there are many derivatives to be evaluated, it is far from practical to use Eq. (7.19) in computing the value Δy_n. To avoid this difficulty, we arbitrarily take

$$\Delta y_n = (\mu_0 z_0 + \mu_1 z_1 + \mu_2 z_2 + \cdots + \mu_m z_m), \qquad (7.20)$$

where

$$z_0 = f(t_n, y_n)\,\Delta t,$$
$$z_1 = f(t_n + \alpha_1\,\Delta t, y_n + \beta_{10}z_0)\,\Delta t,$$
$$z_2 = f(t_n + \alpha_2\,\Delta t, y_n + \beta_{20}z_0 + \beta_{21}z_1)\,\Delta t,$$
$$\vdots$$
$$z_m = f(t_n + \alpha_m\,\Delta t, y_n + \beta_{m0}z_0 + \beta_{m1}z_1 + \cdots)\,\Delta t,$$

and seek to determine all three sets of the constants μ and α, and the β. In Eq. (7.20) the subscript m is used to indicate that the Δy_n-value thus obtained coincides with Eq. (7.19) up to the term involving $(\Delta t)^{m+1}$.

If we take $m = 3$, then Eq. (7.20) becomes

$$\Delta y_n = \mu_0 z_0 + \mu_1 z_1 + \mu_2 z_2 + \mu_3 z_3, \tag{7.21}$$

where

$$z_0 = f(t_n, y_n) \, \Delta t,$$

$$z_1 = f(t_n + \alpha_1 \, \Delta t, y_n + \beta_{10} z_0) \, \Delta t,$$

$$z_2 = f(t_n + \alpha_2 \, \Delta t, y_n + \beta_{20} z_0 + \beta_{21} z_1) \, \Delta t,$$

$$z_3 = f(t_n + \alpha_3 \, \Delta t, y_n + \beta_{30} z_0 + \beta_{31} z_1 + \beta_{32} z_2) \, \Delta t.$$

Since Eq. (7.21) consists of four terms, it is often referred to as a fourth-order formula. We now seek to determine the following thirteen constants:

$$\mu_0, \mu_1, \mu_2, \mu_3;$$

$$\alpha_1, \alpha_2, \alpha_3;$$

$$\beta_{10}, \beta_{20}, \beta_{30}, \beta_{21}, \beta_{31}, \text{ and } \beta_{32}.$$

We recall† that the Taylor series for two independent variables about the point (a, b) is

$$f(a + h, b + k) = f(a, b) + f_x(a, b)h + f_y(a, b)k$$

$$+ \frac{1}{2!} \left[f_{xx}(a, b)h^2 + 2f_{xy}(a, b)hk + f_{yy}(a, b)k^2 \right] + \cdots, \tag{7.22}$$

which is frequently written symbolically as

$$f(a + h, b + k) = f(a, b) + \left(h \frac{\partial}{\partial x} + k \frac{\partial}{\partial y} \right) f(a, b)$$

$$+ \frac{1}{2!} \left(h \frac{\partial}{\partial x} + k \frac{\partial}{\partial y} \right)^2 f(a, b)$$

$$+ \frac{1}{3!} \left(h \frac{\partial}{\partial x} + k \frac{\partial}{\partial y} \right)^3 f(a, b) + \cdots \tag{7.23}$$

Substituting z_0 through z_3 into Eq. (7.23), we have, in the following symbolic form,

$$z_0 = f_n \, \Delta t, \tag{7.24}$$

$$z_1 = \left[f_n + \left(\alpha_1 \, \Delta t \frac{\partial}{\partial t} + \beta_{10} z_0 \frac{\partial}{\partial y} \right) f_n + \frac{1}{2!} \left(\alpha_1 \, \Delta t \frac{\partial}{\partial t} + \beta_{10} z_0 \frac{\partial}{\partial y} \right)^2 f_n \right.$$

$$\left. + \frac{1}{3!} \left(\alpha_1 \, \Delta t \frac{\partial}{\partial t} + \beta_{10} z_0 \frac{\partial}{\partial y} \right)^3 f_n + \cdots \right] \Delta t, \tag{7.25}$$

---

† For example, see Kaplan, W., *Advanced Calculus*. Addison-Wesley, Reading, Mass., p. 370 (1952).

$$z_2 = \left\{ f_n + \left[\alpha_2 \, \Delta t \, \frac{\partial}{\partial t} + (\beta_{20} z_0 + \beta_{21} z_1) \frac{\partial}{\partial y} \right] f_n \right.$$

$$+ \frac{1}{2!} \left[\alpha_2 \, \Delta t \, \frac{\partial}{\partial t} + (\beta_{20} z_0 + \beta_{21} z_1) \frac{\partial}{\partial y} \right]^2 f_n$$

$$\left. + \frac{1}{3!} \left[\alpha_2 \, \Delta t \, \frac{\partial}{\partial t} + (\beta_{20} z_0 + \beta_{21} z_1) \frac{\partial}{\partial y} \right]^3 f_n + \cdots \right\} \Delta t, \quad (7.26)$$

$$z_3 = \left\{ f_n + \left[\alpha_3 \, \Delta t \, \frac{\partial}{\partial t} + (\beta_{30} z_0 + \beta_{31} z_1 + \beta_{32} z_2) \frac{\partial}{\partial y} \right] f_n \right.$$

$$+ \frac{1}{2!} \left[\alpha_3 \, \Delta t \, \frac{\partial}{\partial t} + (\beta_{30} z_0 + \beta_{31} z_1 + \beta_{32} z_2) \frac{\partial}{\partial y} \right]^2 f_n$$

$$\left. + \frac{1}{3!} \left[\alpha_3 \, \Delta t \, \frac{\partial}{\partial t} + (\beta_{30} z_0 + \beta_{31} z_1 + \beta_{32} z_2) \frac{\partial}{\partial y} \right]^3 f_n + \cdots \right\} \Delta t. \quad (7.27)$$

So far, we have obtained two expressions for Δy_n: one from Eqs. (7.11) and (7.23) through (7.27); the other from Eq. (7.19). Equating all of the coefficients of corresponding terms in these two expressions, we have

$$\alpha_1 = \beta_{10},$$
$$\alpha_2 = \beta_{20} + \beta_{21},$$
$$\alpha_3 = \beta_{30} + \beta_{31} + \beta_{32};$$
$$\mu_0 + \mu_1 + \mu_2 + \mu_3 = 1,$$
$$\mu_1 \alpha_1 + \mu_2 \alpha_2 + \mu_3 \alpha_3 = \tfrac{1}{2},$$
$$\mu_1 \alpha_1^2 + \mu_2 \alpha_2^2 + \mu_3 \alpha_3^2 = \tfrac{1}{3},$$
$$\mu_1 \alpha_1^3 + \mu_2 \alpha_2^3 + \mu_3 \alpha_3^3 = \tfrac{1}{4}, \quad (7.28)$$
$$\mu_2 \alpha_1 \beta_{21} + \mu_3 (\alpha_1 \beta_{31} + \alpha_2 \beta_{32}) = \tfrac{1}{6},$$
$$\mu_2 \alpha_1^2 \beta_{21} + \mu_3 (\alpha_1^2 \beta_{31} + \alpha_2^2 \beta_{32}) = \tfrac{1}{12},$$
$$\mu_2 \alpha_1 \alpha_2 \beta_{21} + \mu_3 (\alpha_1 \beta_{31} + \alpha_2 \beta_{32}) \alpha_3 = \tfrac{1}{8},$$
$$\mu_3 \alpha_1 \beta_{21} \beta_{32} = \tfrac{1}{24}.$$

With eleven equations and thirteen unknowns, we arbitrarily take $\mu_1 = \mu_2 = \tfrac{1}{3}$.

It can be shown by simple substitution that Eqs. (7.28) are satisfied by the values shown in the following table:

μ_0	μ_1	μ_2	μ_3	α_1	α_2	α_3	β_{10}	β_{20}	β_{30}	β_{21}	β_{31}	β_{32}
$\tfrac{1}{6}$	$\tfrac{1}{3}$	$\tfrac{1}{3}$	$\tfrac{1}{6}$	$\tfrac{1}{2}$	$\tfrac{1}{2}$	1	$\tfrac{1}{2}$	0	0	$\tfrac{1}{2}$	0	1

Substituting these values into Eq. (7.20), we obtain Eq. (7.7), the Runge-Kutta formula with Runge's coefficients. If, however, a choice is made from

μ_0	μ_1	μ_2	μ_3	α_1	α_2	α_3	β_{10}	β_{20}	β_{30}	β_{21}	β_{31}	β_{32}
$\frac{1}{8}$	$\frac{3}{8}$	$\frac{3}{8}$	$\frac{1}{8}$	$\frac{1}{3}$	$\frac{2}{3}$	1	$\frac{1}{3}$	$-\frac{1}{3}$	1	1	-1	1

it can readily be shown that these values also satisfy Eqs. (7.28). Substituting them into Eq. (7.20), we have the following Runge-Kutta formula with coefficients due to Kutta:

$$y_{n+1} = y_n + \Delta y_n, \tag{7.29}$$

where

$$\Delta y_n = \frac{\Delta t}{8}(k_0 + 3k_1 + 3k_2 + k_3),$$

and

$$k_0 = f(t_n, y_n),$$

$$k_1 = f\left(t_n + \frac{\Delta t}{3}, y_n + \frac{k_0}{3}\Delta t\right),$$

$$k_2 = f\left[t_n + \frac{2\,\Delta t}{3}, y_n + \left(\frac{-k_0 + k_1}{3}\right)\Delta t\right],$$

$$k_3 = f[t_n + \Delta t, y_n + (k_0 - k_1 + k_2)\,\Delta t].$$

Before we turn to the next section, it is interesting to note that the Runge-Kutta equation with Runge coefficients (Eq. 7.7), has an important special case: when $y' = f(t, y)$ is independent of y, Eq. (7.7) becomes

$$y_{n+1} = y_n + \Delta y_n, \tag{7.7a}$$

where

$$\Delta y_n = \frac{\Delta t}{6}(k_0 + 2k_1 + 2k_2 + k_3)$$

and

$$k_0 = f(t_n),$$

$$k_1 = f\left(t_n + \frac{\Delta t}{2}\right),$$

$$k_2 = f\left(t_n + \frac{\Delta t}{2}\right),$$

$$k_3 = f(t_n + \Delta t),$$

or

$$\Delta y_n = \frac{\Delta t}{6}\left[f(t_n) + 4f\left(t_n + \frac{\Delta t}{2}\right) + f(t_n + \Delta t)\right].$$

This is the well-known Simpson rule.

7.4 RUNGE-KUTTA METHOD WITH GILL'S COEFFICIENTS

In the last section, it was mentioned that in Eqs. (6.28) there are eleven equations but thirteen unknowns; there are, therefore, two degrees of freedom in the choice of these thirteen unknowns. In the formula using Runge's coefficients, we arbitrarily take $\mu_1 = \mu_2 = \frac{1}{3}$. This choice leads to Eq. (7.7). In the formula using Kutta's coefficients, μ_1 and μ_2 are taken as $\frac{3}{8}$, and this second choice leads to Eq. (7.29).

A third, and different, choice of two out of the thirteen unknowns in Eqs. (7.28) was suggested by Gill. Here we arbitrarily take

$$\mu_1 = \frac{1}{3}\left(1 - \frac{1}{\sqrt{2}}\right) \quad \text{and} \quad \mu_2 = \frac{1}{3}\left(1 + \frac{1}{\sqrt{2}}\right).$$

By direct substitution it can be shown that Eqs. (7.28) are satisfied by the following values:

μ_0	$\dfrac{1}{6}$
μ_1	$\dfrac{1}{3}\left(1 - \dfrac{1}{\sqrt{2}}\right)$
μ_2	$\dfrac{1}{3}\left(1 + \dfrac{1}{\sqrt{2}}\right)$
μ_3	$\dfrac{1}{6}$

α_1	$\dfrac{1}{2}$
α_2	$\dfrac{1}{2}$
α_3	1

β_{10}	$\dfrac{1}{2}$
β_{20}	$-\dfrac{1}{2} + \dfrac{1}{\sqrt{2}}$
β_{30}	0
β_{21}	$1 - \dfrac{1}{\sqrt{2}}$
β_{31}	$\dfrac{-1}{\sqrt{2}}$
β_{32}	$1 + \dfrac{1}{\sqrt{2}}$

Substituting the above values into Eq. (7.21), we have

$$\Delta y_n = y_{n+1} - y_n$$
$$= \tfrac{1}{6}[z_0 + (2 - \sqrt{2})z_1 + (2 + \sqrt{2})z_2 + z_3], \tag{7.30}$$

where

$$z_0 = f(t_n, y_n)\,\Delta t,$$

$$z_1 = f(t_n + \tfrac{1}{2}\,\Delta t, y_n + \tfrac{1}{2}z_0)\,\Delta t,$$

$$z_2 = f\left[t_n + \tfrac{1}{2}\,\Delta t, y_n + \left(-\frac{1}{2} + \frac{1}{\sqrt{2}}\right)z_0 + \left(1 - \frac{1}{\sqrt{2}}\right)z_1\right]\Delta t,$$

$$z_3 = f\left[t_n + \Delta t, y_n - \frac{1}{\sqrt{2}}z_1 + \left(1 + \frac{1}{\sqrt{2}}\right)z_2\right]\Delta t.$$

The chief advantage of Gill's procedure lies in the requirement of a minimum amount of storage locations. This is important when the digital computer available has limited memory locations. To accomplish this minimization of storage, one may introduce the following three new quantities:

$$q_1 = z_0,$$
$$q_2 = (-2 + 3/\sqrt{2})z_0 + (2 - \sqrt{2})z_1,$$
$$q_3 = -\tfrac{1}{2}z_0 - (1 + \sqrt{2})z_1 + (2 + \sqrt{2})z_2.$$

(7.31)

For ordinary differential equations of first order, the algorithm we use to compute y_{n+1}, when we know t_n, y_n, and Δt, is:

Step 1

$$z_0 = f(t_n, y_n)\,\Delta t,$$
$$y_{n+1}^{(1)} = y_n + \tfrac{1}{2}z_0,$$
$$q_1 = z_0.$$

Step 2

$$z_1 = f(t_n + \tfrac{1}{2}\,\Delta t, y_{n+1}^{(1)})\,\Delta t,$$
$$y_{n+1}^{(2)} = y_{n+1}^{(1)} + (1 - \sqrt{2}/2)(z_1 - q_1),$$
$$q_2 = (-2 + 3/\sqrt{2})q_1 + (2 - \sqrt{2})z_1.$$

Step 3

$$z_2 = f(t_n + \tfrac{1}{2}\,\Delta t, y_{n+1}^{(2)})\,\Delta t,$$
$$y_{n+1}^{(3)} = y_{n+1}^{(2)} + (1 + 1/\sqrt{2})(z_2 - q_2),$$
$$q_3 = -(2 + 3/\sqrt{2})q_2 + (2 + \sqrt{2})z_2.$$

Step 4

$$z_3 = f(t_n + \Delta t, y_{n+1}^{(3)})\,\Delta t,$$
$$y_{n+1}^{(4)} = y_{n+1}^{(3)} + \tfrac{1}{6}z_3 - \tfrac{1}{3}q_3.$$

Here $y_{n+1}^{(4)}$ is the final answer at any stage. It should be noted from the above algorithm that four temporary storage spaces are required throughout the computation, namely, memory for one z-value, one y_{n+1}-value, one q-value, and one Δt-value. One memory space for the initial value t_0 should also be provided.

Steps 1 through 3 may be conveniently written in a shorthand form as follows:

$$y^{(j)} = y^{(j-1)} + a_j[f_{j-1}\,\Delta t - q_{j-1}],$$
$$q_j = (1 - 3a_j)q_{j-1} + 2a_j f_{j-1}\,\Delta t,$$

(7.32)

where

$j = 1, 2, 3$ (indicating the three different computational steps for a new Δy-value),

$t^{(0)}, y^{(0)}$ = values of t and y, respectively, at the beginning of an increment in which y is to be computed,

$$f_0 = f(t^{(0)}, y^{(0)}), \qquad f_1 = f(t^{(0)} + \tfrac{1}{2}\Delta t, y^{(1)}), \qquad f_2 = f(t^{(0)} + \tfrac{1}{2}\Delta t, y^{(2)}),$$

$q_0 = 0,$

$$a_1 = \tfrac{1}{2}, \qquad a_2 = 1 - (\sqrt{2}/2) = 0.2928, \qquad a_3 = 1 + (\sqrt{2}/2) = 1.707,$$

$y^{(1)}, y^{(2)}$ = values of y obtained in the first and second steps, respectively.

Example 7.2

The differential equation associated with the motion of a simple pendulum is

$$\frac{d\omega}{d\phi} = -\frac{g}{L}\frac{\sin \phi}{\omega}, \tag{7.9}$$

and the given initial conditions are

$$\phi = 79°, \qquad \omega = 0.66488546 \text{ rad/sec}, \qquad L = 2.5', \qquad g = 32.2014 \text{ ft/sec}^2.$$

Integrate to $\phi = 0°$; the interval of $\phi = -0.5°$.

Using the Runge-Kutta method with Gill's coefficients, we proceed as follows.

Solution: We shall show the detailed calculation to obtain ω at $\phi = 78.5$. Let f be the derivative of ω. We then have

$$f = \frac{d\omega}{d\phi} = \frac{-g}{L}\frac{\sin \phi}{\omega}.$$

Using Eq. (7.32) with $y^{(j)} = \omega_j$ and $t = \phi$, we obtain

$$\omega_1 = \omega_0 + a_1(f_0 \Delta \phi - q_0)$$

$$= 0.66488546 + \frac{1}{2}\left(\frac{-32.2014}{2.5}\frac{\sin 79°}{0.66488546}\frac{-0.5 \times 3.14159}{180}\right) = 0.747859,$$

where $q_0 = 0$. From Eq. (7.32), we have

$$q_1 = (1 - 3a_1)q_0 + 2a_1 f_0 \Delta \phi$$

$$= 0 + 2 \times 0.5 \left(\frac{-32.2014}{2.5}\frac{\sin 79°}{0.66488546}\frac{-0.5 \times 3.14159}{180}\right) = 0.165947,$$

$$\omega_2 = \omega_1 + a_2(f_1 \Delta \phi - q_1)$$

$$= 0.747859 + \left(1 - \frac{\sqrt{2}}{2}\right)$$

$$\times \left(\frac{-32.2014}{2.5}\frac{\sin 78.75°}{0.747859}\frac{-0.5 \times 3.14159}{180} - 0.165947\right) = 0.74243,$$

$$q_2 = \left(-2 + \frac{3}{\sqrt{2}}\right) z_0 + (2 - \sqrt{2})z_1$$

$$= \left(-2 + \frac{3}{\sqrt{2}}\right) \times 0.16594716 + (2 - \sqrt{2})$$

$$\times \left(\frac{-32.2014}{2.5} \frac{\sin 78.75°}{0.747859} \frac{-0.5 \times 3.14159}{180}\right) = 0.1065042,$$

$$\omega_3 = \omega_2 + a_3(f_2 \, \Delta\phi - q_2)$$

$$= 0.742431 + \left(1 + \frac{\sqrt{2}}{2}\right)$$

$$\times \left(\frac{-32.2014}{2.5} \frac{\sin 78.75°}{0.742431} \frac{-0.5 \times 3.14159}{180} - 0.1065042\right) = 0.814102,$$

$$q_3 = (1 - 3a_3)q_2 + 2a_3 f_2 \, \Delta\phi$$

$$= \left[1 - 3\left(1 + \frac{\sqrt{2}}{2}\right)\right] \times 0.1065042 + 2\left(1 + \frac{\sqrt{2}}{2}\right)$$

$$\times \left(\frac{-32.2014}{2.5} \frac{\sin 78.75°}{0.742431} \frac{-0.5 \times 3.14159}{180}\right) = 0.0680443,$$

$$\omega_4 = \omega_3 + \tfrac{1}{6} f(\phi_n + \Delta\phi, \omega_3) \, \Delta\phi - \tfrac{1}{3}q_3$$

$$= 0.814102 + \frac{1}{6}\left(\frac{-32.2014}{2.5} \frac{\sin 78.5°}{0.814102} \frac{-0.5 \times 3.14159}{180}\right) - \tfrac{1}{3}(0.0680443)$$

$$= 0.8139.$$

This is the value of ω at $\phi = 78.5°$. The process can be repeated for various ω-values. A flow chart and a FORTRAN program for this example are shown in Figs. 7.7 and 7.8, respectively. Table 7.3 lists the correct angular velocities.

7.5　HIGHER-ORDER ORDINARY DIFFERENTIAL EQUATIONS—BOUNCING BALL PROBLEM

In the development of the previous sections in this chapter, computer solutions of differential equations *of first order* were discussed in considerable detail. In this and the next section, we shall examine how to apply these procedures to ordinary differential equations of higher orders.

The standard procedure for the computer solution of higher-order ordinary differential equations is to transform them to systems of simultaneous first-order equations. In this section, the first-order equations will be solved by an approximation similar to Euler's procedure. We shall consider a specific example of tracking a moving ball which has an initial horizontal velocity v_0 (Fig. 7.9).

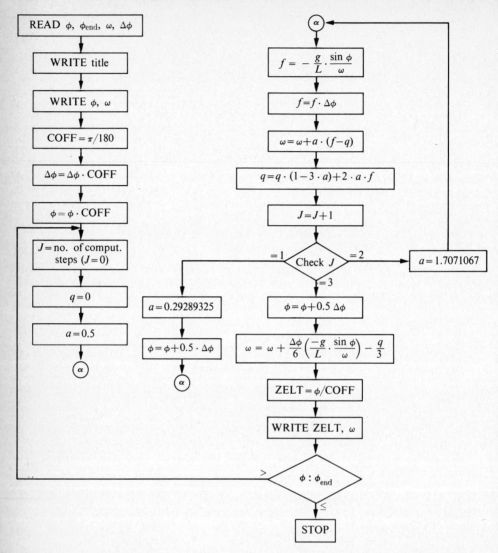

Fig. 7.7. Flow chart for solution of a first-order differential equation by the Runge-Kutta method using Gill's coefficients.

The two governing differential equations are

$$\frac{dx}{dt} = v_0 \tag{7.33}$$

and

$$\frac{d^2y}{dt^2} = -g. \tag{7.34}$$

```
C
C RUNGE-KUTTA-GILL METHOD TO SOLVE SIMPLE PENDULUM PROBLEM
C
C PHI=INITIAL PHI,    PHIN=PHI LAST    W=INITIAL OMEGA    DELTP=DELTA PHI
C
      DATA IREAD,IWRITE/5,6/
      READ(IREAD,7) PHI,PHIN,W,DELTP
    7 FORMAT(F5.1,F5.1,F9.8,F3.1)
      WRITE(IWRITE,100)
  100 FORMAT('1',21X,'TABLE 7.3. ANGULAR VELOCITIES'/,22X,'BY THE
     1 RUNGE-KUTTA-GILL METHOD'//)
      WRITE(IWRITE,500)
  500 FORMAT(24X,'PHI',5X,'ANG. VELOCITY'//)
      WRITE(IWRITE,5) PHI,W
    5 FORMAT(23X,F5.1,5X,F11.7)
      COFF=3.1415926/180.0
      DELTP=DELTP*COFF
      PHI=PHI*COFF
C
C THE FOLLOWING STEPS ARE FINDING GILL SOLUTIONS STEP BY STEP
C INITIALIZE THE COUNTER
C
    3 J=0
C
C INITIALIZE THE VALUE OF Q
C
      Q=0.
      A=0.5
    6 F=-32.2014/2.5*SIN(PHI)/W
      F=F*DELTP
      W=W+A*(F-Q)
      Q=Q*(1.0-3.0*A)+2.0*A*F
C
C INCREMENT
C
      J=J+1
      GO TO(12,13,14),J
   12 A=0.29289325
      PHI=PHI+0.5*DELTP
      GO TO 6
   13 A=1.7071067
      GO TO 6
   14 PHI=PHI+0.5*DELTP
      W=W+(-32.2014/2.5*SIN(PHI)/W*DELTP/6.0)-(Q/3.0)
      ZELT=PHI/COFF
      WRITE(IWRITE,5) ZELT,W
      IF(ZELT.GT.PHIN) GO TO 3
   20 STOP
      END


 79.0 64.0.66488546-.5
```

Fig. 7.8. FORTRAN program.

Equation (7.33) indicates that the horizontal velocity at any time is equal to a constant, v_0; Eq. (7.34) expresses the fact that the gravitational acceleration is equal to $g = 32.2$ ft/sec^2 throughout the motion.

Table 7.3 Angular velocities by the Runge-Kutta-Gill method

PHI	ANG. VELOCITY
79.0	.6648855
78.5	.8139816
78.0	.9395021
77.5	1.0499302
77.0	1.1496163
76.5	1.2411448
76.0	1.3261995
75.5	1.4059500
75.0	1.4812475
74.5	1.5527349
74.0	1.6209110
73.5	1.6861729
73.0	1.7488419
72.5	1.8091831
72.0	1.8674179
71.5	1.9237332
71.0	1.9782886
70.5	2.0312223
70.0	2.0826539
69.5	2.1326884
69.0	2.1814181
68.5	2.2289251
68.0	2.2752824
67.5	2.3205555
67.0	2.3648034
66.5	2.4080793
66.0	2.4504316
65.5	2.4919039
65.0	2.5325367
64.5	2.5723667
64.0	2.6114276

If we let

$$v = \frac{dy}{dt},$$

then

$$\frac{dv}{dt} = \frac{d^2y}{dt^2},$$

and Eqs. (7.33) and (7.34) may be recast in the following system of three *first-order equations*:

$$\frac{dx}{dt} = v_0, \tag{7.35a}$$

$$\frac{dv}{dt} = -32.2, \tag{7.35b}$$

$$\frac{dy}{dt} = v. \tag{7.35c}$$

In the following discussion we shall be concerned only with Eq. (7.35).

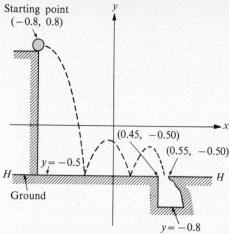

Fig. 7.9   Bouncing ball problem: initial horizontal velocity $= v_0$, initial vertical velocity $= 0$.

As the ball collides with the horizontal plane HH, it either bounces or falls through the hole. If it bounces, the horizontal velocity remains constant as v_0, and the vertical velocity v after one bounce is equal to the negative value of 85% of the vertical velocity before the bounce.

Our problem is to find the v_0 required to ensure that the ball falls into the hole. We first take $v_0 = 0.78$ ft/sec. If the ball misses the hole, we then try $v_0 = 0.775$ ft/sec, and so forth, decreasing v_0 by 0.005 ft/sec each time.

If the ball misses the hole, we continue the computation until $x = +0.58$. On the other hand, if the ball falls into the hole, the computation should go on until $y = -0.8$, and at this point this particular set of computation should be repeated. During this repeated computation, the step-by-step position of the ball, together with the corresponding time in seconds and the v_0-value, will be printed out. We shall use $\Delta t = 0.005$ sec.

Solution: For the sake of simplicity, let us use Euler's approximation as expressed in Eq. (7.8). As a result, Eqs. (7.35) may be replaced by the following equations:

$$x_{j+1} = x_j + v_0(\Delta t), \tag{7.36a}$$

$$v_{j+1} = v_j - 32.2(\Delta t), \tag{7.36b}$$

$$y_{j+1} = y_j + \tfrac{1}{2}(v_j + v_{j+1})\,\Delta t. \tag{7.36c}$$

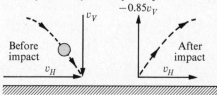

Fig. 7.10

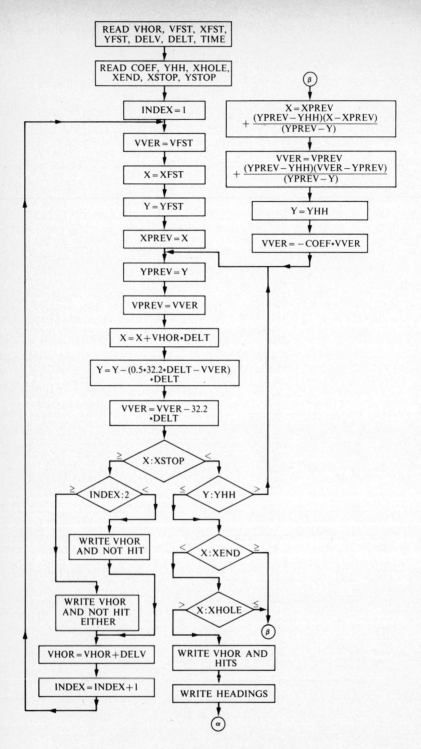

Fig. 7.11. Flow chart for bouncing ball problem.

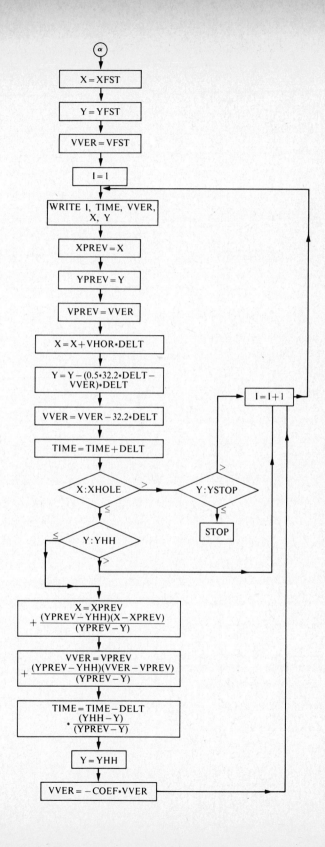

```
C      BOUNCING BALL PROBLEM
C  VHOR=HORIZONTAL VELOCITY, VFST=VERTICAL VELOCITY,
C  XFST=INITIAL X, YFST=INITIAL Y, DELV=DELTA V, DELT=DELTA T
C
       DATA NREAD,NWRITE/5,6/
       WRITE(NWRITE,9)
     9 FORMAT('1')
       READ(NREAD,1) VHOR,VFST,XFST,YFST,DELV,DELT,TIME
     1 FORMAT(7F10.4)
C
C  COEF=COEF.OF RESTITUTION, YHH=Y FOR GROUND,XHOLE=LEFT SIDE OF HOLE,
C  XEND=RIGHT SIDE OF HOLE, XSTOP=MAXIMUM X IF BALL MISSES HOLE,
C  YSTOP=MAXIMUM Y IF BALL HITS HOLE.
C
       READ(NREAD,2) COEF,YHH,XHOLE,XEND,XSTOP,YSTOP
     2 FORMAT(6F10.4)
       INDEX=1
    10 VVER=VFST
       X=XFST
       Y=YFST
    12 XPREV=X
       YPREV=Y
       VPREV=VVER
       X=X+VHOR*DELT
       Y=Y-(0.5*32.2*DELT-VVER)*DELT
       VVER=VVER-32.2*DELT
       IF(X.LT.XSTOP) GO TO 16
    14 IF(INDEX.GE.2) GO TO 42
    40 WRITE(NWRITE,3) INDEX,VHOR
     3 FORMAT(3X,'RUN ',I3,7X,'INITIAL VELOCITY IS ',F5.3)
       WRITE(NWRITE,131)
   131 FORMAT(44X,'DID NOT HIT THE HOLE')
       GO TO 44
    42 WRITE(NWRITE,3) INDEX,VHOR
       WRITE(NWRITE,132)
   132 FORMAT(44X,'DID NOT HIT THE HOLE EITHER')
    44 VHOR=VHOR+DELV
       INDEX=INDEX+1
       GO TO 10
    16 IF(Y.GT.YHH) GO TO 12
    20 IF(X.GE.XEND) GO TO 24
    22 IF(X.GT.XHOLE) GO TO 26
    24 X=XPREV+((YPREV-YHH)*(X-XPREV))/(YPREV-Y)
       VVER=VPREV+((VVER-VPREV)*(YPREV-YHH))/(YPREV-Y)
       Y=YHH
       VVER=-COEF*VVER
       GO TO 12
    26 WRITE(NWRITE,3) INDEX,VHOR
       WRITE(NWRITE,133)
   133 FORMAT(44X,'IT HITS THE HOLE',/)
       WRITE(NWRITE,6)
     6 FORMAT(10X,'THE NEXT TRAJECTORY IS FOR THE SUCCESSFUL RUN',///)
       WRITE(NWRITE,7)
     7 FORMAT(8X,'J',7X,'TIME',5X,'VERT. VEL.',9X,'X',14X,'Y',//)
       X=XFST
```

Fig. 7.12. FORTRAN program.

```
      Y=YFST
      VVER=VFST
      DO 34 I=1,1000
      WRITE(NWRITE,8) I,TIME,VVER,X,Y
  8   FORMAT(6X,I4,F10.3,3F15.6)
      XPREV=X
      YPREV=Y
      VPREV=VVER
      X=X+VHOR*DELT
      Y=Y-(0.5*32.2*DELT-VVER)*DELT
      VVER=VVER-32.2*DELT
      TIME=TIME+DELT
      IF(X.LE.XHOLE) GO TO 30
      IF(Y.GT.YSTOP) GO TO 34
      GO TO 36
  30  IF(Y.GT.YHH) GO TO 34
  32  X=XPREV+((YPREV-YHH)*(X-XPREV))/(YPREV-Y)
      VVER=VPREV+((VVER-VPREV)*(YPREV-YHH))/(YPREV-Y)
      TIME=TIME-DELT*(YHH-Y)/(YPREV-Y)
      Y=YHH
      VVER=-COEF*VVER
  34  CONTINUE
  36  STOP
      END
```

.7800	.0000	-.8000	.8000	-.0050	.0050	.0000
.8500	-.5000	.4500	.5500	.5800	-.8000	

Substituting the value of v_{j+1} in Eq. (7.36b) into Eq. (7.36c), one obtains

$$x_{j+1} = x_j + v_0(\Delta t),$$

$$v_{j+1} = v_j - 32.2(\Delta t),$$

$$y_{j+1} = y_j - [\tfrac{1}{2}g(\Delta t) - v_j]\,\Delta t. \tag{7.37}$$

The actual computational form may be arranged as follows:

$$v_x = 0.775$$

j	t	x	v_y	y
0	0	-0.8	0	0.8
1	0.005	-0.76125	-0.161	0.7987925
2	0.01	$\vdots$	$\vdots$	$\vdots$

    A sample flow chart and a FORTRAN program are shown in Figs. 7.11 and 7.12, respectively. This program is written to determine what initial velocity v_0 will cause the ball to drop into the hole. For this purpose, the following information is read in as data:

$$v_0 = 0.780, \qquad \Delta v = -0.005,$$

$$\Delta t = +0.005, \qquad e = 0.850.$$

Table 7.4   Answers to bouncing ball problem

RUN	1	INITIAL VELOCITY IS	.780

DID NOT HIT THE HOLE

| RUN | 2 | INITIAL VELOCITY IS | .775 |

DID NOT HIT THE HOLE EITHER

| RUN | 3 | INITIAL VELOCITY IS | .770 |

DID NOT HIT THE HOLE EITHER

| RUN | 4 | INITIAL VELOCITY IS | .765 |

DID NOT HIT THE HOLE EITHER

| RUN | 5 | INITIAL VELOCITY IS | .760 |

DID NOT HIT THE HOLE EITHER

| RUN | 6 | INITIAL VELOCITY IS | .755 |

DID NOT HIT THE HOLE EITHER

| RUN | 7 | INITIAL VELOCITY IS | .750 |

DID NOT HIT THE HOLE EITHER

| RUN | 8 | INITIAL VELOCITY IS | .745 |

DID NOT HIT THE HOLE EITHER

| RUN | 9 | INITIAL VELOCITY IS | .740 |

DID NOT HIT THE HOLE EITHER

| RUN | 10 | INITIAL VELOCITY IS | .735 |

IT HITS THE HOLE

THE NEXT TRAJECTORY IS FOR THE SUCCESSFUL RUN

J	TIME	VERT. VEL.	X	Y
+1	.000	.00000000	-.80000000	.80000000
+2	.005	-.16100000	-.79632500	.79959750
+3	.010	-.32200000	-.79265000	.79839000
+4	.015	-.48300000	-.78897500	.79637750
.....				
36	.175	-5.63500000	-.67137500	.30693750
.....				
57	.280	-9.01600000	-.59420000	-.46224000
58	.284	7.77721410	-.59114899	-.50000000
59	.289	7.61621410	-.58747399	-.46151643
.....				
79	.389	4.39621410	-.51397399	.13910497
.....				
105	.519	.21021410	-.41842399	.43852279
106	.524	.04921410	-.41474899	.43917136
107	.529	-.11178590	-.41107399	.43901494
.....				
144	.714	-6.06878590	-.27509899	-.13268785
.....				
154	.764	-7.67878590	-.23834899	-.47637715
155	.767	6.61029490	-.23611132	-.50000000
156	.772	6.44929490	-.23243632	-.46735103
.....				
176	.872	3.22929490	-.15893632	.01657840
.....				
196	.972	.00929490	-.08543632	.17850783
197	.977	-.15170510	-.08176132	.17815181
198	.982	-.31270510	-.07808632	.17699079
.....				

219	1.087	-3.69370510	-.00091132	-.03334567
•••••				
237	1.177	-6.59170510	.06523868	-.49618906
238	1.177	5.61858210	.06565848	-.50000000
239	1.182	5.45758210	.06933348	-.47230959
•••••				
253	1.252	3.20358210	.12078348	-.16916885
•••••				
272	1.347	.14458210	.19060848	-.01013105
273	1.352	-.01641790	.19428348	-.00981064
274	1.357	-.17741790	.19795848	-.01029523
•••••				
289	1.432	-2.59241790	.25308348	-.11416407
•••••				
307	1.522	-5.49041790	.31923348	-.47789169
308	1.526	4.77547370	.32215034	-.50000000
309	1.531	4.61447370	.32582534	-.47652514
•••••				
326	1.616	1.87747370	.38830034	-.20061752
•••••				
337	1.671	.10647370	.42872534	-.14605906
338	1.676	-.05452630	.43240034	-.14592920
339	1.681	-.21552630	.43607534	-.14660433
•••••				
353	1.751	-2.46952630	.48752534	-.24058115
•••••				
370	1.836	-5.20652630	.55000034	-.56681336
•••••				
377	1.871	-6.33352630	.57572534	-.76876427

So far as the problem is concerned, e means that the vertical velocity after the impact is 0.850 times as large as the vertical velocity before the impact (see Fig. 7.10).

If the ball misses, the calculation is carried out to $x = 0.580$ ft. If it drops into the hole, the calculation is carried out until it strikes the bottom. To record the hits and misses, the following format is used (use a new sheet for these results):

```
RUN 1   INITIAL VELOCITY IS .780   DID NOT HIT THE HOLE
RUN 2   INITIAL VELOCITY IS .775   DID NOT HIT THE HOLE EITHER
RUN 3   INITIAL VELOCITY IS .770   DID NOT HIT THE HOLE EITHER

____   _____   _____

RUN X   _____   IT HITS THE HOLE
```

After the correct v_0 for a hit is found, the next step is to print out the trajectory which the ball followed. Note that t is incremented by -0.005 sec, and that j is the number of iterations. (Use a new sheet for printing these results.)

J	T	X	VERT. VEL.	Y
1	0.000	-0.800	0.000	0.800
2	0.005	———	———	———
—	———	———	———	———
X	———	———	———	-.8000

Linear interpolation is used to obtain the vertical velocity, x-distance, and time when the ball goes below the ground (*HH* in Fig. 7.9). The correct printout of these answers appears in Table 7.4.

7.6 HIGHER-ORDER EQUATIONS BY RUNGE-KUTTA METHOD

In the previous section, a higher-order equation was transformed into a system of first-order equations and then solved by an approximation similar to Euler's method. We shall now turn to the solution of higher-order equations using the Runge-Kutta method with Runge's coefficients. Regardless of what method we may choose, the first step is to transform the given higher-order equation to a system of first-order equations.

Consider the second-order differential equation,

$$\frac{d^2y}{dt^2} + c\frac{dy}{dt} + k^2y = 0, \tag{7.38}$$

where c and k are two given constants. To develop a procedure which can be easily used for a computer solution, we set

$$y_1 = y \quad \text{and} \quad y_2 = \frac{dy}{dt} = \frac{dy_1}{dt}.$$

We then let f_1 and f_2 stand for actual functions which are to be used to find the values of Δy_1 and Δy_2, respectively. We see that f_1 and f_2 can be easily obtained by transforming the original differential equation into two simultaneous first-order differential equations. If we set

$$f_1 = \frac{dy_1}{dt} = y_2, \tag{7.39}$$

we can easily solve (7.38) for f_2:

$$f_2 = \frac{dy_2}{dt} = -cy_2 - k^2y_1. \tag{7.40}$$

To obtain a numerical solution for Eq. (7.38), we need a set of initial conditions. Here we take $y = 1$ and $dy/dt = 0$ at $t = 0$. Final values of t and Δt are also needed. These are taken as 3 and 0.2, respectively. The basic problem is, therefore, to determine the values of y_1 and y_2 as t increases by the increment Δt. We have already discussed one simple, but crude, method in Section 7.5. If we simply choose

$$\Delta y_1 = \Delta t f_1, \quad \text{at} \quad t = t_0 \quad \text{and} \quad \Delta y_2 = \Delta t f_2, \quad \text{at} \quad t = t_0,$$

then
$$y_1 \ (\text{at} \quad t = t_0 + \Delta t) = y_1 \ (\text{at} \quad t = t_0) \ + \Delta y_1.$$

$$y_2 \ (\text{at} \quad t = t_0 + \Delta t) = y_2 \ (\text{at} \quad t = t_0) \ + \Delta y_2. \tag{7.41}$$

This equation will directly give us approximations for y and $y'(y_1$ and $y_2)$ at $t + \Delta t$. These, in turn, will enable us to find new Δy_1 and Δy_2, which will give us y_1 and y_2 at $t = t_0 + 2\,\Delta t$. The procedure can be carried on indefinitely.

Up to now, all that has been done toward developing the Runge-Kutta method is the transformation of Eq. (7.38) into two simultaneous first-order equations. We shall now

1) further develop the notation,

2) present the basic Runge-Kutta formula for solving simultaneous first-order equations,

3) and, finally, use this for solving Eqs. (7.39) and (7.40).

Let us develop these steps.

1) The notation used in Eq. (7.41) is awkward. We can simplify it by letting

t_0 = initial value of t,

$t_n = n \, \Delta t + t_0$,

$y_n^{(2)} = y_2 \Big|_{t=t_n}$ (note that the superscript 2 is used to indicate the second ordinary differential equation of first order),

m = the total number of first-order simultaneous equations that must be solved,

$y_n^{(i)}$ ($i = 1, \ldots, m$) = the value of any one of the dependent variables at $t = t_n$.

$$f^{(i)}(t_n, y_n^{(1)}, y_n^{(2)}, \ldots, y_n^{(m)}) = \frac{dy_n^{(i)}}{dt}.$$

2) If the values of the dependent variables in the m first-order differential equations are given at $t = t_n$, we can use the following algorithm to find their numerical values at $t_{n+1} = t_n + \Delta t$:

$$y_{n+1}^{(i)} = y_n^{(i)} + \Delta y_n^{(i)}, \tag{7.42}$$

where

$$\Delta y_n^{(i)} = \frac{\Delta t}{6} (k_0^{(i)} + 2k_1^{(i)} + 2k_2^{(i)} + k_3^{(i)})$$

and

$$k_0^{(i)} = f^{(i)}(t_n, y_n^{(i)}),$$

$$k_1^{(i)} = f^{(i)} \left(t_n + \frac{\Delta t}{2}, \ y_n^{(i)} + \frac{k_0^{(i)}}{2} \, \Delta t \right),$$

$$k_2^{(i)} = f^{(i)} \left(t_n + \frac{\Delta t}{2}, \ y_n^{(i)} + \frac{k_1^{(i)}}{2} \, \Delta t \right),$$

$$k_3^{(i)} = f^{(i)}(t_n + \Delta t, \ y_n^{(i)} + k_2^{(i)} \, \Delta t),$$

and the superscript $i = 1, 2, 3, \ldots, m$ serves to indicate the m first-order differential equations.

If, for example, we have two simultaneous equations of first order, $m = 2$, then

$$\dot{y} = f(t, y, v), \qquad \dot{v} = g(t, y, v). \tag{7.43}$$

```
      SUBROUTINE RUNGE(T,DT,N,Y,DY,F,L,M,J)
      DIMENSION DY(2),Y(2),F(14)
      GO TO (100,110,300),L
100   GO TO (101,110) ,IG
101   J = 1
      L = 2
      DO 106 K = 1,N
      K1 = K+3*N
      K2 = K1+N
      K3 = N + K
      F(K1) = Y(K)
      F(K3) = F(K1)
106   F(K2) = DY(K)
      GO TO 406
110   DO 140 K=1,N
      K1 = K
      K2 = K+5*N
      K3 = K2+N
      K4 = K + N
      GO TO (111,112,113,114),J
111   F(K1) = DY(K)*DT
      Y(K) = F(K4)+.5*F(K1)
      GO TO 140
112   F(K2) = DY(K)*DT
      GO TO 124
113   F(K3) = DY(K)*DT
      GO TO 134
114   Y(K) = F(K4) +(F(K1)+2.*(F(K2)+F(K3))+DY(K)*DT)/6.
      GO TO 140
124   Y(K) = .5*F(K2)
      Y(K) = Y(K)+F(K4)
      GO TO 140
134   Y(K) = F(K4)+F(K3)
140   CONTINUE
      GO TO (170,180,170,180),J
170   T = T + .5*DT
180   J = J+1
      IF(J-4)404,404,299
299   M=1
      GO TO 406
300   IG=1
      GO TO 405
404   IG=2
405   L=1
406   RETURN
      END
```

Fig. 7.13. Subprogram RUNGE.

Equation (7.42) now takes the form

$$y_{n+1} = y_n + \tfrac{1}{6}(k_0 + 2k_1 + 2k_2 + k_3)\, \Delta t,$$
$$v_{n+1} = v_n + \tfrac{1}{6}(m_0 + 2m_1 + 2m_2 + m_3)\, \Delta t, \qquad (7.44)$$

where

$$k_0 = f(t_n, y_n, v_n),$$

$$k_1 = f(t_n + \Delta t/2, y_n + k_0 \Delta t/2, v_n + m_0 \Delta t/2),$$

$$k_2 = f(t_n + \Delta t/2, y_n + k_1 \Delta t/2, v_n + m_1 \Delta t/2),$$

$$k_3 = f(t_n + \Delta t, y_n + k_2 \Delta t, v_n + m_2 \Delta t),$$

$$m_0 = g(t_n, y_n, v_n),$$

$$m_1 = g(t_n + \Delta t/2, y_n + k_0 \Delta t/2, v_n + m_0 \Delta t/2),$$

$$m_2 = g(t_n + \Delta t/2, y_n + k_1 \Delta t/2, v_n + m_1 \Delta t/2),$$

$$m_3 = g(t_n + \Delta t, y_n + k_2 \Delta t, v_n + m_2 \Delta t).$$

3) We now turn to the evaluation of Eq. (7.42), the basic Runge-Kutta procedure. A subprogram RUNGE is provided for this purpose (see Fig. 7.13), in which the given initial conditions for the N dependent variables are used as solutions at $t = t_0$. RUNGE performs the computation necessary to find the solution for the new value of the independent variable T. The following symbols are used:

T = independent variable,

DT = step size,

N = number of dependent variables,

Y = first dependent variable followed by the remaining N − 1 dependent variables (must be dimensioned),

DY = derivative of first dependent variable followed by the remaining N − 1 derivatives in the same order as the corresponding dependent variables (must be dimensioned),

F = storages (must be dimensioned as 7N),

L = a fixed-point variable controlled by subroutine RUNGE to exit to the proper section of the program calling RUNGE,

J = same as L,

M = a fixed-point variable used to return from subprogram RUNGE to the main program.

A flow chart for the subroutine RUNGE is shown in Fig. 7.14.

Example

Write a main program, together with the required data cards, to numerically solve Eqs. (7.39) and (7.40), where

$$c = 0.1, \qquad k = 3.2, \qquad \Delta t = 0.2,$$

the initial conditions are $t = 0$, $y_1 = 1$, $y_2 = 0$, and the range of t is $0 \le t \le 3$.

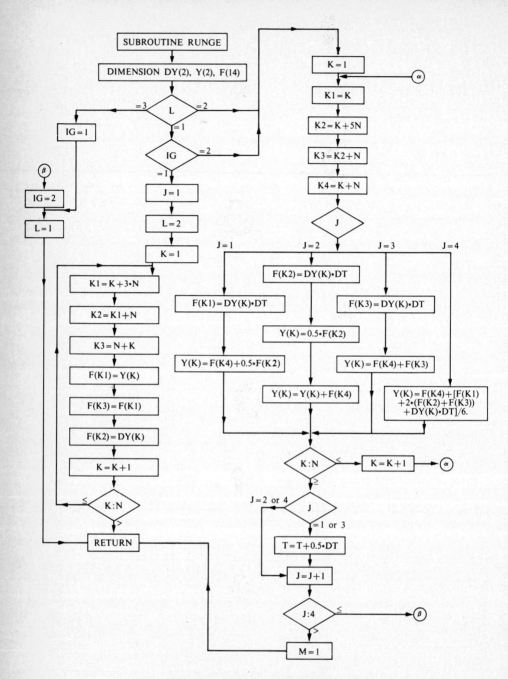

Fig. 7.14. Flow chart for RUNGE.

```
C HIGHER ORDER EQUATIONS BY RUNGE-KUTTA METHOD
C
C SOLVE THE EQUATION   D2Y+C*(DY)+(K*K)*Y=0
C C=0.1     K=3.2     DT=0.2
C INITIAL CONTIONS
C      T=0.      YI=1.      Y2=0
C RANGE OF T      0 TO 3
      DATA IREAD,IWRITE/5,6/
      DIMENSION DY(2),Y(2),F(14)
C
C READ IN NUMBER OF DEPENDENT VARIABLES, INDEPENDENT VARIABLE,ETC
C
      READ(IREAD,1) N,DT,T,Y(1),Y(2)
    1 FORMAT(I2,F5.2,F5.2,F5.2,F5.2)
      WRITE(IWRITE,3)
    3 FORMAT('1',10X,'TABLE 7.5. COMPUTER SOLUTIONS'//)
      WRITE(IWRITE,59)
   59 FORMAT(16X,'T',20X,'Y',15X,'DY/DT'//)
C
C DEFINE TWO FIXED-POINT VARIABLES
C
   10 L=3
      M=0
   50 CALL RUNGE(T,DT,N,Y,DY,F,L,M,J)
      IF(M.EQ.1) GO TO 10
   75 GO TO (100,200,999),L
  100 DY(1)=Y(2)
      DY(2)=-0.1*Y(2)-3.2*3.2*Y(1)
      GO TO 50
C
C PRINT OUT RESULTS
C
  200 WRITE(IWRITE,800) T,Y(1),Y(2)
C
C TEST FOR TERMINATION
C
  250 IF(T.GE.3.) GO TO 999
  260 GO TO 50
  800 FORMAT(10X,E15.8,5X,E15.8,5X,E15.8)
  999 STOP
      END
```

Fig. 7.15. Main program.

This main program is used to (a) call the subroutine RUNGE; (b) read in Eqs. (7.39) and (7.40); (c) write out answers; and (d) terminate the calculations. Figures 7.15 and 7.16 show a possible main program and a flow chart, respectively. The calling sequence in the main program to call subroutine RUNGE consists of the three statements

```
      L = 3
S     CALL RUNGE (T, DT, N, Y, DY, F, L, M, J)
      GO TO (BOXA, BOXB, BOXC), L
```

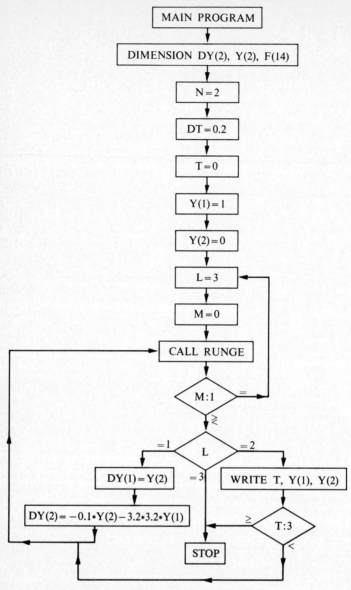

Fig. 7.16. Main program flow chart.

where

S = a statement number,

BOXA = statement number for derivative computations,

BOXB = statement number for print routine and check for end of program,

BOXC = statement number for STOP.

The statements related with BOXA, BOXB, and BOXC are:

BOXA	... Evaluate N derivatives
	⋮
	GO TO S
BOXB	PRINT
	⋮
	GO TO S
BOXC	STOP

The computer solutions are shown in Table 7.5

Table 7.5 Computer solutions

T	Y	DY/DT
.00000000E-99	.10000000E+01	.00000000E-99
.20000000E+00	.80354910E+00	-.18892436E+01
.40000000E+00	.29713235E+00	-.30013440E+01
.60000000E+00	-.31497695E+00	-.29177088E+01
.80000000E+00	-.79140636E+00	-.16956232E+01
.10000000E+01	-.94877030E+00	.16392680E+00
.12000000E+01	-.73213954E+00	.19211571E+01
.14000000E+01	-.23386334E+00	.28914892E+01
.16000000E+01	.34554886E+00	.27119313E+01
.18000000E+01	.77800717E+00	.14763096E+01
.20000000E+01	.89754080E+00	-.31079540E+00
.22000000E+01	.66387740E+00	-.19396786E+01
.24000000E+01	.17559422E+00	-.27770667E+01
.26000000E+01	-.37126041E+00	-.25120137E+01
.28000000E+01	-.76178357E+00	-.12707790E+01
.30000000E+01	-.84658468E+00	.44150700E+00

BIBLIOGRAPHY

General

Bennett, A. A., W. E. Milne, and H. Bateman, *Numerical Integration of Differential Equations.* Dover, New York, 1956.

Clenshaw, C. W., and F. W. J. Oliver, "Solution of differential equations by recurrence relations," *Math. Tab., Wash.*, **5,** 34–39 (1951).

Collatz, L., *The Numerical Treatment of Differential Equations*, 3rd ed. Translated by P. G. Williams. Springer, Berlin, 1960.

Fox, L , "A note on the numerical integration of first-order differential equations," *Quart. J. Mech.*, **7,** 367–378 (1954).

Fox, L., *The Numerical Solution of Two-Point Boundary Problems in Ordinary Differential Equations.* Oxford University Press, London, 1957.

Henrici, P., *Discrete Variable Methods in Ordinary Differential Equations.* Wiley, New York, 1962.

Hull, T. E., and A. C. R. Newberry, "Integration procedures which minimize propagated errors," *J. Soc. Ind. Appl. Math.*, **9**, 31–47 (1961).

Levy, H., and E. A. Baggott, *Numerical Studies in Differential Equations.* Dover, New York, 1950.

Lindelof, E., "Remarques sur l'intégration numérique des équations différentielles ordinaires," *Acta Soc. Sci. Fennic. Nova Ser. A.*, **2**, 1–21 (1938).

Milne, W. E., *Numerical Solution of Differential Equations.* Wiley, New York, 1953.

National Physical Laboratory, *Modern Computing Methods, Notes on Applied Science,* No. 16, 2nd ed. H.M. Stationery Office, London, 1961.

Ralston, A., "Numerical integration methods for the solution of ordinary differential equations," *Mathematical Methods for Digital Computers*, pp. 95–109, A. Ralston and H. S. Wilf, eds. Wiley, New York, 1960.

Todd, J., "Notes on modern numerical analysis, I, solution of differential functions by recurrence relations," *Math. Tables Aids Comput.*, **4**, 39–44 (1950).

Voight, S. J., "Bibliography on the numerical solution on integral and differential equations and related topics," Report 2423, Naval Ship Res. and Dev. Center, Wash. D.C., 1967.

Runge-Kutta Method

Conte, S. D., and R. F. Reeves, "A Kutta third-order procedure for solving differential equations requiring minimum storage," *J. Assoc. Comput. Mach.*, **3**, 22–25 (1956).

Fehlberg, E., "Classical Fifth-, Sixth-, Seventh-, and Eighth-Order Runge-Kutta Formulas with Stepsize Control." Clearinghouse, U.S. Dept. Commerce, Springfield, Va., NASA TR T-287, Oct. 1968.

Gill, S., "A process for the step-by-step integration of differential equations in an automatic digital computing machine," *Proc. Cambridge Phil. Soc.*, **47**, 96–108 (1951).

Heun, K., "Neue Methode zur Approximativen Integration der Differentialgleichungen einer Unabhängigen Variabel," *Z. Math. Phys.*, **45**, 23–38 (1900).

Kutta, W., "Beitrag zur Näherungsweisen Integration Totaler Differentialgleichungen," *Z. Math. Phys.*, **46**, 435–453 (1901).

Martin, D. W., "Runge-Kutta methods for integrating differential equations on high-speed digital computers," *Comput. J.*, **1**, 118–123 (1958).

Ralston, A., "Runge-Kutta methods with minimum error bounds," *MTAC*, **16**, no. 80, 431–437 (1962).

Rice, J. R., "Split Runge-Kutta method for simultaneous equations," *J. Res. Natn. Bur. Stand.*, **64B**, 151–170 (1960).

Romanelli, M. J., "Runge-Kutta methods for the solution of ordinary differential equations," *Mathematical Methods for Digital Computers*, pp. 110–120, A. Ralston and H. S. Wilf, eds. Wiley, New York, 1960.

Runge, C., "Über die numerische Auflösung von Differentialgleichungen," *Math. Ann.*, **46**, 167–178 (1895).

Runge, C., "Über die numerische Auflösung totaler Differentialgleichungen," *Nachr. K. Ges. Wiss. Göttingen*, 252–257 (1905).

Adams' Method

Alonso, R., "A starting method for the three-point Adams predictor-corrector method," *J. Assoc. Comput. Mach.*, **7,** 176–180 (1960).

Predictor-Corrector Procedures

Bulirsch, R., and J. Stoer, "Numerical treatment of ordinary differential equations by extrapolation methods," *Numerische Mathematik*, **8,** 1–13 (1966).

Crane, R. L., and R. W. Klopfenstein, "A predictor-corrector algorithm with an increased range of absolute stability," *J. Assoc. Comput. Mach.*, **12,** no. 2, 227–241 (1965).

Forrington, C. V. D., "Extensions of the predictor-corrector method for the solution of systems of ordinary differential equations," *Comput. J.*, **4,** 80–84 (1961).

Hamming, R. W., "Stable predictor-corrector methods for ordinary differential equations," *J. Assoc. Comput. Mach.*, **6,** 37–47 (1959).

Ralston, A., "Some theoretical and computational matters relating to predictor-corrector methods of numerical integration," *Comput. J.*, **4,** 64–67 (1961).

Miscellaneous Methods

Brock, P., and F. J. Murray, "The use of exponential sums in step-by-step integration," *Math. Tables Aids Comput.*, **6,** 63–78 (1952).

Clenshaw, C. W., "The numerical solution of linear differential equations in Chebyshev series," *Proc. Cambridge Phil. Soc.*, **53,** 134–149 (1957).

Curtiss, J. H., "Sampling methods applied to differential and integral equations," *Proceedings, Seminar on Scientific Computation.* IBM Corporation, New York, 1949.

Dennis, S. C., "The numerical integration of ordinary differential equations possessing exponential type solutions," *Proc. Cambridge Phil. Soc.*, **56,** 240–246 (1960).

de Vogelaere, R., "A method for the numerical integration of differential equations of second order without explicit first derivatives," *J. Res. Natn. Bur. Stand.*, **54,** 119–125 (1955).

Fox, L., "The solution by relaxation methods of ordinary differential equations," *Proc. Cambridge Phil. Soc.*, **45,** 50–68 (1949).

Fox, L., and E. T. Goodwin, "Some new methods for the numerical integration of ordinary differential equations," *Proc. Cambridge Phil. Soc.*, **45,** 373–388 (1949).

Norton, H. J., "The iterative solution of non-linear ordinary differential equations in Chebyshev series," *Comput. J.*, **7,** 76–85 (1964).

Rahme, H. S., "Stability analysis of a new algorithm used for integrating a system of ordinary differential equations," *J. Assoc. Comput. Mach.*, **17,** no. 2, 284–293 (1970).

Richards, P. L., W. D. Lanning, and M. D. Torrey, "Numerical integration of large, highly damped nonlinear systems," *SIAM Rev.*, **7,** 376–380 (1965).

Sandberg, I. W., and H. Shichman, "Numerical integration of systems of stiff nonlinear differential equations," *Bell Systems Tech. J.*, **47,** 511–527 (1968).

Salzer, H. E., "Osculatory extrapolation and a new method for the numerical integration of differential equations," *J. Franklin Inst.*, **262,** 111–119 (1956).

Stoller, L., and D. Morrison, "A method for the numerical integration of ordinary differential equations," *Math. Tables Aids Comput.*, **12**, 269–272 (1958).

Wilf, H. S., "An open formula for the numerical integration of first-order differential equations," *Math. Tables Aids Comput.*, **11**, 201–203 (1957); **12**, 55–58 (1958).

PROBLEMS

In the following problems compare the Runge-Kutta solution with the closed form solution; or compare the Runge-Kutta solution with the Euler's solution, when the analytic solution is not available. In many of these problems, the choice of increment size and interval length in the Runge-Kutta and Euler procedures has been left to the reader.

1. A cylindrical tank with a small hole at its bottom is full of water (Fig. 7.17). Water is flowing out through the hole due to gravity. The differential equation expressing the depth in time is

$$\frac{-dh}{dt} = \frac{d^2}{D^2}\sqrt{2gh}.$$

   Find the *h-t* relation by Euler and Runge-Kutta procedures. Given: $D = 5$ ft, $d = 2$ in., and $h_0 = 10$ ft. Compare your results with the analytical solution

$$t = (2/c_1)(\sqrt{h_0} - \sqrt{h}),$$

   where $h_0 = 10$ ft, $c_1 = (d^2/D^2)\sqrt{2g}$.

2. Let x be the temperature at time t of a body immersed in a medium of temperature described by the expression

$$G(t) = -100(t - 4).$$

   The temperature satisfies the differential equation

$$dx/dt + kx = k\dot{G}(t),$$

   where k is a proportionality constant equal to 0.005.

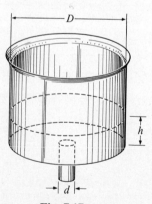

Fig. 7.17

and the initial condition is that at

$$t = 0, \qquad x = 1000°F.$$

Obtain the x-t relation by the Runge-Kutta procedure and compare the result with the analytical solution

$$x = 400 - \left(\frac{100}{k}\right)(kt - 1) + \left(600 - \frac{100}{k}\right)e^{-kt}.$$

3. Find the relation between velocity v and altitude r for a projectile that has been fired with initial velocity v_0 from the surface of the earth. The equation is

$$v\frac{dv}{dr} = -g\frac{R^2}{r^2},$$

where $R = 20,908,800$ ft, $g = 32.2$ ft/sec^2, and the initial condition is $v_0 = 10,000$ ft/sec at $r = R$.

Use the Runge-Kutta procedure and compare your answers with the analytical solution

$$v = \sqrt{(2gR^2/r) + v_0^2 - 2gR}.$$

4. The current i in the driven RL circuit at any time t after a switch is thrown at $t = 0$ can be expressed by the equation

$$di/dt = (E \sin \omega t - Ri)/L,$$

where $E = 100$ volts, $L = 1$ henry, $\omega = 600$, $R = 100$ ohms and the initial condition is that at $t = 0$, $i = 0$.

Solve the differential equation numerically using the Runge-Kutta method and compare your answers with the analytical solution.

$$i = \frac{E}{Z^2}(R \sin \omega t - \omega L \cos \omega t + \omega L e^{-Rt/L}),$$

where

$$Z = \sqrt{R^2 + \omega^2 L^2}.$$

5. A quantity of 20 lb of sugar is dumped into a vessel containing 120 lb of water. The concentration of the solution, c, in percentage at any time t, is expressed as

$$(120 - 1.212c)\,dc/dt = (k/3)(200 - 14c)(100 - 4c),$$

where k, the mass transfer coefficient, is equal to 0.05889. The initial condition is that at $t = 0$, $c = 0$.

Find the c-t relation by the Runge-Kutta and Euler methods.

6. The reaction of NO with H_2 is known as a third-order reaction, and its rate is given by

$$\frac{dx}{dt} = k(P_{NO} - 2x)^2(P_{H_2} - x),$$

where x is the partial pressure of the product; k is the velocity constant; P_{NO} and P_{H_2} denote the partial pressures of NO and H_2, respectively, in the reaction mixture before the reaction takes place; $P_{NO} - 2x$ is the partial pressure of NO in the reactor; and

$P_{H_2} - x$ is the partial pressure of H_2 in the reactor at any time after the reaction has been started. In this problem

$$P_{NO} = 359 \text{ mm Hg},$$

$$P_{H_2} = 400 \text{ mm Hg},$$

$$k = 1.12 \times 10^{-7} \text{ mm}^{-2}\text{sec}^{-1},$$

and the initial condition is that $x_0 = 0, 0 \le t \le 3$ min.

Find the x-t relation by the Runge-Kutta procedure and compare the result with the analytical solution.

7. A rod protrudes from a satellite into a solar radiation field (see Fig. 7.18). The differential equation based on a one-dimensional heat-condition model is

$$\frac{dT}{dx} = \sqrt{\frac{2\sigma\epsilon CT^4}{5kA} - \frac{2S\alpha DT \sin \theta}{kA}} + K,$$

where

$\alpha = \text{absorptivity} = 0.4,$

$\epsilon = \text{emissivity} = 0.4,$

$\sigma = \text{Stefan-Boltzmann constant}$

$\quad = 0.173 \times 10^{-8} \text{ Btu/hr} \cdot \text{ft}^2 \cdot {}^\circ\text{R}^4,$

$A = \text{cross-sectional area of the rod},$

$C = \text{circumference of the rod},$

$D = \text{diameter of the rod} = 1 \text{ in.},$

$K = \text{constant of integration},$

$L = \text{length of rod} = 3.0 \text{ ft},$

$S = \text{solar radiation constant} = 425 \text{ Btu/hr} \cdot \text{ft}^2,$

$T = \text{temperature of the rod } {}^\circ\text{R},$

$k = \text{thermal conductivity of rod} = 100 \text{ Btu/hr} \cdot \text{ft} \cdot {}^\circ\text{R},$

$\theta = 30^\circ.$

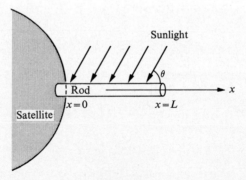

Fig. 7.18

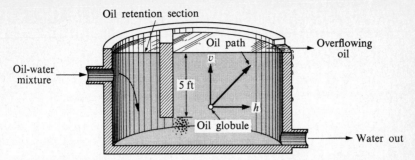

Fig. 7.19

The initial condition is that at $x = 3.0$ ft, $T = 637.6285$ °R, and $dT/dx = 0$.

Using the Runge-Kutta and Euler procedures, find the T-x relation.

8. A standard oil separator is used to separate an oil-water mixture. The overflow level of oil is 5 ft above the lower end of the retention wall (see Fig. 7.19). Obtain the v-t relation by using the equation

$$\frac{dv}{dt} = \left[\frac{(\rho_w - \rho_o)}{\rho_o}\right] g - \left[\frac{3 f \rho_w}{4 D \rho_o}\right] v^2,$$

where

$D = $ average diameter of oil globule $= \frac{1}{16}$ in.,

$v = $ vertical velocity of oil globule,

$\rho_o = $ density of oil $= 0.6 \rho_w$,

$\rho_w = $ density of water,

$g = $ acceleration due to gravity $= 32.2$ ft/sec^2,

$f = $ coefficient of viscous friction $= 0.06$.

The initial condition is that at

$$t = 0, \qquad v = 0 \qquad (0 \le t \le 0.03 \text{ sec}).$$

Use Euler and Runge-Kutta procedures to solve numerically the following first order differential equations. Compare the results with the analytical solutions.

9. $dy/dx = x + \sin x + y$. Initial condition: $x = 0$, $y = 0.5$.
Analytical solution: $y = 2e^x - x - 1 - (\cos x + \sin x)/2$.

10. $dy/dx = yx/(y^2 + x^2)$. Initial condition: $x = 0.1223$, $y = 0.05$.
Analytical solution: $y = e^{x^2/2y^2}$.

11. $dv/dx = (3x^2 + 2x)/8v$. Initial condition: $x = 3$, $v = 3$.
Analytical solution: $v = (x/2)\sqrt{x + 1}$.

12. The differential equation that describes a sphere in which heat is generated by a heat source at its center is expressed as

$$d^2T/dr^2 + (2/r)(dT/dr) + w/k = 0,$$

where T indicates the temperature inside the sphere and k is the thermal conductivity of the sphere in Btu/ft · °F · hr. The temperature of the heat source is kept at a constant temperature T_c, while the temperature gradient dT/dr is zero at the center. Let the diameter of the sphere equal 2 ft. Other values are

$$w = 100 \text{ Btu/hr} \cdot \text{ft}^3,$$

$$k = 212 \text{ Btu/ft} \cdot °F \cdot \text{hr},$$

$$T_c = 900° \text{ F.}$$

What is the axial temperature distribution of the sphere? Use the Runge-Kutta procedure and $\Delta r = 0.1$ ft.

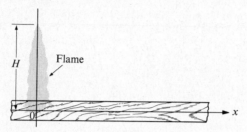

Fig. 7.20

13. A long, horizontal strip of wood is burned at one end as shown in Fig. 7.20. The differential equation of temperature profile ahead of the flame front is

$$\frac{d^2T}{dx^2} + \frac{Vs\rho}{k} \cdot \frac{dT}{dx} - \frac{2U}{kL}(T - T_a) - \frac{2\sigma\epsilon}{kL}(T^4 - T_a^4)$$

$$+ \frac{1}{2}\frac{\sigma\epsilon\epsilon_f}{kL}(T_f^4 - T_a^4)\left(1 - \frac{x}{\sqrt{x^2 + H^2}}\right) = 0,$$

where

$V = $ velocity of the flame propagation along the x-axis $= 11.010$ ft/hr,

$s = $ specific heat of wood $= 0.55$ Btu/°R $\cdot$ lb$_m$,

$\rho = $ density of wood $= 20{,}362$ lb$_m$/ft^3,

$k = $ thermal conductivity of wood $= 0.2$ Btu/ft $\cdot$ hr $\cdot$ °R,

$U = $ mean heat transfer coefficient between the wood and air $= 2.0$ Btu/ft$^2 \cdot$ hr $\cdot$ °R,

$L = $ thickness of the wooden strip $= 0.00526$ ft,

$T_a = $ temperature of surrounding air $= 530$ °R,

$T_f = $ temperature of the flame $= 2160$ °R,

$\sigma = $ Stefan-Boltzmann constant $= 0.173 \times 10^{-8}$ Btu/°R$^4 \cdot$ hr $\cdot$ ft^2,

$\epsilon = $ radiant emissivity of wood $= 0.9$,

$\epsilon_f = $ radiant emissivity of flame $= 0.85$,

$H = $ height of flame $= 0.87$ in. (0.0725 ft).

The experimental data show that

$$T = 1200 \text{ °R} \quad \text{at} \quad x = 0.5 \text{ in.}$$

and

$$dT/dx = -280761.72 \quad \text{at} \quad x = 0.5 \text{ in.}$$

Integrate the equation from $x = 0.5$ in. into $x = 1.0$ in. by use of the Runge-Kutta procedure.

14. Use the Runge-Kutta procedure to integrate the equation

$$0.3 \frac{d^2Q}{dt^2} + 5000 \frac{dQ}{dt} + \frac{Q}{0.08 \times 10^{-6}} = 1,$$

which describes the electric circuit shown in Fig. 7.21. The charge Q is initially zero ($t = 0$); that is, $Q_0 = Q'_0 = 0$.

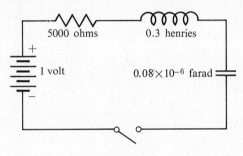

Fig. 7.21

Obtain also the current $i = dQ/dt$ of the circuit at any time $t (0 \le t \le 1.5 \times 10^{-3}$ sec).

15. A gate is hung on a frictionless inclined support as shown in Fig. 7.22. The differential equation expressing the oscillation of the gate is

$$\frac{d^2\phi}{dt^2} + \frac{3g}{2b} \sin \alpha \sin \phi = 0,$$

where $g = 32.2$ ft/sec^2. Let the initial values be $\phi_0 = 90°$, $(d\phi/dt)_{t=0} = 4\pi$ rad/sec. Integrate the equation over the interval $0 \le t \le 3$ sec.

16. The differential equation of a simple pendulum is given by

$$\frac{d^2\theta}{dt^2} + \frac{g}{L} \sin \theta = 0.$$

Let the initial values be

$$L = 3.0 \text{ ft,}$$
$$g = 32.2 \text{ ft/sec}^2,$$
$$\theta_0 = 80°,$$
$$\dot{\theta}_0 = (d\theta/dt)_{t=0} = 2\pi \text{ rad/sec.}$$

Evaluate numerically the quantities θ and $\dot{\theta} (0 \le t \le 3$ min).

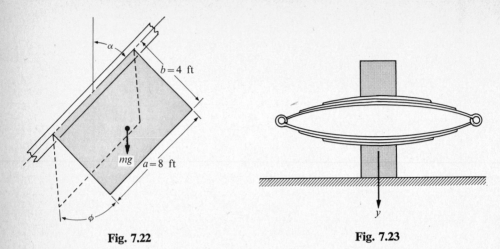

Fig. 7.22   Fig. 7.23

17. The motion of a nonlinear spring (Fig. 7.23) is expressed by the equation

$$\frac{d^2y}{dt^2} = -\frac{a}{4k^2}[(1 - k^2)y + 2k^2y^3].$$

Solve the equation by taking $a = 4$, $k = 5$, $y_0' = 0.04$, and $y_0 = 0$.

18. In a hemispherical bowl of radius R, a body with mass M starts from rest on its rim and slides down under the influence of gravity (see Fig. 7.24). The differential equation of the motion of the body is given by

$$\frac{d^2\theta}{dt^2} + \frac{g}{R}(f \sin \theta - \cos \theta) = 0,$$

where f is the coefficient of friction. Let $f = 0.006$, $R = 7$ in.; at $t = 0$, $d\theta/dt = 0$ and $\theta = 0$.

Integrate the equation by the Runge-Kutta procedure ($0 \le t \le 6$ sec).

19. The Bessel equation

$$x\frac{d^2y}{dx^2} + \frac{dy}{dx} - \left(2hx_w\frac{\sec \theta}{kw}\right)y = 0$$

expresses the temperature distribution of a triangular fin of an air heater (Fig. 7.25),

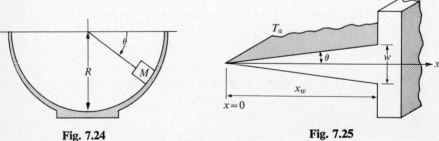

Fig. 7.24   Fig. 7.25

where

$y = T - T_a$ = temperature distribution of fin temperature of air,

h = heat transfer coefficient between the fin and air = 2.5 Btu/ft$^2 \cdot$ hr $\cdot$ °F,

k = heat conductivity of the fin = 212 Btu/ft $\cdot$ hr $\cdot$ °F.

Initial conditions are that at $x = 1$ ft, $T = 170°$ and $dy/dx = 20$. Also, $\theta = 10°$.

Integrate the above equation by letting $x_w = 3$ ft and $T_a = 70$ °F. Use the Runge-Kutta procedure.

20. In the study of bubble motion, the second-order nonlinear differential equation

$$\frac{d^2y}{dt^2} = \frac{c_1 y}{(1 + y^2)^3} - c_2 \frac{dy}{dt}$$

is used. The initial conditions are that at $t = 0$, $dy/dt = 0$ and $y = 0.00001$.

Find the y-t and $(dy/dt) - t$ relations, using the Runge-Kutta procedure from $t = 0$ to $t = 10$ with $\Delta t = 0.2$. Assume that $c_1 = c_2 = 1$.

21. The following nonlinear differential equation is the well-known van der Pol equation

$$\ddot{x} - (1 - x^2)\dot{x} + x = 0.$$

Initial conditions are as follows: $t = 0$, $x = 0.5$, $t = 0$, $\dot{x} = 0$.

Write a FORTRAN main program to find the x vs. t, $\dot{x}$ vs. t, and $\ddot{x}$ vs. t relations from $t = 0$ to $t = 5$. The increment of t, Δt, equals 0.1. Use the Runge-Kutta method.

22. Obtain the relation between radial velocity $\overline{V}$ of a particle and radial distance r from the center of a disk used in a spray drier. The differential equation is

$$\overline{V}(d\overline{V}/dr) + A\overline{V}^3 - Br = 0,$$

where $A = 1.66 \times 10^{-4}$, $B = 3.6 \times 10^7$, and $r = 0$ to 4 in.

23. The differential equations of motion of a single wheel of an automobile suspension system (Fig. 7.26) are

$$\frac{d^2x_1}{dt^2} = -\frac{D_1}{m_1}\left(\frac{dx_1}{dt} - \frac{dx_2}{dt}\right) - \frac{K_1}{m_1}(x_1 - x_2).$$

$$\frac{d^2x_2}{dt^2} = -\frac{D_1}{m_2}\left(\frac{dx_2}{dt} - \frac{dx_1}{dt}\right) - \frac{K_1}{m_2}(x_2 - x_1) - \frac{K_2}{m_2}(x_2 - x_3),$$

where

m_1 = one-fourth the mass of the automobile = 25 slugs,

m_2 = mass of the wheel and axle = 2 slugs,

K_1 = spring constant of the main auto spring = 1000 lb/ft,

K_2 = spring constant of the tire (assumed linear) = 4500 lb/ft,

D_1 = shock-absorber damping constant = 20 lb $\cdot$ sec/ft,

x_1 = displacement of the auto body,

x_2 = displacement of the wheel,

x_3 = roadway profile displacement.

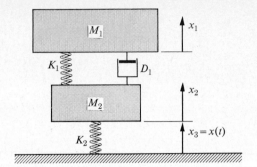

Fig. 7.26

The initial conditions and forcing functions are as follows:

$$x_1 = x_2 = \frac{dx_1}{dt} = \frac{dx_2}{dt} = 0 \quad \text{at} \quad t = 0;$$

$$x_3 = 1, \quad 0 \le t \le 35 \text{ msec};$$

$$= 0, \quad t > 35 \text{ msec}.$$

Obtain the x_1-t and x_2-t relations by the Runge-Kutta method. Use $\Delta t = 0.1$ sec and $0 \le t \le 4$ sec.

24. In the three-body system consisting of a projectile, the earth, and the moon, we can calculate the path of the projectile by using the following position, velocity, and acceleration relationships:

$$\frac{dx}{dt} = u + \omega y,$$

$$\frac{dy}{dt} = v - \omega x,$$

$$\frac{du}{dt} = \omega v - \frac{m_1 G(x + J)}{r^3} - \frac{m_2 G(x - K)}{s^3},$$

$$\frac{dv}{dt} = -\omega u - \frac{m_1 G y}{r^3} - \frac{m_2 G y}{s^3}.$$

The coordinate system which defines the values of x and y has as its origin the centers of mass of the earth and moon, with the earth and moon along the x-axis at $-J$ and K, respectively (see Fig. 7.27). The constants ω, m_1, m_2, G, K, and J are defined as

$$m_1 = \text{mass of the earth} = 5.975 \times 10^{27} \text{ gm},$$

$$m_2 = \text{mass of the moon} = 7.343 \times 10^{25} \text{ gm},$$

$$G = \text{gravitational constant} = 6.67 \times 10^{-8} \text{ dyne-cm}^2/\text{gm}^2,$$

$$\omega = \text{angular velocity of rotation of the moon and earth} \atop \text{about each other} = 2.662 \times 10^{-6} \text{ rad/sec},$$

$$K = 3.797 \times 10^{10} \text{ cm},$$

$$J = 4.667 \times 10^8 \text{ cm}.$$

The quantities r and s

$$r = \sqrt{(x + J)^2 + y^2}, \qquad s = \sqrt{(x - K)^2 + y^2}$$

are used to indicate the projectile's distance from the earth and moon, respectively. Given that at $t = 0$, $x = -J$, $y = 0$, $u = 2.3 \times 10^5$ cm/sec and $v = 2.0 \times 10^5$ cm/sec, find the position and velocity of the projectile after 20 sec.

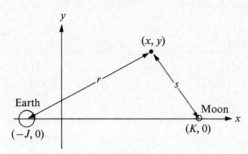

Fig. 7.27

CHAPTER 8

Matrix Algebra and Simultaneous Equations

8.1 INTRODUCTION

In this chapter we discuss the solution of simultaneous linear equations and the related topics of matrices and determinants. Many physical problems are expressed in terms of simultaneous linear equations. Many other problems are expressed in the form of simultaneous linear differential equations which may be readily transformed into simultaneous algebraic equations. We shall limit our discussion to two widely used procedures, the Gauss-Jordan method and the Gauss-Seidel method. The elements of matrix algebra will also be introduced.

8.2 ELEMENTARY OPERATIONS OF MATRICES

As mentioned in Section 3.4 and illustrated in Fig. 3.8(b), a matrix is an array of numbers which may be either rectangular or square. A matrix is indicated by brackets and written in the following form:

$$[A] = \begin{bmatrix} a_{11} & a_{12} & a_{13} \\ a_{21} & a_{22} & a_{23} \\ a_{31} & a_{32} & a_{33} \end{bmatrix}. \tag{8.1}$$

The first subscript of the elements of the matrix corresponds to the row containing the element and the second subscript corresponds to the column. The rules for addition, subtraction, and multiplication of two matrices are stated below.

Consider matrix $[A]$ in Eq. (8.1) and matrix $[B]$,

$$[B] = \begin{bmatrix} b_{11} & b_{12} & b_{13} \\ b_{21} & b_{22} & b_{23} \\ b_{31} & b_{32} & b_{33} \end{bmatrix}. \tag{8.2}$$

If matrices $[A]$ and $[B]$ are added, a matrix $[C]$ is obtained in a 3×3 array of numbers:

$$[C] = \begin{bmatrix} c_{11} & c_{12} & c_{13} \\ c_{21} & c_{22} & c_{23} \\ c_{31} & c_{32} & c_{33} \end{bmatrix}. \tag{8.3}$$

Each element of $[C]$ is obtained by adding the corresponding elements of $[A]$ and $[B]$. Thus, the elements of $[C]$ are computed as follows:

$$\begin{aligned} c_{11} &= a_{11} + b_{11}, \\ c_{12} &= a_{12} + b_{12}, \\ &\vdots \\ c_{21} &= a_{21} + b_{21}, \\ &\vdots \\ c_{32} &= a_{32} + b_{32}, \\ c_{33} &= a_{33} + b_{33}. \end{aligned} \tag{8.4}$$

Subtraction of matrices is done in the same manner.

The following numerical examples illustrate the processes of addition and subtraction. Given

$$[E] = \begin{bmatrix} 1 & 4 \\ 3 & 2 \end{bmatrix} \quad \text{and} \quad [F] = \begin{bmatrix} 2 & 1 \\ 3 & 6 \end{bmatrix},$$

then

$$[G] = [E] + [F] = \begin{bmatrix} 3 & 5 \\ 6 & 8 \end{bmatrix},$$

and

$$[H] = [F] - [E] = \begin{bmatrix} 1 & -3 \\ 0 & 4 \end{bmatrix}.$$

If matrices $[A]$ and $[B]$ are multiplied, a matrix $[P]$ is obtained in a 3×3 array of numbers:

$$[A][B] = [P],$$

in which

$$[P] = \begin{bmatrix} p_{11} & p_{12} & p_{13} \\ p_{21} & p_{22} & p_{23} \\ p_{31} & p_{32} & p_{33} \end{bmatrix}. \tag{8.5}$$

The elements of $[P]$ are computed from the elements of $[A]$ and $[B]$ by the following definition:

$$\begin{aligned} p_{11} &= a_{11}b_{11} + a_{12}b_{21} + a_{13}b_{31}, \\ p_{12} &= a_{11}b_{12} + a_{12}b_{22} + a_{13}b_{32}, \\ p_{13} &= a_{11}b_{13} + a_{12}b_{23} + a_{13}b_{33}, \\ p_{21} &= a_{21}b_{11} + a_{22}b_{21} + a_{23}b_{31}, \end{aligned} \tag{8.6}$$

and in general,

$$p_{ij} = a_{i1}b_{1j} + a_{i2}b_{2j} + a_{i3}b_{3j}.$$

The following numerical examples illustrate the process of multiplication, where $[E]$ and $[F]$ are as given above:

$$[E][F] = \begin{bmatrix} 1 & 4 \\ 3 & 2 \end{bmatrix} \begin{bmatrix} 2 & 1 \\ 3 & 6 \end{bmatrix} = \begin{bmatrix} 14 & 25 \\ 12 & 15 \end{bmatrix},$$

$$[F][E] = \begin{bmatrix} 2 & 1 \\ 3 & 6 \end{bmatrix} \begin{bmatrix} 1 & 4 \\ 3 & 2 \end{bmatrix} = \begin{bmatrix} 5 & 10 \\ 21 & 24 \end{bmatrix}.$$

These examples show the importance of the order of multiplication. The two examples below illustrate the process of multiplying rectangular matrices:

$$\begin{bmatrix} 1 & 4 & -1 \\ 3 & 2 & 2 \end{bmatrix} \begin{bmatrix} 2 & 1 \\ 3 & 6 \\ 4 & 5 \end{bmatrix} = \begin{bmatrix} 10 & 20 \\ 20 & 25 \end{bmatrix},$$

$$\begin{bmatrix} 1 & 4 & -1 \\ 3 & 2 & 2 \end{bmatrix} \begin{bmatrix} 2 \\ 3 \\ 4 \end{bmatrix} = \begin{bmatrix} 10 \\ 20 \end{bmatrix}.$$

It is important that the first matrix in a multiplication have the same number of columns as there are rows in the second matrix. If this number is designated as m, the multiplication of matrices can be defined as

$$[P] = [A][B], \tag{8.7}$$

where

$$p_{ij} = \sum_{k=1}^{k=m} a_{ik} b_{kj},$$

and the orders of the matrices $[A]$, $[B]$, and $[P]$ are (n, m), (m, l), and (n, l), respectively.

The multiplication of two matrices may readily be carried out by the following FORTRAN subprogram MATMPY:

```
SUBROUTINE MATMPY(A,N,M,B,L,C)
DIMENSION A(12,12),B(12,12),C(12,12)
DO 5 I = 1,N
DO 5 J = 1,L
C(I,J) = 0.
DO 5 K = 1,M
5 C(I,J) = C(I,J) + A(I,K)*B(K,J)
RETURN
END
```

In this subroutine, the three matrices are dimensioned as 12×12. These dimensions may be changed but they must agree with the dimensions used in the main program.

If a matrix is to be multiplied by a constant (*scalar* multiplication), then each element of the matrix must be multiplied by that constant. Thus

$$3 \begin{bmatrix} 1 & 4 \\ 3 & 2 \end{bmatrix} = \begin{bmatrix} 3 & 12 \\ 9 & 6 \end{bmatrix}.$$

In the remainder of this section we shall introduce several definitions which are related to a square matrix.

a) The *principal diagonal* of a square matrix $[A]$ of order n consists of the n elements $a_{11}, a_{22}, \ldots, a_{nn}$.

b) The *transpose* of $[A]$ is obtained by interchanging the rows and columns of $[A]$. A transpose is indicated by a prime or a T. Thus,

$$[A] = \begin{bmatrix} 1 & 2 & 5 \\ 3 & 4 & 6 \\ 7 & 8 & 9 \end{bmatrix} \quad \text{and} \quad [A]^T = \begin{bmatrix} 1 & 3 & 7 \\ 2 & 4 & 8 \\ 5 & 6 & 9 \end{bmatrix}.$$

If $[B] = [A]$, then $b_{ij} = a_{ji}$. Note that the principal diagonal is not altered in the above example.

c) A *symmetric matrix* is one in which the elements $a_{ij} = a_{ji}$, or $[A] = [A]^T$.

d) A *diagonal matrix* is one in which all the elements other than those in the principal diagonal are zero; thus

$$[D] = \begin{bmatrix} 1 & 0 & 0 \\ 0 & 4 & 0 \\ 0 & 0 & 2 \end{bmatrix}.$$

e) The *unit matrix*, or identity matrix $[I]$, is a matrix in which the elements of the principal diagonal are ones and the remaining elements are zeros; thus

$$[I] = \begin{bmatrix} 1 & 0 & 0 \\ 0 & 1 & 0 \\ 0 & 0 & 1 \end{bmatrix}.$$

8.3 GAUSS-JORDAN ELIMINATION METHOD—DIRECT METHOD

The usefulness of matrix algebra will be explained in the remainder of this chapter.†
Consider the following system of three equations in three unknowns:

$$\begin{aligned} 3x_1 - 6x_2 + 7x_3 &= 3, \\ 9x_1 \qquad\quad - 5x_3 &= 3, \\ 5x_1 - 8x_2 + 6x_3 &= -4. \end{aligned} \qquad (8.8)$$

---

† Matrix algebra also plays an important part in eigenvalue problems (Chapter 9) and in linear programming (Chapter 15).

The above equation can be expressed by using matrices in the following manner:

$$\begin{bmatrix} 3 & -6 & 7 \\ 9 & 0 & -5 \\ 5 & -8 & 6 \end{bmatrix} \begin{bmatrix} x_1 \\ x_2 \\ x_3 \end{bmatrix} = \begin{bmatrix} 3 \\ 3 \\ -4 \end{bmatrix}, \tag{8.9}$$

where the leftmost square matrix is commonly known as the *coefficient matrix*.

If the column matrix, representing the constant terms on the right-hand side, is combined with the coefficient matrix, then the resulting matrix is called the *augmented matrix*. Thus the augmented matrix in our example is

$$\begin{bmatrix} 3 & -6 & 7 & 3 \\ 9 & 0 & -5 & 3 \\ 5 & -8 & 6 & -4 \end{bmatrix}. \tag{8.10}$$

The Gauss-Jordan elimination method is used to reduce the augmented matrix to the form

$$\begin{bmatrix} a_{11} & 0 & 0 & c_1 \\ 0 & a_{22} & 0 & c_2 \\ 0 & 0 & a_{33} & c_3 \end{bmatrix}. \tag{8.11}$$

The solution is

$$x_1 = c_1/a_{11}, \qquad x_2 = c_2/a_{22}, \qquad x_3 = c_3/a_{33}.$$

The procedure involved is essentially a simple, direct process, one which is easily learned by a beginner and readily adapted to digital computers.

The reduction from Eq. (8.10) to Eq. (8.11) is based on three elementary operations which are well known for any system of linear equations:

a) interchange any two equations;

b) multiply any equation by a nonzero number;

c) add to one equation d times a second equation, where d is any real number.

The new system of linear equations, resulting from any of these three operations, has exactly the same solution as the original system of linear equations. These two systems are known as *equivalent*.

We will now demonstrate the step-by-step operations as they are used to reduce Eq. (8.10) to Eq. (8.11). In this procedure, the given augmented matrix is first written

$$\begin{bmatrix} 3 & -6 & 7 & 3 \\ 9 & 0 & -5 & 3 \\ 5 & -8 & 6 & -4 \end{bmatrix}.$$

Next, we eliminate the *first* element in each row, except the *first* row. In other words, we wish to eliminate a_{21} and a_{31}. To reduce a_{21} from 9 to zero, we obtain a new second row by using the following expression:

The new second row $= a_{11} \cdot$ (the present second row) $- a_{21} \cdot$ (the first row).

Our matrix, after this first step, looks like this:

$$\begin{bmatrix} 3 & -6 & 7 & 3 \\ 0 & 54 & -78 & -18 \\ 5 & -8 & 6 & -4 \end{bmatrix}.$$

To reduce a_{31} to zero, a new third row is obtained by the following expression:

The new third row = $a_{11} \cdot$ (the present third row) $- a_{31} \cdot$ (the first row).

This completes the second step and yields the matrix

$$\begin{bmatrix} 3 & -6 & 7 & 3 \\ 0 & 54 & -78 & -18 \\ 0 & 6 & -17 & -27 \end{bmatrix}.$$

For the sake of discussion, the above two steps of eliminating a_{21} and a_{31}, but not a_{11}, are referred to as the first pass. In this pass, the element a_{11} plays such a prominent role that we shall call it the pivot element for this pass. The first row is called the pivot row. It should be noted that the above matrix, representing a system of simultaneous linear equations, is completely equivalent to the given augmented matrix, as shown in Eq. (8.10).

Let us now return to our problem and initiate the second pass, consisting of two steps, whose purpose is to eliminate the *second* element in each row except the *second* row. This pass is shown in Table 8.1, where the pivot element is circled, and the row involved in the elimination is shaded. A third and final pass eliminates the *third* element in each row except the *third* row. This pass is also shown in Table 8.1. Our original system of equations, (8.8), is now seen to be equivalent to the following new set of equations:

$$-72{,}900x_1 = -145{,}800,$$
$$-24{,}300x_2 = -97{,}200,$$
$$-450x_3 = -1350.$$

The values of x_1, x_2, and x_3 can readily be obtained.

In recapitulation, the Gauss-Jordan elimination method as illustrated in this example has three passes. In each of the three passes, two elements are reduced to zero. Thus in pass one ($k = 1$), a_{21} and a_{31} are reduced to zero. In pass two ($k = 2$), the elements eliminated are a_{12} and a'_{32}, and in the final pass ($k = 3$), a'_{13} and a'_{23} are eliminated.

We note that in each pass ($k = 1, 2,$ or 3) a pivot element a_{kk} is first selected. The kth row, or the pivot row, is then used to eliminate the coefficients in the kth column except the pivot element itself. In general, if we take a_{kk} as the pivot element, then the kth row, or the pivot row, is used to eliminate the elements a_{ij} where $i = 1, 2, 3, \ldots, n$ but $i \neq k(j = 1, 2, 3, \ldots, n)$. When the last pass has been completed, the values of the x's in the system of linear equations can be easily determined, as shown in the last step of Table 8.1.

Table 8.1 Gauss-Jordan elimination method

Given Matrix	$\begin{matrix} a_{11} & a_{12} & a_{13} & a_{14} \\ a_{21} & a_{22} & a_{23} & a_{24} \\ a_{31} & a_{32} & a_{33} & a_{34} \end{matrix}$	$\begin{matrix} 3 & -6 & 7 & 3 \\ 9 & 0 & -5 & 3 \\ 5 & -8 & 6 & -4 \end{matrix}$

PASS 1 $k = 1$, i.e., pivot row = row 1; pivot element = a_{11}		
$i = 2$ $j = 1, 2, 3, 4$	New row 2 = $a_{11} \times$ row 2 $- a_{21} \times$ row 1 $\begin{matrix} \textcircled{a_{11}} & a_{12} & a_{13} & a_{14} \\ 0 & a'_{22} & a'_{23} & a'_{24} \\ a_{31} & a_{32} & a_{33} & a_{34} \end{matrix}$	$\begin{matrix} 3 & -6 & 7 & 3 \\ 0 & 54 & -78 & -18 \\ 5 & -8 & 6 & -4 \end{matrix}$
$i = 3$ $j = 1, 2, 3, 4$	New row 3 = $a_{11} \times$ row 3 $- a_{31} \times$ row 1 $\begin{matrix} \textcircled{a_{11}} & a_{12} & a_{13} & a_{14} \\ 0 & a'_{22} & a'_{23} & a'_{24} \\ 0 & a'_{32} & a'_{33} & a'_{34} \end{matrix}$	$\begin{matrix} 3 & -6 & 7 & 3 \\ 0 & 54 & -78 & -18 \\ 0 & 6 & -17 & -27 \end{matrix}$

PASS 2 $k = 2$, i.e., pivot row = row 2; pivot element = a'_{22}		
$i = 1$ $j = 1, 2, 3, 4$	New row 1 = $a'_{22} \times$ row 1 $- a_{12} \times$ row 2 $\begin{matrix} a'_{11} & 0 & a'_{13} & a'_{14} \\ 0 & \textcircled{a'_{22}} & a'_{23} & a'_{24} \\ 0 & a'_{32} & a'_{33} & a'_{34} \end{matrix}$	$\begin{matrix} 162 & 0 & -90 & 54 \\ 0 & 54 & -78 & -18 \\ 0 & 6 & -17 & -27 \end{matrix}$
$i = 3$ $j = 1, 2, 3, 4$	New row 3 = $a'_{22} \times$ row 3 $- a'_{32} \times$ row 2 $\begin{matrix} a'_{11} & 0 & a'_{13} & a'_{14} \\ 0 & \textcircled{a'_{22}} & a'_{23} & a'_{24} \\ 0 & 0 & a''_{33} & a''_{34} \end{matrix}$	$\begin{matrix} 162 & 0 & -90 & 54 \\ 0 & 54 & -78 & -18 \\ 0 & 0 & -450 & -1350 \end{matrix}$

PASS 3 $k = 3$, i.e., pivot row = row 3; pivot element = a''_{33}		
$i = 1$ $j = 1, 2, 3, 4$	New row 1 = $a''_{33} \times$ row 1 $- a'_{13} \times$ row 3 $\begin{matrix} a''_{11} & 0 & 0 & a''_{14} \\ 0 & a'_{22} & a'_{23} & a'_{24} \\ 0 & 0 & \textcircled{a''_{33}} & a''_{34} \end{matrix}$	$\begin{matrix} 72{,}900 & 0 & 0 & -145{,}800 \\ 0 & 54 & -78 & -18 \\ 0 & 0 & -450 & -1350 \end{matrix}$
$i = 2$ $j = 1, 2, 3, 4$	New row 2 = $a''_{33} \times$ row 2 $- a'_{23} \times$ row 3 $\begin{matrix} a''_{11} & 0 & 0 & a''_{14} \\ 0 & a''_{22} & 0 & a''_{24} \\ 0 & 0 & \textcircled{a''_{33}} & a''_{34} \end{matrix}$	$\begin{matrix} -72{,}900 & 0 & 0 & -145{,}800 \\ 0 & -24{,}300 & 0 & -97{,}200 \\ 0 & 0 & -450 & -1350 \end{matrix}$

Solutions	$\begin{matrix} a''_{11}x_1 = a''_{14} \\ a''_{22}x_2 = a''_{24} \\ a''_{33}x_3 = a''_{34} \end{matrix}$	$\begin{matrix} x_1 = 2 \\ x_2 = 4 \\ x_3 = 3 \end{matrix}$

8.4 NECESSITY OF NORMALIZATION

In the numerical example shown in the previous section (Table 8.1) the numbers increased greatly as we proceeded through each step. If the given system of simultaneous equations begins with large coefficients, overflow may occur. This difficulty can be solved if a slight modification is made on the Gauss-Jordan method by simply dividing the pivot row by the value of the pivot element prior to the elimination in each pass. Each pivot element is then equal to one. This process is commonly known as *normalization*. It should be noted that normalization does not alter the equivalence of the system, since an elementary operation, division, is used.

Table 8.2 shows a new tabulation for the Gauss-Jordan method with normalization. In each of the three passes ($k = 1, 2, 3,$) there are now three operations. The first is a normalization step and the second and third operations are elimination steps.

Before we turn to the next section, let us note two important cases for diagonal element a_{ii}. First, $a_{ii} = 0$. In this case, the normalization process breaks down completely.

Second, the value of a_{ii} is nearly equal to zero. When this a_{ii} is used as the divisor in the division a_{ij}/a_{ii}, $(j \neq i)$ as required in an elimination step, it causes substantial magnification of error†. In other words, the result of a_{ij}/a_{ii} may become arbitrarily large. To avoid difficulties associated with these two cases, we shall describe a technique known as *positioning for size* in the next section.

8.5 ZERO DIAGONAL ELEMENT AND POSITIONING FOR SIZE

Consider the following system of simultaneous equations:

$$- 6y + 9z = -3, \tag{8.12a}$$

$$7x \qquad - 5z = \quad 3, \tag{8.12b}$$

$$5x - 8y + 6z = -4. \tag{8.12c}$$

If the Gauss-Jordan elimination method with normalization is to be used for Eqs. (8.12), the augmented matrix appears as

$$[A] = \begin{bmatrix} 0 & -6 & 9 & -3 \\ 7 & 0 & -5 & 3 \\ 5 & -8 & 6 & -4 \end{bmatrix}, \tag{8.13}$$

and we proceed to operate on it as discussed in Section 8.3. However, with zero as the first diagonal element, we would run into difficulty with the normalization process. This situation can be remedied by the elementary operation of interchanging two equations, for example, Eq. (8.12a) with Eqs. (8.12b) or (8.12c).

---

† Roundoff errors are discussed in Chapter 13.

Table 8.2 Elimination method with normalization

Given augmented matrix	$a_{11}\ a_{12}\ a_{13}\ a_{14}$ $a_{21}\ a_{22}\ a_{23}\ a_{24}$ $a_{31}\ a_{32}\ a_{33}\ a_{34}$	3 $\quad$ −6 $\quad$ 7 $\quad$ 3 9 $\quad$ 0 $\quad$ −5 $\quad$ 3 5 $\quad$ −8 $\quad$ 6 $\quad$ −4

$$\text{Pass 1}$$
$$k = 1, \text{ i.e., pivot row} = \text{row 1; pivot element} = a'_{11} = 1$$

	New row 1 = row 1/a_{11}	
$i = 1$ $j = 1, 2, 3, 4$	$1 \quad a'_{12}\ a'_{13}\ a'_{14}$ $a_{21}\ a_{22}\ a_{23}\ a_{24}$ $a_{31}\ a_{32}\ a_{33}\ a_{34}$	1 $\quad$ −2 $\quad$ 2.33 $\quad$ 1 9 $\quad$ 0 $\quad$ −5 $\quad$ 3 5 $\quad$ −8 $\quad$ 6 $\quad$ −4
	New row 2 = 1 × row 2 − a_{21} × row 1	
$i = 2$ $j = 2, 3, 4$	① $\quad a'_{12}\ a'_{13}\ a'_{14}$ 0 $\quad a'_{22}\ a'_{23}\ a'_{24}$ $a_{31}\ a_{32}\ a_{33}\ a_{34}$	1 $\quad$ −2 $\quad$ 2.33 $\quad$ 1 0 $\quad$ 18 $\quad$ −26 $\quad$ −6 5 $\quad$ −8 $\quad$ 6 $\quad$ −4
	New row 3 = 1 × row 3 − a_{31} × row 1	
$i = 3$ $j = 2, 3, 4$	① $\quad a'_{12}\ a'_{13}\ a'_{14}$ 0 $\quad a'_{22}\ a'_{23}\ a'_{24}$ 0 $\quad a'_{32}\ a'_{33}\ a'_{34}$	1 $\quad$ −2 $\quad$ 2.33 $\quad$ 1 0 $\quad$ 18 $\quad$ −26 $\quad$ −6 0 $\quad$ 2 $\quad$ −5.65 $\quad$ −9

$$\text{Pass 2}$$
$$k = 2, \text{ i.e., pivot row} = \text{row 2; pivot element} = a''_{22} = 1$$

	New row 2 = row 2/a'_{22}	
$i = 2$ $j = 2, 3, 4$	$1 \quad a'_{12}\ a'_{13}\ a'_{14}$ 0 $\quad 1 \quad a''_{23}\ a''_{24}$ 0 $\quad a'_{32}\ a'_{33}\ a'_{34}$	1 $\quad$ −2 $\quad$ 2.33 $\quad$ 1 0 $\quad$ 1 $\quad$ −1.44 $\quad$ −0.33 0 $\quad$ 2 $\quad$ −5.65 $\quad$ −9
	New row 1 = 1 × row 1 − a'_{12} × row 2	
$i = 1$ $j = 3, 4$	$1 \quad 0 \quad a''_{13}\ a''_{14}$ 0 $\quad$ ① $\quad a''_{23}\ a''_{24}$ 0 $\quad a'_{32}\ a'_{33}\ a'_{34}$	1 $\quad$ 0 $\quad$ −0.55 $\quad$ 0.33 0 $\quad$ 1 $\quad$ −1.44 $\quad$ −0.33 0 $\quad$ 2 $\quad$ −5.65 $\quad$ −9
	New row 3 = 1 × row 3 − a'_{32} × row 2	
$i = 3$ $j = 3, 4$	$1 \quad 0 \quad a''_{13}\ a''_{14}$ 0 $\quad$ ① $\quad a''_{23}\ a''_{24}$ 0 $\quad 0 \quad a''_{33}\ a''_{34}$	1 $\quad$ 0 $\quad$ −0.55 $\quad$ 0.33 0 $\quad$ 1 $\quad$ −1.44 $\quad$ −0.33 0 $\quad$ 0 $\quad$ −2.77 $\quad$ −8.33

	PASS 3 $k = 3$, i.e., pivot row = row 3; pivot element = $a_{33}''' = 1$	
$i = 3$ $j = 3, 4$	New row 3 = row 3$/a_{33}''$ $\begin{matrix} 1 & 0 & a_{13}'' & a_{14}'' \\ 0 & 1 & a_{23}'' & a_{24}'' \\ 0 & 0 & 1 & a_{34}''' \end{matrix}$	$\begin{matrix} 1 & 0 & -0.55 & 0.33 \\ 0 & 1 & -1.44 & -0.33 \\ 0 & 0 & 1 & 3.01 \end{matrix}$
$i = 1$ $j = 4$	New row 1 = $1 \times$ row 1 $- a_{13}'' \times$ row 3 $\begin{matrix} 1 & 0 & 0 & a_{14}''' \\ 0 & 1 & a_{23}'' & a_{24}'' \\ 0 & 0 & 1 & a_{34}''' \end{matrix}$	$\begin{matrix} 1 & 0 & 0 & 1.99 \\ 0 & 1 & -1.44 & -0.33 \\ 0 & 0 & 1 & 3.01 \end{matrix}$
$i = 1$ $j = 4$	New row 2 = $1 \times$ row 2 $- a_{23}'' \times$ row 3 $\begin{matrix} 1 & 0 & 0 & a_{14}''' \\ 0 & 1 & 0 & a_{24}''' \\ 0 & 0 & 1 & a_{34}''' \end{matrix}$	$\begin{matrix} 1 & 0 & 0 & 1.99 \\ 0 & 1 & 0 & 4.00 \\ 0 & 0 & 1 & 3.01 \end{matrix}$
Solutions	$1(x_1) = a_{14}'''$ $1(x_2) = a_{24}'''$ $1(x_3) = a_{34}'''$	$1(x_1) = 1.99$ $1(x_2) = 4.00$ $1(x_3) = 3.01$

A practical way to alleviate the difficulty arising from the presence of zero diagonal elements is to find the largest absolute number $|a_{ij}|$ in the given coefficient matrix. Since this is found to be the element a_{13}, or 9, row 3 and row 1 are interchanged. In general, we interchange rows i and j if the element a_{ij} has the largest absolute value. After the interchange of two rows, we normalize the original row i (or the new row j). When the two elements a_{ij} and a_{kl} are equal to each other and have the largest absolute value, we use a_{ij}, where $j \leq l$. This technique is known as *positioning for size* or *using pivotal elements*.

Table 8.3 shows in detail the interchange of rows before each of three regular passes. As mentioned in Table 8.2, each regular pass consists of one normalization and two elimination steps. This procedure is also used to alleviate serious errors incurred when any of the diagonal elements a_{ii} is small in magnitude compared with the element $a_{ji}, j > i$.

8.6 DETERMINANTS

We shall now discuss the evaluation of the value of a given determinant. It is known†
that the value of a determinant is unchanged if each element of any row is added c

---

† For example, see Thomas, G. B., *Calculus and Analytic Geometry*, 3rd ed. Addison-Wesley, Reading, Mass., 1960, p. 422.

Table 8.3 Elimination method with interchange of rows

Given augmented matrix	$\begin{array}{llll} a_{11} & a_{12} & a_{13} & a_{14} \\ a_{21} & a_{22} & a_{23} & a_{24} \\ a_{31} & a_{32} & a_{33} & a_{34} \end{array}$	$\begin{array}{rrrr} 0 & -6 & 9 & -3 \\ 7 & 0 & -5 & 3 \\ 5 & -8 & 6 & -4 \end{array}$

	PASS 1 Pivot row = row 3	

Interchange of rows	Find the element a_{ij} which has the largest absolute value $(i = 1, 2, 3; j = 1, 2, 3)$. Interchange rows i and j.	$i = 1, \quad j = 3$ Interchange rows 1 and 3
	$\begin{array}{llll} a'_{11} & a'_{12} & a'_{13} & a'_{14} \\ a_{21} & a_{22} & a_{23} & a_{24} \\ a'_{31} & a'_{32} & a'_{33} & a'_{34} \end{array}$	$\begin{array}{rrrr} 5 & -8 & 6 & -4 \\ 7 & 0 & -5 & 3 \\ 0 & -6 & 9 & -3 \end{array}$
$i = 3$ $j = 1, 2, 3, 4$	New row 3 = row $3/a'_{33}$ $\begin{array}{llll} a'_{11} & a'_{12} & a'_{13} & a'_{14} \\ a_{21} & a_{22} & a_{23} & a_{24} \\ a''_{31} & a''_{32} & 1 & a''_{34} \end{array}$	Normalize row j = row 3 $\begin{array}{rrrr} 5 & -8 & 6 & -4 \\ 7 & 0 & -5 & 3 \\ 0 & -0.66 & 1 & -0.33 \end{array}$
$i = 1$ $j = 1, 2, 4$	New row 1 = row 1 $- a'_{13}$ row 3 $\begin{array}{llll} a''_{11} & a''_{12} & 0 & a''_{14} \\ a_{21} & a_{22} & a_{23} & a_{24} \\ a''_{31} & a''_{32} & 1 & a''_{34} \end{array}$	Elimination $\begin{array}{rrrr} 5 & -4.04 & 0 & -2.02 \\ 7 & 0 & -5 & 3 \\ 0 & -0.66 & 1 & -0.33 \end{array}$
$i = 2$ $j = 1, 2, 4$	New row 2 = row 2 $- a_{23}$ row 3 $\begin{array}{llll} a''_{11} & a''_{12} & 0 & a''_{14} \\ a'_{21} & a'_{22} & 0 & a'_{24} \\ a''_{31} & a''_{32} & 1 & a''_{34} \end{array}$	Elimination $\begin{array}{rrrr} 5 & -4.04 & 0 & -2.02 \\ 7 & -3.30 & 0 & 1.35 \\ 0 & -0.66 & 1 & -0.33 \end{array}$

	PASS 2 Pivot row = row 1	

Interchange of rows	Find max $|a_{ij}|$ $\quad i = 1, 2$ $\qquad\qquad\qquad j = 1, 2$ Interchange rows i and j	$i = 2 \quad j = 1$ Interchange rows 1 and 2
	$\begin{array}{llll} a'''_{11} & a'''_{12} & 0 & a'''_{14} \\ a''_{21} & a''_{22} & 0 & a''_{24} \\ a''_{31} & a''_{32} & 1 & a''_{34} \end{array}$	$\begin{array}{rrrr} 7 & -3.30 & 0 & 1.35 \\ 5 & -4.04 & 0 & -2.02 \\ 0 & -0.66 & 1 & -0.33 \end{array}$
$i = 1$ $j = 1, 2, 4$	New row 1 = row $1/a'''_{11}$ $\begin{array}{llll} 1 & a^{iv}_{12} & 0 & a^{iv}_{14} \\ a''_{21} & a''_{22} & 0 & a''_{24} \\ a''_{31} & a''_{32} & 1 & a''_{34} \end{array}$	Normalize row j = row 1 $\begin{array}{rrrr} 1 & -0.47 & 0 & 0.19 \\ 5 & -4.04 & 0 & -2.02 \\ 0 & -0.66 & 1 & -0.33 \end{array}$

	New row 2 = row 2 − a''_{21} Row 1	Elimination
$i = 2$ $j = 2, 4$	$\begin{array}{cccc} 1 & a^{iv}_{12} & 0 & a^{iv}_{14} \\ 0 & a'''_{22} & 0 & a'''_{24} \\ a''_{31} & a''_{32} & 1 & a''_{34} \end{array}$	$\begin{array}{cccc} 1 & -0.47 & 0 & 0.19 \\ 0 & -1.69 & 0 & -2.97 \\ 0 & -0.66 & 1 & -0.33 \end{array}$
	New row 3 = row 3 − a''_{31} row 1	Elimination
$i = 3$ $j = 2, 4$	$\begin{array}{cccc} 1 & a^{iv}_{12} & 0 & a^{iv}_{14} \\ 0 & a'''_{22} & 0 & a'''_{24} \\ 0 & a''_{32} & 1 & a''_{34} \end{array}$	$\begin{array}{cccc} 1 & -0.47 & 0 & 0.19 \\ 0 & -1.69 & 0 & -2.97 \\ 0 & -0.66 & 1 & -0.33 \end{array}$
	PASS 3 Pivot row = row 2	
	Find max $\lvert a_{ij} \rvert$ $i = 2, j = 2$	No interchange needed
	New row 2 = row $2/a''_{22}$	Normalization
$i = 2$ $j = 2, 4$	$\begin{array}{cccc} 1 & a^{iv}_{12} & 0 & a^{iv}_{14} \\ 0 & 1 & 0 & a'''_{24} \\ 0 & a''_{32} & 1 & a''_{34} \end{array}$	$\begin{array}{cccc} 1 & -0.47 & 0 & 0.19 \\ 0 & 1 & 0 & 1.76 \\ 0 & -0.66 & 1 & -0.33 \end{array}$
	New row 1 = row 1 − a''_{12} row 2	Elimination
$i = 1$ $j = 4$	$\begin{array}{cccc} 1 & 0 & 0 & a^{v}_{14} \\ 0 & 1 & 0 & a^{iv}_{24} \\ 0 & a''_{32} & 1 & a''_{34} \end{array}$	$\begin{array}{cccc} 1 & 0 & 0 & 1.03 \\ 0 & 1 & 0 & 1.76 \\ 0 & -0.66 & 1 & -0.33 \end{array}$
	New row 3 = row 3 − a''_{32} row 2	Elimination
$i = 3$ $j = 4$	$\begin{array}{cccc} 1 & 0 & 0 & a^{v}_{14} \\ 0 & 1 & 0 & a^{iv}_{24} \\ 0 & 0 & 1 & a'''_{34} \end{array}$	$\begin{array}{cccc} 1 & 0 & 0 & 1.03 \\ 0 & 1 & 0 & 1.76 \\ 0 & 0 & 1 & 0.83 \end{array}$
Solutions	$\begin{array}{l} x_1 = a^{v}_{14} \\ x_2 = a^{iv}_{24} \\ x_3 = a'''_{34} \end{array}$	$\begin{array}{l} x_1 = 1.03 \\ x_2 = 1.76 \\ x_3 = 0.83 \end{array}$

times to the corresponding element of some other row, where c is a known constant. In other words, the following relation is valid:

$$\begin{vmatrix} a_1 & b_1 \\ a_2 & b_2 \end{vmatrix} = \begin{vmatrix} a_1 & b_1 \\ (a_2 + ca_1) & (b_2 + cb_1) \end{vmatrix}. \tag{8.14}$$

Using this relation, we will show that the value of the determinant of the coefficient matrix

$$\begin{bmatrix} 3 & -6 & 7 \\ 9 & 0 & -5 \\ 5 & -8 & 6 \end{bmatrix},$$

which has the same elements as those in the coefficient matrix (not the augmented matrix) in Table 8.2 or in Eq. (8.9), is simply equal to $(3)(18)(-2.77) = -149.58$†.

In Table 8.2 each pass consists of two different operations, normalization and elimination. The combined effect of these is to reduce the coefficient matrix to a unit matrix.

We now wish to examine what effect each of the two operations has on the value of the determinant of the coefficient matrix. The elimination process does not alter the value of the determinant, as can be seen from Eq. (8.14). However, with each normalization process, the value of the given determinant is reduced by its normalization factor, for example, a_{11}, a'_{22}, or a''_{33} in Table 8.2. Hence the true value of the determinant is the product of the normalizing factors a_{11}, a'_{22}, and a''_{33}.

Before we begin the next section, it is important to note that the zero determinant, or a determinant having zero value, plays an important role in matrix inversion (see Section 8.7). For example, it is known that the following determinant‡ has zero value:

$$\begin{vmatrix} 1 & 4 & 13 & 16 \\ 15 & 14 & 3 & 2 \\ 12 & 9 & 8 & 5 \\ 6 & 7 & 10 & 11 \end{vmatrix}.$$

A square matrix $[A]$ having a zero determinant is called *singular*. If $|A| \neq 0$, then $[A]$ is called *nonsingular*.

When a determinant of order two is zero, we have one of the following two cases:

Case 1: There is no solution of the associated simultaneous equations if the equations are *incompatible* or *inconsistent*. For example, the following given linear equations are incompatible:

$$x_1 + 2x_2 = 6,$$
$$2x_1 + 4x_2 = 11.$$

Here, the lines are parallel, and no pair of numbers x_1 and x_2 can satisfy both equations.

Case 2: There are infinitely many solutions if the given linear equations coincide. For example, the following simultaneous equations represent only one straight line:

$$x_1 + 2x_2 = 6,$$
$$2x_1 + 4x_2 = 12.$$

---

† The calculation of this determinant which would, in the usual way, yield -150.0 involves round-off errors. They will be discussed in Chapter 13.

‡ This is a 4×4 magic square.

Similarly, when a determinant of order three is zero, we have one of the following four cases:

1) any two of the three planes are parallel,
2) any two of the three planes coincide,
3) the intersecting line of any two of the planes is parallel to the third plane, or
4) the intersecting line lies in the third plane.

8.7 MATRIX INVERSION

As mentioned in Section 8.2, a unit matrix $[I]$ is one in which the elements on the principal diagonal are ones and the remaining elements are zeros. We shall now define the inverse of a matrix $[A]$. The inverse of $[A]$, or $[A]^{-1}$, is a matrix which has the property

$$[A][A]^{-1} = [A]^{-1}[A] = [I].$$

The basic problem in this section is to find $[A]^{-1}$ when $[A]$ is given. The method we discuss below is essentially the Gauss-Jordan elimination method (Sections 8.3 through 8.5). When applied to a system of n simultaneous equations, the method consists of n passes, each having the following three steps:

1) interchange of rows (discussed in Section 8.5),
2) normalization (discussed mainly in Section 8.4), and
3) elimination (discussed mainly in Section 8.3).

We shall now consider the inversion process using the Gauss-Jordan reduction. For the sake of clarity we shall temporarily ignore the interchange of rows.

If the matrix $[A]$ is known and its inverse is to be found, we append a unit matrix of the same order, forming a double square matrix as follows:

$$\begin{bmatrix} a_{11} & a_{12} & a_{13} & 1 & 0 & 0 \\ a_{21} & a_{22} & a_{23} & 0 & 1 & 0 \\ a_{31} & a_{32} & a_{33} & 0 & 0 & 1 \end{bmatrix}. \tag{8.15}$$

We proceed to normalize row 1 and eliminate a_{21} and a_{31} as shown in pass 1 of Table 8.2. The result of the first elimination is

$$\begin{bmatrix} 1 & a_{12}/a_{11} & a_{13}/a_{11} & 1/a_{11} & 0 & 0 \\ 0 & a_{22} - a_{21}(a_{12}/a_{11}) & a_{23} - a_{21}(a_{13}/a_{11}) & -a_{21}/a_{11} & 1 & 0 \\ 0 & a_{32} - a_{31}(a_{12}/a_{11}) & a_{33} - a_{31}(a_{13}/a_{11}) & -a_{31}/a_{11} & 0 & 1 \end{bmatrix}.$$

By continuing the process to eliminate the other variables, we obtain the following form:

$$\begin{bmatrix} 1 & 0 & 0 & b_{11} & b_{12} & b_{13} \\ 0 & 1 & 0 & b_{21} & b_{22} & b_{23} \\ 0 & 0 & 1 & b_{31} & b_{32} & b_{33} \end{bmatrix}.$$

Table 8.4 Matrix inversion

Given coefficient matrix and appended unit matrix	3	−6	7		1	0	0
	9	0	−5		0	1	0
	5	−8	6		0	0	1

PASS 1
$k = 1$, pivot row = row 1

New row 1 = (row 1)/3

$i = 1$	1	−2	2.33		0.33	0	0
$j = 1, 2, 3, 4$	9	0	−5		0	1	0
	5	−8	6		0	0	1

New row 2 = row 2 − 9 × row 1

$i = 2$	1	−2	2.33		0.33	0	0
$j = 2, 3, 4$	0	18	−26		−3	1	0
	5	−8	6		0	0	1

New row 3 = row 3 − 5 × row 1

$i = 3$	1	−2	2.33		0.33	0	0
$j = 2, 3, 4$	0	18	−26		−3	1	0
	0	2	−5.65		−1.65	0	1

PASS 2
$K = 2$, pivot row = row 2

New row 2 = (row 2)/18

$i = 2$	1	−2	2.33		0.33	0	0
$j = 2, 3, 4$	0	1	−1.44		−0.17	0.06	0
	0	2	−5.65		−1.65	0	1

New row 1 = row 1 − (−2) × row 2

$i = 1$	1	0	−0.55		−0.01	0.12	0
$j = 3, 4$	0	1	−1.44		−0.17	0.06	0
	0	2	−5.65		−1.65	0	1

New row 3 = row 3 − 2 × row 2

$i = 3$	1	0	−0.55		−0.01	0.12	0
$j = 3, 4$	0	1	−1.44		−0.17	0.06	0
	0	0	−2.77		−1.31	−0.12	1

		PASS 3 $K = 3$, pivot row = row 3				
		New row 3 = (row 3)/-2.77				
$i = 3$	1	0	-0.55	-0.01	0.12	0
$j = 3, 4$	0	1	-1.44	-0.17	0.06	0
	0	0	1	0.48	0.04	-0.36
		New row 2 = row 2 $-$ (-1.44) $\times$ row 3				
$i = 2$	1	0	-0.55	-0.01	0.12	0
$j = 4$	0	1	0	0.52	0.12	-0.52
	0	0	1	0.48	0.04	-0.36
		New row 1 = row 1 $-$ (-0.55) $\times$ row 3				
$i = 1$	1	0	0	0.26	0.14	-0.2
$j = 4$	0	1	0	0.52	0.12	-0.52
	0	0	1	0.48	0.04	-0.36

In this resultant matrix, the matrix $[B]$ is the inverse of $[A]$. Table 8.4 shows a step-by-step numerical computation for obtaining the matrix $[A]^{-1}$, where

$$[A] = \begin{bmatrix} 3 & -6 & 7 \\ 9 & 0 & -5 \\ 5 & -8 & 6 \end{bmatrix}.$$

We note that the answer is

$$[A]^{-1} = \begin{bmatrix} 0.26 & 0.14 & -0.2 \\ 0.52 & 0.12 & -0.52 \\ 0.48 & 0.04 & -0.36 \end{bmatrix}.$$

Turning to the relation between the solution of simultaneous equations and matrix inversion, let us consider Eq. (8.9) which, in matrix form, becomes

$$[A][X] = [C]. \tag{8.16}$$

This equation is solved by multiplying through by $[A]^{-1}$. Thus

$$[A]^{-1}[A][X] = [A]^{-1}[C]$$

or

$$[X] = [A]^{-1}[C] \tag{8.17}$$

Equation (8.17) gives the values of all the unknown X's by a simple multiplication of matrices, provided that $[A]^{-1}$ is determined. In the example shown in Tables 8.1 and 8.2, Eq. (8.17) becomes

$$\begin{bmatrix} x_1 \\ x_2 \\ x_3 \end{bmatrix} = \begin{bmatrix} 0.26 & 0.14 & -0.2 \\ 0.52 & 0.12 & -0.52 \\ 0.48 & 0.04 & -0.36 \end{bmatrix} \begin{bmatrix} 3 \\ 3 \\ -4 \end{bmatrix} = \begin{bmatrix} 2 \\ 4 \\ 3 \end{bmatrix},$$

or

$$x_1 = 2, \qquad x_2 = 4, \qquad x_3 = 3,$$

which agree with the solutions obtained in Table 8.1.

It should be noted that $[A]^{-1}$ does not always exist. This is indeed the case if $[A]$ is singular,† or $|A| = 0$. In addition, when the determinant $|A|$ has a very small value, the overflow problem usually occurs in computing $[A]^{-1}$. Suppose, for example, the absolute value of an element in floating point turns out to be larger than 16^{63} (or approximately 10^{75}) used in a typical FORTRAN IV compiler; we then have an overflow condition.

8.8 MATRIX INVERSION IN PLACE

In the last section we discussed matrix inversion through the use of an appended unit matrix. We shall now proceed to show how this powerful method can be modified to save a portion of computer storage. In this modified method each element of $[A]^{-1}$ is to be stored, as computed, in a location in $[A]$, and hence the appended matrix is not needed. To understand this procedure, it is necessary to inspect Table 8.4. We note that (1) at the end of each step, one of the elements in the left-hand column of the given matrix $[A]$ is made either zero or one, and (2) this element, once reduced to zero or one remains unaffected by later operations.

In pass 1, for example, the elements a_{11} in Step 1, a_{21} in Step 2, and a_{31} in Step 3 are reduced to a value of one, zero, and zero, respectively, and these values are kept unchanged. At each step it is therefore possible to store the value of the corresponding element of the appended matrix in these locations. Thus the element b_{11} of the appended matrix, now having the value 0.33, can be stored in the location of the element a_{11} in place of the value of one. Similarly, the element b_{21}, now having the value -3.0, is stored in a_{21} in place of the value zero, etc.

Table 8.5 shows a step-by-step storage map of $[A]$ and compares this modified method with the method discussed in Table 8.4. In the left column of Table 8.5, matrix $[A]$ at each of the nine steps is reproduced from Table 8.4. These values may be compared with their counterparts, which are shown in the right column. We note that at each step, each of the nine elements of $[A]$ is replaced, one by one, by an element of the required $[A]^{-1}$-matrix. In other words, the inversion is done within $[A]$ itself, and no storages for the appended unit matrix are needed. This modified method is frequently referred to as matrix inversion in place.

Before we turn to the next section for a FORTRAN program we shall define an orthogonal matrix. A square matrix $[T]$ is called *orthogonal* if $[T][T]^T = [T]^T[T] = [I]$. In other words, if $[T]^T = [T]^{-1}$, then $[T]$ is said to be orthogonal. For example, the matrix

$$[T] = \begin{bmatrix} \cos\theta & \sin\theta \\ -\sin\theta & \cos\theta \end{bmatrix}$$

---

† For example, see Fox, L., *An Introduction to Numerical Linear Algebra.* Oxford Univ. Press, New York, 1965, p. 37.

Table 8.5 Matrix inversion in place

Column 1 [A] reproduced from Table 8.4			Column 2 Inversion in place		
Pass 1, Step 1					
1	−2	2.33	0.33	−2	2.33
9	0	−5	9	0	−5
5	−8	6	5	−8	6
Step 2					
1	−2	2.33	0.33	−2	2.33
0	18	−26	−3	18	−26
5	−8	6	5	−8	6
Step 3					
1	−2	2.33	0.33	−2	2.33
0	18	−26	−3	18	−26
0	2	−5.65	−1.65	2	−5.65
Pass 2, Step 1					
1	−2	2.33	0.33	−2	2.33
0	1	−1.44	−0.17	0.06	−1.44
0	2	−5.65	−1.65	2	−5.65
Step 2					
1	0	−0.55	−0.01	0.12	−0.55
0	1	−1.44	−0.17	0.06	−1.44
0	2	−5.65	−1.65	2	−5.65
Step 3					
1	0	−0.55	−0.01	0.12	−0.55
0	1	−1.44	−0.17	0.06	−1.44
0	0	−2.77	−1.31	−0.12	−2.77
Pass 3, Step 1					
1	0	−0.55	−0.01	0.12	−0.55
0	1	−1.44	−0.17	0.06	−1.44
0	0	1	0.48	0.04	−0.36
Step 2					
1	0	−0.55	−0.01	0.12	−0.55
0	1	0	0.52	0.12	−0.52
0	0	1	0.48	0.04	−0.36
Step 3					
1	0	0	0.26	0.14	−0.20
0	1	0	0.52	0.12	−0.52
0	0	1	0.48	0.04	−0.36

is an orthogonal matrix because it has the following property:

$$[T]^T = [T]^{-1} = \begin{bmatrix} \cos \theta & -\sin \theta \\ \sin \theta & \cos \theta \end{bmatrix}.$$

The orthogonal matrix plays an important role in Chapter 9. However, we impart a word of warning: the inverse of a nonsingular matrix is not always equal to the transpose of the given matrix.

8.9 FORTRAN PROGRAM FOR MATRIX INVERSION AND SIMULTANEOUS LINEAR EQUATIONS

A tested FORTRAN subprogram to carry out the matrix inversion with accompanying solution of simultaneous linear equations $[A][X] = [B]$ appears in Fig. 8.1. The procedure used is similar to the one discussed in Sections 8.5, 8.6, and 8.8. The arguments in the subprogram CHAP8 are as follows:

A = the given coefficient matrix A; A^{-1} will be stored in this matrix at return to the main program;

N = order of A; N ≥ 1;

B = matrix of constant vector, used for solution of simultaneous equation only;

M = the number of column vectors in the matrix of constant vectors (M = 0 if inversion is the sole aim; M = 1, 2, ... for solution of simultaneous equations);

DET = value of the determinant |A|.

```
C
      SUBROUTINE CHAP8 (A,N,B,M,DET)
C     THIS SUBPROGRAM IS FOR MATRIX INVERSION AND SIMULT.LINEAR EQS.
      DIMENSION A(30,30),B(30,30),IPVOT(30),INDEX(30,2),PIVOT(30)
      COMMON IPVOT,INDEX,PIVOT
      EQUIVALENCE (IROW,JROW),(ICOL,JCOL)
C
C     FOLLOWING 3 STATEMENTS FOR INITIALIZATION
C
   57 DET=1.
      DO 17 J=1,N
   17 IPVOT(J)=0
      DO 135 I=1,N
C
C     FOLLOWING 12 STATEMENTS FOR SEARCH FOR PIVOT ELEMENT
C
      T=0.
      DO 9 J=1,N
      IF(IPVOT(J).EQ.1) GO TO 9
   13 DO 23 K=1,N
      IF(IPVOT(K)-1) 43,23,81
   43 IF(ABS(T).GE.ABS(A(J,K))) GO TO 23
   83 IROW=J
```

Fig. 8.1. Subprogram CHAP8.

```
        ICOL=K
        T=A(J,K)
     23 CONTINUE
      9 CONTINUE
        IPVOT(ICOL)=IPVOT(ICOL)+1
C
C       FOLLOWING 15 STATEMENTS TO PUT PIVOT ELEMENT ON DIAGONAL
C
        IF(IROW.EQ.ICOL) GO TO 109
     73 DET=-DET
        DO 12 L=1,N
        T=A(IROW,L)
        A(IROW,L)=A(ICOL,L)
     12 A(ICOL,L)=T
        IF(M.LE.0) GO TO 109
     33 DO 2 L=1,M
        T=B(IROW,L)
        B(IROW,L)=B(ICOL,L)
      2 B(ICOL,L)=T
    109 INDEX(I,1)=IROW
        INDEX(I,2)=ICOL
        PIVOT(I)=A(ICOL,ICOL)
        DET=DET*PIVOT(I)
C
C       FOLLOWING 6 STATEMENTS TO DIVIDE PIVOT ROW BY PIVOT ELEMENT
C
        A(ICOL,ICOL)=1.
        DO 205 L=1,N
    205 A(ICOL,L)=A(ICOL,L)/PIVOT(I)
        IF(M.LE.0) GO TO 347
     66 DO 52 L=1,M
     52 B(ICOL,L)=B(ICOL,L)/PIVOT(I)
C
C       FOLLOWING 10 STATEMENTS TO REDUCE NON-PIVOT ROWS
C
    347 DO 135 LI=1,N
        IF(LI.EQ.ICOL) GO TO 135
     21 T=A(LI,ICOL)
        A(LI,ICOL)=0.
        DO 89 L=1,N
     89 A(LI,L)=A(LI,L)-A(ICOL,L)*T
        IF(M.LE.0) GO TO 135
     18 DO 68 L=1,M
     68 B(LI,L)=B(LI,L)-B(ICOL,L)*T
    135 CONTINUE
C
C       FOLLOWING 11 STATEMENTS TO INTERCHANGE COLUMNS
C
    222 DO 3 I=1,N
        L=N-I+1
        IF(INDEX(L,1).EQ.INDEX(L,2)) GO TO 3
     19 JROW=INDEX(L,1)
        JCOL=INDEX(L,2)
        DO 549 K=1,N
        T=A(K,JROW)
        A(K,JROW)=A(K,JCOL)
        A(K,JCOL)=T
    549 CONTINUE
      3 CONTINUE
     81 RETURN
        END
```

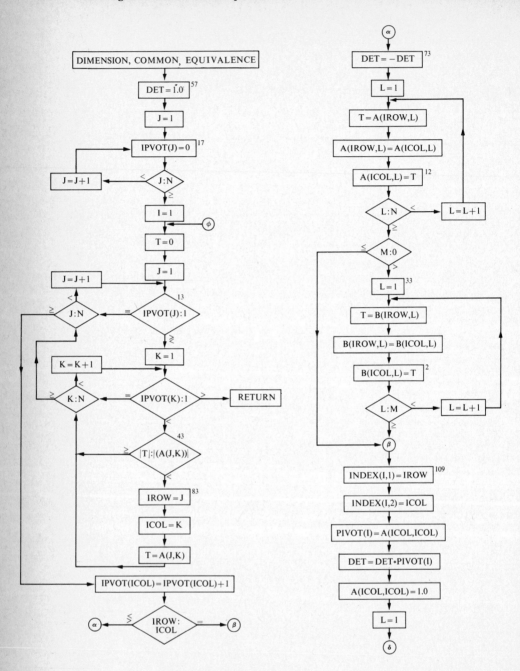

Fig. 8.2 Flowchart for matrix inversion and solutions of simultaneous equations

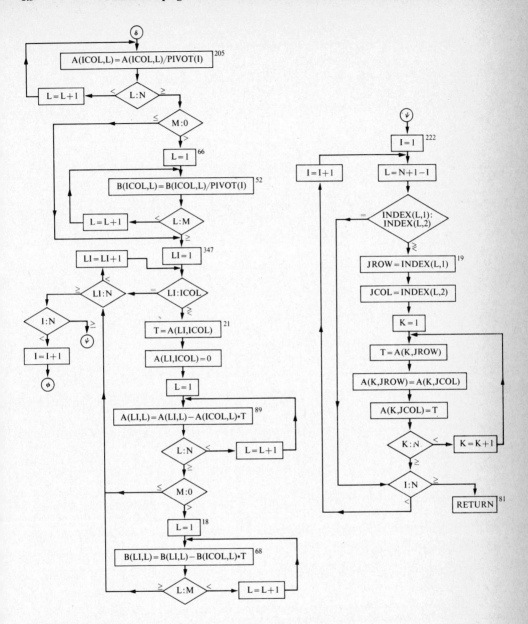

It should be noted that the column index for the argument INDEX is 2. IPVOT is an array used to prevent duplicate pivotings on any single row. Figure 8.2 shows the corresponding flow chart.

The EQUIVALENCE and COMMON statements used in this subprogram have not been discussed before and need some explanation.

The EQUIVALENCE statement conserves storage locations. This statement allows two or more variables which are never needed at the same time in the program to share the *same* storage location. For example, the statement

EQUIVALENCE (IROW, JROW)

means that the variables IROW and JROW are to share the same storage location. In general, the statement

$$\text{EQUIVALENCE } (a_1, a_2, \ldots, a_n), (b_1, b_2, \ldots, b_m)$$

will cause variables a_1 through a_n to share one location, and variables b_1 through b_m to share *another* storage location.

The COMMON statement is also used to conserve locations. As mentioned in Sections 5.7 and 5.8, if a variable V appears in both the main program and a subprogram, these two V's are not at all related. In fact, a separate storage area is set up for each one. It is clear that storage locations are wasted if these two V's are actually the same variable. To eliminate this waste, COMMON statements can be used in a FORTRAN program. For example, in Figs. 8.2 and 8.3, the FORTRAN statement

COMMON IPVOT, INDEX, PIVOT

is put in both the subprogram CHAP8 and its main program. The variables are now the same for both programs and therefore storages are not wasted.

```
C        MATRIX INVERSION. MAIN PROGRAM FOR EXAMPLE 8.1
         DIMENSION C(15.15).D(15.15)
         COMMON IPVOT.INDEX.PIVOT
         DATA IREAD.IWRITE/5.6/
         WRITE(IWRITE.1)
    1 FORMAT('1'.10X.'GIVEN MATRIX C'/)
         READ(IREAD.5) N.M
    5 FORMAT(I2.I2)
         DO 10 I=1.N
   10 READ(IREAD.15) (C(I.J).J=1.N)
   15 FORMAT(7F10.4)
         DO 20 I=1.N
   20 WRITE(IWRITE.15) (C(I.J).J=1.N)
   30 CALL CHAP8(C.N.D.M.DETERM)
         WRITE(IWRITE.35)
   35 FORMAT(//10X.'INVERSE OF MATRIX C'/)
         DO 40 I=1.N
   40 WRITE(IWRITE.45) (C(I.J).J=1.N)
   45 FORMAT(7F10.4)
   50 STOP
         END


    0300
    0.0        -6.0         9.0
    7.0         0.0        -5.0
    5.0        -8.0         6.0
```

Fig. 8.3. Main program and data for matrix inversion.

In this example, the three subscripted variables,

IPVOT, INDEX, and PIVOT,

are assigned to the same locations in a continuous special section of the storage which is set aside by the compiler. This is the special common area in the storage for the main program and its required subprograms. In this common area, the arrays IPVOT, INDEX, and PIVOT are assigned sequentially and by columns for each of the three arrays.

As shown in Fig. 8.2, the COMMON and EQUIVALENCE usually follow immediately the DIMENSION statement in the subroutine. They should precede all executable statements.

A COMMON statement also shows whether there is any correspondence between variables in the main program and the subprogram; it stores these variables in the same storage location even if they do not have the same name. For example, in the subroutine:

COMMON (A, B)

in the main program:

COMMON (D, E)

Then A and D, as well as B and E, have the same storage location.

Example 8.1

Write a main program, together with the necessary data cards, to compute $[C]^{-1}$, where $[C]$ is the coefficient matrix shown in Table 8.4. This main program will be used to call the subroutine CHAP8 shown in in Fig. 8.1; it will also serve to read in $[C]$ and print out $[C]^{-1}$.

Figure 8.3 illustrates a possible main program which calls for the subroutine CHAP8. After the first transfer (statement 30 in the main program) to the subprogram CHAP8, values in matrix C and matrix D will be substituted for the dummy variables A and B, respectively, 3 will be substituted for N, and 0 for M.

An echo print and the answers are shown in Table 8.6.

Example 8.2

Write a FORTRAN main program to solve a set of simultaneous linear equations by calling on the subprogram CHAP8 shown in Fig. 8.1. The first input card after the heading print out should be N. If N is zero, terminate the computation; if it is not zero, read in the given matrices.

Table 8.6 Answers for Example 8.1

```
                GIVEN   MATRIX   C
        0.0000      -6.0000       9.0000
        7.0000       0.0000      -5.0000
        5.0000      -8.0000       6.0000


                INVERSE   OF   MATRIX   C
         .3921        .3529      -.2941
         .6568        .4411      -.6176
         .5490        .2941      -.4117
```

```
C        SOLUTION OF SIMULTANEOUS LINEAR EQUATIONS, MAIN PROGRAM
         DIMENSION C(15,15),D(15,15)
         COMMON IPVOT,INDEX,PIVOT
         DATA IREAD,IWRITE/5,6/
   71 READ(IREAD,5) N,M
    5 FORMAT(I2,I2)
         IF(N.EQ.0) GO TO 60
   77 WRITE(IWRITE,1)
    1 FORMAT('1',10X,'GIVEN MATRIX C')
         DO 10 I=1,N
   10 READ(IREAD,15) (C(I,J),J=1,N)
   15 FORMAT(7F10.4)
         DO 20 I=1,N
   20 WRITE(IWRITE,15) (C(I,J),J=1,N)
         WRITE(IWRITE,25)
   25 FORMAT(10X,'GIVEN MATRIX D')
         DO  30 I=1,N
   30 READ(IREAD,15) (D(I,J),J=1,M)
         DO 35 I=1,N
   35 WRITE(IWRITE,15) (D(I,J),J=1,M)
         CALL CHAP8(C,N,D,M,DET)
         WRITE(IWRITE,40)
   40 FORMAT(10X,'SOLUTION OF GIVEN EQUATIONS')
         DO 45 I=1,N
   45 WRITE(IWRITE,15) (D(I,J),J=1,M)
         WRITE(IWRITE,50)
   50 FORMAT(10X,'VALUE OF DETERMINANT')
         WRITE(IWRITE,55) DET
   55 FORMAT(F10.4//)
         GO TO 71
   60 STOP
         END
```

```
0401
        .5000        .3300        .2500        .5000
        .5000        .2500        .2000       1.0000
       1.0000       1.0000       1.0000       1.0000
       2.0000       3.0000       1.0000       5.0000
      13.0000
      11.5000
      40.0000
      75.0000
0301
       2.0000       -7.0000       4.0000
       1.0000        9.0000      -6.0000
      -3.0000        8.0000       5.0000
       9.0000
       1.0000
       6.0000
0000
```

Fig. 8.4. Main program and data for solutions of simultaneous equations.

The output should include:

1) name and date (this is an echo print of an input card);
2) an echo print of the given matrices;
3) answers—a field of F7.3 will suffice;
4) value of the determinants.

The following two sets of equations should be solved:

$$\begin{array}{ll} \text{Set 1} & \text{Set 2} \\ \end{array}$$

$$\begin{aligned} x/2 + y/3 + z/4 + w/2 &= 13.0, & 2x - 7y + 4z &= 9.0, \\ x/2 + y/4 + z/5 + w &= 11.5, & x + 9y - 6z &= 1.0, \\ x + y + z + w &= 40.0, & -3x + 8y + 5z &= 6.0. \\ 2x + 3y + z + 5w &= 75.0; \end{aligned}$$

A possible main program in FORTRAN is shown in Fig. 8.4, and the corresponding results are tabulated in Table 8.7.

Table 8.7 Answers for Example 8.2

```
            GIVEN   MATRIX   C
  .5000        .3300       .2500        .5000
  .5000        .2500       .2000       1.0000
 1.0000       1.0000      1.0000       1.0000
 2.0000       3.0000      1.0000       5.0000
            GIVEN   MATRIX   D
 13.0000
 11.5000
 40.0000
 75.0000
        SOLUTION   OF   GIVEN   EQUATIONS
   7.2290
  12.0654
  19.7954
    .9100
        VALUE   OF   DETERMINANT
 -.2445



            GIVEN   MATRIX   C
 2.0000       -7.0000       4.0000
 1.0000        9.0000      -6.0000
-3.0000        8.0000       5.0000
            GIVEN   MATRIX   D
 9.0000
 1.0000
 6.0000
        SOLUTION   OF   GIVEN   EQUATIONS
   3.9999
    .9999
   1.9999
        VALUE   OF   DETERMINANT
 234.9999
```

```
C        MATRIX INVERSION ACCOMPANYING WITH DETERMINANT AND
C        SOLVING FOR SIMULTANEOUS EQUATIONS
C        FOR INVERSION ONLY M=0, LDET=0, INV=0
C        FOR DETERMINANT ONLY  M=-1, LDET=0, INV=0
C        FOR SOLUTION ONLY M=1 OR MORE, LDET=0, INV=-1
C        FOR INVERSION AND DETERMINANT  M=0, LDET=1, INV=0
C        FOR SOLUTION AND DETERMINANT  M=1 OR MORE, LDET=0, INV=0
C        FOR INVERSION AND SOLUTION  M=1 OR MORE, LDET=0, INV=1
C        FOR ALL VALUE  M=1 OR MORE, LDET=1, INV=1
C        TO TERMINATE THE COMPUTATION N=0, M=0, LDET=0, INV=0
         DIMENSION A(15,15),B(15,15)
         COMMON IPVOT,INDEX,PIVOT
         DATA IREAD,IWRITE/5,6/
  100 READ(IREAD,1) N,M,LDET,INV
    1 FORMAT(4I2)
      IF(N.EQ.0) GO TO 9000
      WRITE(IWRITE,7)
    7 FORMAT('1',' GIVEN MATRIX A'/)
      DO 14 I=1,N
      READ(IREAD,5) (A(I,J),J=1,N)
      WRITE(IWRITE,5) (A(I,J),J=1,N)
    5 FORMAT(7F10.4)
   14 CONTINUE
      IF(M.LE.0) GO TO 57
   10 WRITE(IWRITE,8)
    8 FORMAT(//10X,' GIVEN MATRIX OF CONSTANT VECTOR'/)
      DO 4 I=1,N
      READ(IREAD,5) (B(I,J),J=1,M)
      WRITE(IWRITE,5) (B(I,J),J=1,M)
    4 CONTINUE
   57 DET=1.
      .....................
      .....................
      .....................                    SAME AS SHOWN IN FIG. 8.1.
      .....................
      .....................
    3 CONTINUE
      IF(M) 1003,1001,1004
 1001 WRITE(IWRITE,1011)
 1011 FORMAT(//10X,' INVERSE OF MATRIX A'/)
      DO 1002 I=1,N
      WRITE(IWRITE,5) (A(I,J),J=1,N)
 1002 CONTINUE
      IF(LDET.LE.0) GO TO 9009
 1003 WRITE(IWRITE,1033)
 1033 FORMAT(//10X,' VALUE OF DETERMINANT'/)
      WRITE(IWRITE,5) DET
      GO TO 9009
 1004 WRITE(IWRITE,1044)
 1044 FORMAT(//10X,' SOLUTION OF GIVEN EQUATIONS'/)
      DO 1005 I=1,N
      WRITE(IWRITE,5) (B(I,J),J=1,M)
 1005 CONTINUE
      IF(INV) 9009,1003,1001
 9009 GO TO 100
 9000 STOP
      END
```

Fig. 8.5. Main program for matrix inversion and linear equations.

Example 8.3

Modify subroutine CHAP8 (Fig. 8.1) to a *main* program and find the inverse of matrix

$$A = \begin{bmatrix} 2 & 1 & 1 \\ 1 & 2 & 3 \\ 3 & 2 & 1 \end{bmatrix}.$$

We combine Figs. 8.1 and 8.3 to make a main program, shown in Fig. 8.5. Statements 57 through 3 inclusive are taken directly from subroutine CHAP8. The answers for $[A]^{-1}$ are shown in Table 8.8.

Table 8.8 Answers for Example 8.3

```
                        INPUT
    03010101
       0.0000      -6.0000       9.0000
       7.0000       0.0000      -5.0000
       5.0000      -8.0000       6.0000
      -4.0000
       7.0000
       3.0000
                        OUTPUT
                GIVEN   MATRIX   A
       0.0000      -6.0000       9.0000
       7.0000       0.0000      -5.0000
       5.0000      -8.0000       6.0000
            GIVEN   MATRIX   OF   CONSTANT   VECTOR
      -4.0000
       7.0000
       3.0000
            SOLUTION   OF   GIVEN   EQUATIONS
        .0196
      -1.3922
      -1.3725
            INVERSE   OF   MATRIX   A
        .3922        .3529       -.2941
        .6569        .4412       -.6176
        .5490        .2941       -.4118
            VALUE   OF   DETERMINANT
    -102.0000
                        INPUT
    03000000
       2.0000       1.0000       1.0000
       1.0000       2.0000       3.0000
       3.0000       2.0000       1.0000

                        OUTPUT
                GIVEN   MATRIX   A
       2.0000       1.0000       1.0000
       1.0000       2.0000       3.0000
       3.0000       2.0000       1.0000
            INVERSE   OF   MATRIX   A
       1.0000      -0.2500      -0.2500
      -2.0000       0.2500       1.2500
       1.0000       0.2500      -0.7500
```

8.10 GAUSS-SEIDEL ITERATIVE METHOD

The elimination method discussed in Sections 8.3 through 8.5 is a direct method, not an iterative procedure. When the numerical values of diagonal elements in the given coefficient matrix are large compared with those of the remaining elements, an iterative procedure, the Gauss-Seidel method, is generally favored. This type of coefficient matrix arises frequently in numerical or partial differential equations of scientific and engineering problems. Let us consider the linear system

$$
\begin{aligned}
a_{11}x_1 + a_{12}x_2 + a_{13}x_3 + \cdots + a_{1n}x_n &= c_1, \\
a_{21}x_1 + a_{22}x_2 + a_{23}x_3 + \cdots + a_{2n}x_n &= c_2, \\
&\vdots \\
a_{n1}x_1 + a_{n2}x_2 + a_{n3}x_3 + \cdots + a_{nn}x_n &= c_n.
\end{aligned}
\tag{8.18}
$$

Equation (8.18) can be rewritten so that the ith equation is solved for x_i:

$$
x_1 = -\frac{a_{12}x_2 + a_{13}x_3 + \cdots + a_{1n}x_n - c_1}{a_{11}},
$$

$$
x_2 = -\frac{a_{21}x_1 + a_{23}x_3 + \cdots + a_{2n}x_n - c_2}{a_{22}},
$$

$$
x_3 = -\frac{a_{31}x_1 + a_{32}x_2 + \cdots + a_{3n}x_n - c_3}{a_{33}},
\tag{8.19}
$$

$$
\vdots
$$

$$
x_n = -\frac{a_{n1}x_1 + a_{n2}x_2 + \cdots + a_{nn}x_n - c_n}{a_{nn}}.
$$

The procedure is to assume a set of starting values for x_i ($i = 1, 2, \ldots, n$) and then to calculate new values by substitution into Eq. (8.19) for $i = 1, 2, \ldots, n$. These values are used as new estimates and iterations are continued until some criterion is met.

Consider the following numerical example of two simultaneous linear equations

$$
\begin{aligned}
3x_1 + x_2 &= 5, \\
x_1 + 2x_2 &= 5,
\end{aligned}
\tag{8.20}
$$

which may be rewritten as follows:

$$
x_1 = \frac{-x_2}{3} + \frac{5}{3},
$$

$$
x_2 = \frac{-x_1}{2} + \frac{5}{2}.
$$

Assuming as an initial approximation that

$$
x_1 = 0, \qquad x_2 = 0,
$$

we obtain the iterative solutions shown in the following table:

Step number	x_1	x_2
1	0	0
2	$\frac{5}{3}$	$\frac{5}{2}$
3	$\frac{5}{6}$	$\frac{5}{3}$
4	$\frac{10}{9}$	$\frac{25}{12}$
⋮		
(True solution)	1	2

If the above iteration is modified so that the most recently computed values are used, then the iterative procedure will converge more rapidly, as can be seen:

Step number	x_1	x_2
1	0	0
2	$\frac{5}{3}$	$-(\frac{1}{2})(\frac{5}{3}) + \frac{5}{2} = \frac{5}{3}$
3	$-(\frac{1}{3})(\frac{5}{3}) + \frac{5}{3} = \frac{10}{9}$	$-(\frac{1}{2})(9) + \frac{5}{2} = \frac{35}{18}$

The Gauss-Seidel method suffers from its slow convergence for ill-conditioned systems. An ill-conditioned system is very sensitive to small variations in the values of the coefficients; such systems may be interpreted geometrically in terms of the two straight lines in the upper portion of Fig. 8.6. Here the intersecting angle is very acute, and the intersecting point P is not sharply defined.

To illustrate slow convergence, let us take the following singular system (the parallel lines shown in Fig. 8.6):

$$4x_1 + 3x_2 = 7,$$
$$8x_1 + 6x_2 = 14. \tag{8.21}$$

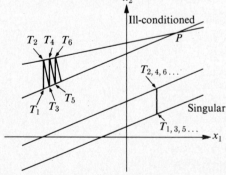

Fig. 8.6. Ill-conditioned system and singular system.

Using the Gauss-Seidel Method, we have at first

$$x_1 = \frac{7 - 3x_2}{4},$$

$$x_2 = \frac{14 - 8x_1}{6} \qquad (8.22)$$

$$= \frac{7 - 4x_1}{3}.$$

The table below shows the iterative processes using the most recently computed values:

Step number	x_1	x_2
1	$\frac{7}{4}$	0
2	$\frac{7}{4}$	0
3	$\frac{7}{4}$	0
4	$\frac{7}{4}$	0

It is clear that no improvement is expected. Similarly for other ill-conditioned systems (see Fig. 8.6), the rate of convergence is very slow indeed.

In this connection, we might use the following convergence criterion:

$$\left| \frac{x_i^{(k+1)}}{x_i^{(k)}} \right| < \delta, \qquad \text{if} \qquad |x_i^{(k+1)}| < |x_i^{(k)}|,$$

where k = step number,

$\quad i$ = number of variables ($i = 1, 2, \ldots, n$),

$\quad \delta$ = tolerance which is less than 1, say, between 0.9 and 0.999.

A FORTRAN program for the Gauss-Seidel method is shown in Fig. 8.7. While this program works for many cases, its success depends upon a number of conditions, including the arrangement of the given equations and the assumed initial values. From a practical point of view, convergence is normally assured† if each diagonal element of $[A]$ in Eq. (8.18) satisfies the following condition:

$$|a_{ii}| > \sum_{\substack{j=1 \\ j \neq i}}^{n} |a_{ij}| \qquad (i = 1, 2, \ldots, n)$$

Take Eq. (8.20) as an example. It must be verified that

$$|3| > |1|$$

and

$$|2| > |1|.$$

Since these two conditions are true, the Gauss-Seidel method does converge.

---

† See also Problem 4 in Chapter 9.

```
C        THE GAUSS-SEIDEL ITERATION METHOD FOR SOLVING SIMULTANEOUS
C        LINEAR EQUATIONS
C        THE EQUATIONS MUST BE PUT IN THE FORM A*X=B
C        THE FOLLOWING DEFINE SOME OF THE VARIABLES USED
C        N-THE NUMBER OF UNKNOWNS,KOUNT-NUMBER OF ITERATIONS
C        DELTA -THE TOLERANCE WITHIN WHICH THE RATIO OF TWO SUCCESSIVE
C        ITERATIONS MUST LIE.
C        DELTA IS LESS THAN 1. NORMALLY USE DELTA=0.9 OR 0.99
C        THE NUMBER OF EQNS.=NUMBER OF UNKNOWNS IN THE METHOD
C        A REPRESENTS THE COEFFICIENT MATRIX
C        B THE COLUMN MATRIX OF CONSTANTS ON THE RIGHT HAND SIDE OF EQNS.
C
         DIMENSION A(20,20),B(20),X(20),OLDX(20)
C        READ IN N,DELTA,A,B
         READ(5,5)N,DELTA
    5 FORMAT(5X,I5,2X,F7.2)
   63 DO 6 I=1,N
    6 READ(5,7)(A(I,J),J=1,N)
    7 FORMAT(7F10.4)
      READ(5,7)(B(I),I=1,N)
C        BEGIN GAUSS SEIDEL METHOD
C        FIND MAX.NUMBER OF ITERATIONS
      ITRTE=N*20
C        INITIALIZATION
      KOUNT=0
      DO 8 I=1,N
      OLDX(I)=0.0
C        USUAL FIRST APPROXIMATION
    8 X(I)=0.0
C
C
C        COMPUTE NEW VALUES FOR X(I)
C
   16 DO 9 I=1,N
      DIL=0.0
      DO 11 J=1,N
      IF(I-J)10,11,10
   10 DIL=DIL+A(I,J)*X(J)
   11 CONTINUE
    9 X(I)=(B(I)-DIL)/A(I,I)
      KOUNT=KOUNT+1
C
C        TEST TO SEE IF ITRTE HAS BEEN EXCEEDED
      IF(ITRTE-KOUNT)18,15,15
C        COMPARE EACH X TO ITS PREDECESSOR OLDX
C        TEST FOR CONVERGENCE
   15 DO 12 I=1,N
      IF(ABS(X(I))-ABS(OLDX(I)))25,12,26
   25 IF(DELTA-ABS(X(I)/OLDX(I)))12,13,13
   26 IF(DELTA-ABS(OLDX(I)/X(I)))12,13,13
   12 CONTINUE
      GO TO 31
   13 DO 17 I=1,N
   17 OLDX(I)=X(I)
      GO TO 16
C        PRINT OUT NUMBER OF ITERATIONS AND X VALUES
   18 WRITE(6,19)ITRTE
   19 FORMAT(//,15X,5HAFTER,I3,32H ITERATIONS THE VALUES OF X ARE )
      GO TO 14
   31 WRITE(6,19)KOUNT
   14 DO 20 I=1,N
      IF(I-10)21,24,24
   24 WRITE(6,22)I,X(I)
   22 FORMAT(25X,2HX(,I2,3H) =,E14.7)
      GO TO 20
   21 WRITE(6,23)I,X(I)
   23 FORMAT(25X,2HX(,I1,3H) =,E14.7)
   20 CONTINUE
C
      CALL EXIT
      END
```

Fig. 8.7. FORTRAN program for the Gauss-Seidel method.

BIBLIOGRAPHY

Determinants and Matrices

Aitken, A. C., *Determinants and Matrices*, 9th ed. Oliver and Boyd, Edinburgh; Interscience, New York, 1956.

Beckenbach, E. F. (ed.), *Modern Mathematics for the Engineer*. McGraw-Hill, New York, 1956.

Bellman, R., *Introduction to Matrix Algebra*. McGraw-Hill, New York, 1960.

Bodewig, E., *Matrix Calculus*. North Holland, Amsterdam; Interscience, New York, 1956. "Conference on Matrix Computations," *J. Assoc. Comput. Mach.*, **4**, 100–115 (1958).

Frazer, R. A., W. J. Duncan, and A. R. Collar, *Elementary Matrices*. Cambridge Univ. Press, London, 1947.

Householder, A. S., "The approximate solution of matrix problems," *J. Assoc. Comp. Mach.*, **5**, 205–243 (1958).

Marcus, M., "Basic theorems in matrix theory," *Natn. Bur. Stand. Appl. Math. Ser.*, **57** (1960).

Matrix Inversion

Fox, L., "Practical solution of linear equations and inversion of matrices," *U.S. Bur. Stand. Appl. Math. Ser.*, **39**, 1–54 (1954).

Fox, L., and J. G. Hayes, "More practical methods for the inversion of matrices," *J. Roy. Stat. Soc. B*, **13**, 83–91 (1951).

Goldstine, H. H., and J. von Neumann, "Numerical inverting of matrices of high order," *Proc. Am. Math. Soc.*, **2**, 188–202 (1951).

Householder, A. S., "A survey of some closed methods for inverting matrices," *J. Soc. Ind. Appl. Math.*, **5**, 155–169 (1957).

Newman, M., and J. Todd, ' The evaluation of matrix inversion programs," *J. Soc. Ind. Appl. Math.*, **6**, 466–476 (1958).

von Neumann, J., and H. H. Goldstine, "Numerical inverting of matrices of high order," *Bull. Am. Math. Soc.*, **53**, 1021–1099 (1947), *Proc. Am. Math. Soc.*, **2**, 188–202 (1951).

Linear Algebra and Simultaneous Equations

Bowdler, H. J., R. S. Martin, G. Peters, and J. H. Wilkinson, "Solution of real and complex systems of linear equations," *Num. Math.*, **8**, no. 3 (1966).

Carré, B. A., "The partitioning of network equations for block iteration," *Comput. J.*, **9**, 84–97 (1966).

Dwyer, P. S., *Linear Computations*. Wiley, New York, 1951.

Faddeeva, V. N., *Computational Methods in Linear Algebra*. Translated by C. D. Benster. Dover, New York, 1959.

Forsythe, G. E., "Solving linear equations can be interesting," *Bull. Am. Math. Soc.*, **59**, 299–329 (1953).

Forsythe, G. E., "Tentative classification of methods and bibliography on solving systems of linear equations," *Inst. Num. Anal. Rpt.* 52–7, Natn. Bur. Stand., Los Angeles, Calif. (Multilith Typescript, 1951); *Appl. Math. Ser.*, **29**, Natn. Bur. Stand., Washington, D.C., pp. 1–28 (1953).

Forsythe, G. E., and C. B. Moler, *Computer Solution of Linear Algebraic Systems.* Prentice-Hall, New York, 1967.

Fox, L., *Introduction to Numerical Linear Algebra.* Clarendon Press, Oxford, 1964.

Gloudeman, J. F., R. A. Rosenoff, and S. Levy, "Numerical conditioning of stiffness matrix formulation for frame structures." Proc. 2nd Conf. on Matrix Methods in Struct. Mech., U.S. Air Force Flight Dynamics Lab., Wright-Patterson A.F.B., Ohio, Dec. 1969, pp. 1029–1060.

Hestenes, M. R., and E. L. Stiefel, "Methods of conjugate gradients for solving linear systems," *J. Res. Natn. Bur. Stand.*, **49**, no. 6, 409–436 (1952).

Householder, A. S., *The Theory of Matrices in Numerical Analysis.* Ginn, 1964.

Martin, R. S., G. Peters, and J. H. Wilkinson, "Iterative refinement of the solution of a positive definite system of equations," *Num. Math.*, **8**, 203–216 (1966).

Ostrowski, A. M., "Über näherungsweise Auflösung von Systemen homogener linearer Gleichungen," *Z. Angew. Math. Phys.*, **8**, 280–285 (1957).

Paige, L. J., and O. Taussky (eds.), "Simultaneous linear equations and the determination of eigenvalues," *Natn. Bur. Stand. Appl. Math. Ser.*, **29** (1953).

Stiefel, E. L., P. Henrici, and H. Rutishauser, "Further contributions to the solution of simultaneous linear equations and the determination of eigenvalues," *Natn. Bur. Stand. Appl. Math. Ser.*, **49** (1958).

Taussky, O. (ed.), "Contributions to the solution of systems of linear equations and the determination of eigenvalues," *Natn. Bur. Stand. Appl. Math. Ser.*, **39** (1954).

Varga, R. S., *Iterative Numerical Analysis.* Prentice-Hall, Englewood Cliffs, N.J., 1962.

Westlake, J. A., *A Handbook of Numerical Matrix Inversion and Solution of Linear Equations.* Wiley, New York, 1968.

System of Non-Linear Equations

Mancino, O. G., "Resolution by iteration of some nonlinear systems," *J. Assoc. Comput. Mach.*, **14**, no. 2, 341–350 (1967).

Remmler, K. L., D. W. Cawood, J. A. Stanton, and R. Hill, "Solutions of systems of nonlinear equations," *Report LMSC/HREC A783333.* Lockheed Missiles and Space Co., Huntsville, Ala., 1966.

Robinson, S. M., "Interpolative solution of systems of nonlinear equations," *SIAM J. Num. Anal.*, **3**, no. 4, 650 (1966).

Wolfe, P., "The secant method for simultaneous nonlinear equations," *Commun. Assoc. Comp. Mach.*, **2**, no. 12, 12–13 (1959).

PROBLEMS

1. Euler rotations of variables require three stages, corresponding to the factored plane rotations:

$$
\begin{bmatrix} \bar{X} \\ \bar{Y} \\ \bar{Z} \end{bmatrix} = \begin{bmatrix} 1 & 0 & 0 \\ 0 & \cos\phi & \sin\phi \\ 0 & -\sin\phi & \cos\phi \end{bmatrix} \begin{bmatrix} \cos\theta & \sin\theta & 0 \\ -\sin\theta & \cos\theta & 0 \\ 0 & 0 & 1 \end{bmatrix} \begin{bmatrix} 1 & 0 & 0 \\ 0 & \cos\psi & \sin\psi \\ 0 & -\sin\psi & \cos\psi \end{bmatrix} \begin{bmatrix} X \\ Y \\ Z \end{bmatrix}.
$$

Write a main program to call the subroutine MATMPY to find $\bar{X}, \bar{Y}, \bar{Z}$ if $\phi = 34°$, $\theta = 22.5°$, $\psi = 18°$, and $X = 1$, $Y = 2$, and $Z = 4$.

2. Use Gauss-Jordan procedure and consider the interchange of rows by normalization as well as elimination to find the inverse of each of the following two matrices:

$$
[A] = \begin{bmatrix} 0 & 1 & 0 \\ -9 & 0 & 0 \\ 0 & 4 & 3 \end{bmatrix}, \quad [B] = \begin{bmatrix} 0 & -4 & 8 \\ -4 & 8 & -4 \\ 8 & -4 & 0 \end{bmatrix}.
$$

3. a) Use the Gauss-Jordan procedure to find the inversion of the following matrix:

$$
\begin{bmatrix} 1 & 3 \\ 4 & 2 \end{bmatrix}.
$$

Both normalization and possible zero element on diagonal should be considered.
b) Do (a) by matrix inversion in place.
c) Find the value of the related determinant by using the steps in (a).

4. In structural analysis, one may use the slope-deflection method to solve for moments in the various parts of a frame. In a typical skew frame problem one obtains

$$
\begin{bmatrix} 3 & 1 & 2.4 \\ 1 & 5 & 0.3 \\ -41 & 52 & -209.1 \end{bmatrix} \begin{bmatrix} \theta_b \\ \theta_c \\ \theta_a \end{bmatrix} = \begin{bmatrix} 0 \\ 0 \\ -500 \end{bmatrix}.
$$

What is the value of θ_a?

5. Kirchhoff's law may be used to establish a set of linear equations for the circuit shown in Fig. 8.8. We have

$$
\begin{aligned}
I_1 + I_2 + I_3 &= A_1, \\
I_8 + I_9 + I_{10} &= A_2, \\
-I_1 + I_4 - I_6 &= 0, \\
-I_3 + I_5 - I_9 &= 0, \\
I_6 + I_7 - I_{10} &= 0, \\
-R_7 I_7 + R_8 I_8 - R_{10} I_{10} &= 0, \\
-R_5 I_5 + R_8 I_8 - R_9 I_9 &= 0, \\
R_2 I_2 - R_3 I_3 - R_5 I_5 &= 0, \\
-R_1 I_1 + R_2 I_2 - R_4 I_4 &= 0, \\
-R_4 I_4 - R_6 I_6 + R_7 I_7 &= 0,
\end{aligned}
$$

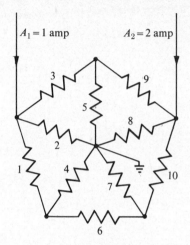

$A_1 = 1$ amp $A_2 = 2$ amp

Fig. 8.8

where $R_1 = 1$, $R_2 = 10$, $R_3 = 35$, $R_4 = 23$, $R_5 = 100$, $R_6 = 25$, $R_7 = 50$, $R_8 = 75$, $R_9 = 5$, and $R_{10} = 50$ ohms. Solve for the current I_1 through I_{10}.

6. Two sets of equations are given as follows:

$$2x - 2y - 2z - 3w = 11.0, \qquad x/5 - y/2 - z/4 = 15.0,$$
$$x - y - z - 2w = 4.0, \qquad x - y - z = 57.0,$$
$$-x - y - 6w = -3.0, \qquad z/5 - x/2 - 3y = 2.5.$$
$$9x - 9y - 7z - 23w = 43.0,$$

Write a main program to solve the two given sets of equations by calling the subroutine CHAP8. Make the main program flexible enough to handle any number of sets of equations. The required format for output is:

1. Name, edition, date

2. Echo print of the first coefficient matrix:

$$
\begin{array}{ccccc}
2.00 & -2.00 & 2.00 & -3.00 & 11.00 \\
1.00 & \cdots & \cdots & \cdots & 4.00 \\
\vdots & & & & \vdots \\
9.00 & \cdots & \cdots & \cdots & 43.00
\end{array}
$$

Echo print of the second coefficient matrix:

$$
\begin{array}{ccc}
0.20 & 0.50 & \cdots \\
\vdots & &
\end{array}
$$

3. Answers to the first set of equations:

X = XXXX.XXX,   Y = XXXX.XXX,   Z = XXXX.XXX,   W = XXXX.XXX.

Answers to the second set of equations: same.

7. Find the inverse and determinant of the following matrix (this is a 3 × 3 magic square):

$$\begin{bmatrix} 2 & 7 & 6 \\ 9 & 5 & 1 \\ 4 & 3 & 8 \end{bmatrix}.$$

8. Use the Gauss-Seidel method to solve

$$
\begin{aligned}
10x + 2y + 3z - w &= 80, \\
x - 20y - z + 3w &= 40, \\
x + y - 10z + 2w &= 40, \\
2x - y - z + 30w &= 120.
\end{aligned}
$$

CHAPTER 9

Eigenvalues and Eigenvectors of a Real Symmetric Matrix

9.1 INTRODUCTION

In studying vibration or buckling problems, it is often necessary to solve linear equations having the form

$$\begin{bmatrix} a_{11} - \lambda & a_{12} & a_{13} \\ a_{21} & a_{22} - \lambda & a_{23} \\ a_{31} & a_{32} & a_{33} - \lambda \end{bmatrix} \begin{bmatrix} x_1 \\ x_2 \\ x_3 \end{bmatrix} = 0 \tag{9.1}$$

or the form

$$\begin{bmatrix} a_{11} & a_{12} & a_{13} \\ a_{21} & a_{22} & a_{23} \\ a_{31} & a_{32} & a_{33} \end{bmatrix} \begin{bmatrix} x_1 \\ x_2 \\ x_3 \end{bmatrix} = \begin{bmatrix} \lambda & 0 & 0 \\ 0 & \lambda & 0 \\ 0 & 0 & \lambda \end{bmatrix} \begin{bmatrix} x_1 \\ x_2 \\ x_3 \end{bmatrix}. \tag{9.2}$$

In general, Eq. (9.2) is expressed in the following matrix form,

$$[A][x] = \lambda[I][x], \tag{9.3}$$

where $[A]$ is a real symmetric matrix, $[x]$ is a column matrix of independent variables, and λ is a scalar parameter known as a *characteristic value* or *eigenvalue*. As examples of the parameter λ, we have the *natural frequencies* in vibration systems and the *buckling loads* in buckling problems.

Our problem is to find λ and its corresponding $[x]$, which is known as an *eigenvector*.

In this chapter only Jacobi's method will be discussed. The basic aim of the method is to transform Eqs. (9.2) or (9.3) into the following identity:

$$\begin{bmatrix} a'_{11} & 0 & 0 \\ 0 & a'_{22} & 0 \\ 0 & 0 & a'_{33} \end{bmatrix} \begin{bmatrix} \bar{x}_1 \\ \bar{x}_2 \\ \bar{x}_3 \end{bmatrix} \equiv \begin{bmatrix} \lambda_1 & 0 & 0 \\ 0 & \lambda_2 & 0 \\ 0 & 0 & \lambda_3 \end{bmatrix} \begin{bmatrix} \bar{x}_1 \\ \bar{x}_2 \\ \bar{x}_3 \end{bmatrix}, \tag{9.4}$$

where both $[a']$ and $[\lambda]$ are diagonal matrices. Once the transformation is completed, the eigenvalues and eigenvectors can readily be found. This will be discussed in the following three sections.

9.2 TRANSFORMATION OF COORDINATES

Before we discuss a step-by-step procedure of the Jacobi method, we will briefly review the basic concept of transformation of coordinates. For this purpose, we shall consider the column matrix $[x]$ in Eq. (9.3) as a column vector having two components x_1 and x_2, as shown in Fig. 9.1. If a new set of axes $\bar{x}_1$ and $\bar{x}_2$ with the same origin is introduced (Fig. 9.1), then the vector $\overline{OD}$ has two different components $\bar{x}_1$ and $\bar{x}_2$. By trigonometry, we have

$$x_1 = \bar{x}_1 \cos \theta - \bar{x}_2 \sin \theta,$$
$$x_2 = \bar{x}_1 \sin \theta + \bar{x}_2 \cos \theta \tag{9.5}$$

or, in matrix form,

$$\begin{bmatrix} x_1 \\ x_2 \end{bmatrix} = \begin{bmatrix} \cos \theta & -\sin \theta \\ \sin \theta & \cos \theta \end{bmatrix} \begin{bmatrix} \bar{x}_1 \\ \bar{x}_2 \end{bmatrix}. \tag{9.6}$$

This is the transformation relation

$$[x] = [T][\bar{x}], \tag{9.7}$$

which connects the two systems in x and $\bar{x}$.

The significance of matrix $[T]$ in Eq. (9.7) can be shown only after a general discussion on matrix operations in Eq. (9.3). Substituting Eq. (9.7) into Eq. (9.3), we have

$$[A][T][\bar{x}] = \lambda[T][\bar{x}], \tag{9.8}$$

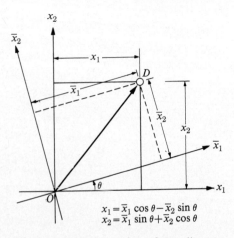

$$x_1 = \bar{x}_1 \cos \theta - \bar{x}_2 \sin \theta$$
$$x_2 = \bar{x}_1 \sin \theta + \bar{x}_2 \cos \theta$$

Fig. 9.1. Transformation of coordinates.

or

$$[T]^T[A][T][\bar{x}] = \lambda[T]^T[T][\bar{x}], \tag{9.9}$$

where $[T]^T$ is the transpose of $[T]$,

$$[T]^T[A][T][\bar{x}] = \lambda[I][\bar{x}].$$
$$= [\lambda][\bar{x}] \tag{9.10}$$

The relation $[T]^T[T] = [I]$ can be verified by the following multiplication of the two matrices:

$$\begin{bmatrix} \cos\theta & \sin\theta \\ -\sin\theta & \cos\theta \end{bmatrix} \begin{bmatrix} \cos\theta & -\sin\theta \\ \sin\theta & \cos\theta \end{bmatrix} = \begin{bmatrix} 1 & 0 \\ 0 & 1 \end{bmatrix}.$$

We recall, from Section 9.1, that our basic aim is to alter the given $[A]$ so that all off-diagonal terms become zero. At this point, $[A]$ in Eq. (9.3) is already transformed to $[T]^T[A][T]$, called $[B]$ for simplicity. We ask, can this transformation possibly be used to eliminate one of the off-diagonal terms in $[A]$? The answer is affirmative provided that a proper angle of rotation θ is selected. To select θ properly, let us write out the matrix $[T]^T[A][T]$, or $[B]$:

$$[B] = \begin{bmatrix} \cos\theta & \sin\theta \\ -\sin\theta & \cos\theta \end{bmatrix} \begin{bmatrix} a_{11} & a_{12} \\ a_{21} & a_{22} \end{bmatrix} \begin{bmatrix} \cos\theta & -\sin\theta \\ \sin\theta & \cos\theta \end{bmatrix}$$

$$= \begin{bmatrix} a_{11}\cos^2\theta + 2a_{12}\sin\theta\cos\theta + a_{22}\sin^2\theta \\ a_{12}(\cos^2\theta - \sin^2\theta) + \cos\theta\sin\theta(a_{22} - a_{11}) \end{bmatrix}$$

$$\begin{matrix} a_{12}(\cos^2\theta - \sin^2\theta) + \sin\theta\cos\theta(a_{22} - a_{11}) \\ a_{11}\sin^2\theta - 2a_{12}\sin\theta\cos\theta + a_{22}\cos^2\theta \end{matrix} \Bigg]. \tag{9.11}$$

We now wish to set the off-diagonal element b_{12} in Eq. (9.11) to zero. The necessary value is then found to be

$$a_{12}(\cos^2\theta - \sin^2\theta) + \cos\theta\sin\theta(a_{22} - a_{11}) = 0, \tag{9.12}$$

or

$$\tan 2\theta = \frac{2a_{12}}{a_{11} - a_{22}}. \tag{9.13}$$

For a symmetric $[A]$, both off-diagonal elements b_{12} and b_{21} in $[B]$ are reduced to zero by using the relation (9.13) and we find that

$$[B] = \begin{bmatrix} a_{11}\cos^2\theta + 2a_{12}\sin\theta\cos\theta + a_{22}\sin^2\theta & 0 \\ 0 & a_{11}\sin^2\theta - 2a_{12}\sin\theta\cos\theta + a_{22}\cos^2\theta \end{bmatrix}. \tag{9.14}$$

Hence the desired eigenvalues are the diagonal elements b_{11} and b_{22}, respectively.

Example 9.1

$$[A] = \begin{bmatrix} 4 & 1 \\ 1 & 2 \end{bmatrix}.$$

From Eq. (9.13) we have

$$\tan 2\theta = \frac{2(1)}{4 - 2} = 1, \qquad \theta = 22.5°$$

Since $[A]$ is symmetric, we can use Eq. (9.14), which establishes $[B]$ or $[T]^T[A][T]$ as follows:

$$[B] = \begin{bmatrix} 4\cos^2\theta + 2\sin\theta\cos\theta + 2\sin^2\theta & 0 \\ 0 & 4\sin^2\theta - 2\sin\theta\cos\theta + 2\cos^2\theta \end{bmatrix}$$

$$= \begin{bmatrix} 4.414 & 0 \\ 0 & 1.5868 \end{bmatrix}.$$

From Eq. (9.10) we know that

$$[T]^T[A][T] = [B] = [\lambda];$$

therefore, the eigenvalues are found to be

$$\lambda_1 = 4.414,$$

$$\lambda_2 = 1.586.$$

It is interesting to note that this example is closely related to the Mohr circle approach used in studying the strength of materials. Here, $\sigma_x = 4$, $\sigma_y = 2$, and $\tau_{xy} = \tau_{yx} = 1$ are the elements in $[A]$; the eigenvalues λ_1 and λ_2 are two principal stresses 4.414 and 1.586. The orientation of the principal axes is related to the eigenvectors to be considered in Section 9.4.

We shall now extend our discussion to the situation in which both $[A]$ and $[x]$ are of third order. Specifically, the column vector $[x]$ now has three components which are either x_1, x_2, and x_3 or $\bar{x}_1$, $\bar{x}_2$, and $\bar{x}_3$. As seen in Fig. 9.2, we have the following transformation relation for the three components in x and $\bar{x}$:

$$\begin{bmatrix} x_1 \\ x_2 \\ x_3 \end{bmatrix} = \begin{bmatrix} \cos\theta & -\sin\theta & 0 \\ \sin\theta & \cos\theta & 0 \\ 0 & 0 & 1 \end{bmatrix} \begin{bmatrix} \bar{x}_1 \\ \bar{x}_2 \\ \bar{x}_3 \end{bmatrix}, \tag{9.15}$$

or

$$[x] = [T][\bar{x}], \tag{9.7}$$

where $[T]$ is the transformation matrix, a square matrix.

Equation (9.10) can now be used to eliminate one of the off-diagonal elements. It should be emphasized that one such rotation is not sufficient to alter $[A]$ such that all off-diagonal terms become zero. In order to eliminate the remaining off-diagonal elements, a series of m plane rotations is necessary to transform $[A]$ successively to

$$[T_m]^T \cdots [T_3]^T[T_2]^T[T_1]^T[A][T_1][T_2][T_3] \cdots [T_m], \tag{9.16}$$

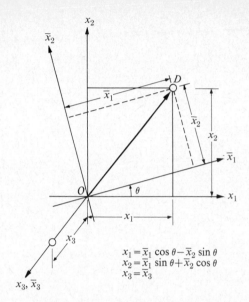

$$x_1 = \bar{x}_1 \cos \theta - \bar{x}_2 \sin \theta$$
$$x_2 = \bar{x}_1 \sin \theta + \bar{x}_2 \cos \theta$$
$$x_3 = \bar{x}_3$$

Fig. 9.2. Transformation of coordinates in a plane.

which converges to a diagonal matrix, the principal diagonal representing the eigenvalues λ. In other words, one off-diagonal element at a time is reduced to zero. The
sole purpose for this reduction to zero and the many rotations of coordinates is to
make *all* nondiagonal elements approach and eventually reach zero. The application
of this concept will be discussed in the next section.

9.3 JACOBI'S METHOD FOR EIGENVALUES

We shall now consider the application of the concept of rotation of axes developed in
Section 9.2. A practical procedure is to select the largest (in absolute value) off-
diagonal element in $[A]$; the position of this element, which is to be eliminated, fixes
the term $-\sin \theta$ in $[T]$. We develop $[T]$ by placing the ($\sin \theta$)-term symmetrically
opposite the term $-\sin \theta$, the ($\cos \theta$)-terms on the diagonal so as to form a square
with the ($\sin \theta$)-terms, and finally by placing 1 on the remainder of the diagonal, and
0 elsewhere. For example, let us take

$$[A] = \begin{bmatrix} 2 & -1 & 0 \\ -1 & 2 & -1 \\ 0 & -1 & 2 \end{bmatrix}, \tag{9.17}$$

where a_{12} is selected as the largest off-diagonal element, since its absolute value is 1.
We could have selected a_{23} as the largest off-diagonal element, since $|a_{23}| = |a_{12}|$;

this would not affect the final answers. Following the above rule, $[T_1]$ will have the form

$$[T_1] = \begin{bmatrix} \cos\theta & -\sin\theta & 0 \\ \sin\theta & \cos\theta & 0 \\ 0 & 0 & 1 \end{bmatrix}. \tag{9.18}$$

Next we use Eq. (9.13) to determine θ and to evaluate the terms in Eq. (9.18). Thus

$$\tan 2\theta = \frac{-2}{2-2} = \frac{-2}{0}, \quad \text{or} \quad \theta = -45°;$$

and

$$[T_1] = \begin{bmatrix} \sqrt{2}/2 & \sqrt{2}/2 & 0 \\ -\sqrt{2}/2 & \sqrt{2}/2 & 0 \\ 0 & 0 & 1 \end{bmatrix}. \tag{9.19}$$

The matrix $[A_1] = [T_1]^T[A][T_1]$ is thus determined. We repeat this procedure, assuming that $[A_1]$ is now the $[A]$, and continue until a matrix $[A_m]$ is created such that all nondiagonal terms are zero. The diagonal terms of this $[A_m]$ will then be equal to the eigenvalues.

It can be seen from Table 9.1 that five successive rotations ($m = 5$) are performed. In each rotation first $[A]$ is given and then the element to be eliminated from $[A]$ is selected. Next, Eq. (9.13) is used to compute the angle of rotation θ, and $[T]$ is written based on the rule mentioned above (see Eq. 9.19). A new $[A]$ is then obtained from $[T]^T[A][T]$.

The given $[A]$ in Eq. (9.17) is symmetric and therefore only two off-diagonal elements need be eliminated. We note that $a_{13} = a_{31} = 0$, and hence no elimination is required. However, it takes five rotations, not two, to obtain the answers shown in Table 9.1, due to the fact that although any one rotation may set an off-diagonal element to zero, it does cause an element previously reduced to zero to attain a nonzero value. In other words, the number of rotations is by necessity greater than the number of off-diagonal nonzero terms. Fortunately, each rotation makes the maximum-valued off-diagonal element decrease in absolute value and $[A]$ generally converges to a diagonal matrix. Thus $[A]$ is diagonalized.

Before concluding this section, we shall examine Eq. (9.13), in which the angle of rotation θ is determined. If the components x_i and x_j are connected with $\bar{x}_i$ and $\bar{x}_j$, then Eq. (9.13) has the form

$$\tan 2\theta = \frac{2a_{ij}}{a_{ii} - a_{jj}}. \tag{9.20}$$

We recall from trigonometry that

$$\tan 2\theta = \frac{2\tan\theta}{1 - \tan^2\theta} = \frac{2a_{ij}}{a_{ii} - a_{jj}}$$

or

$$2a_{ij}\tan^2\theta + 2(a_{ii} - a_{jj})\tan\theta - 2a_{ij} = 0.$$

Table 9.1 Jacobi's method for eigenvalues

$[A_0]$ = given real symmetric matrix; $[A_m] = [T_m]'[A_{m-1}][T_m]$, m = 1,2,3,4,5

m	$[A_m]$			Element to be eliminated	$\tan 2\theta$	$\sin \theta$	$\cos \theta$	$[T_{m+1}]$		
0	2	−1	0	(1,2)	−∞	−0.7071	0.7071	$\cos \theta$	$-\sin \theta$	0
	−1	2	−1					$\sin \theta$	$\cos \theta$	0
	0	−1	2					0	0	1
1	3	0	0.7071	(1,3)	1.4142	0.4597	0.8880	$\cos \theta$	0	$-\sin \theta$
	0	1	−0.7071					0	1	0
	0.7071	−0.7071	2					$\sin \theta$	0	$\cos \theta$
2	3.3660	−0.3250	0	(2,3)	1.9811	0.5241	0.8516	1	0	0
	−0.3250	1	−0.6279					0	$\cos \theta$	$-\sin \theta$
	0	−0.6279	1.6339					0	$\sin \theta$	$\cos \theta$
3	3.3660	−0.1703	−0.2768	(1,3)	−0.2011	−0.0990	0.9950	$\cos \theta$	0	$-\sin \theta$
	−0.1703	2.0204	0					0	1	0
	−0.2768	0	0.6135					$\sin \theta$	0	$\cos \theta$
4	3.3935	−0.1695	0	(1,2)	−0.2469	−0.1207	0.9926	$\cos \theta$	$-\sin \theta$	0
	−0.1695	2.0204	−0.0168					$\sin \theta$	$\cos \theta$	0
	0	−0.0168	0.5859					0	0	1
5	3.4142	0	0.0020							
	0	1.9998	−0.0167							
	0.0020	−0.0167	0.5859							

Solving, we have

$$\tan \theta = \frac{-(a_{ii} - a_{jj}) \pm \sqrt{(a_{ii} - a_{jj})^2 + 4a^2_{ij}}}{2a_{ij}}. \tag{9.21}$$

Multiplying both the denominator and the numerator of the right-hand side of Eq. (9.21) by

$$-(a_{ii} - a_{jj}) \mp \sqrt{(a_{ii} - a_{jj})^2 + 4a^2_{ij}},$$

we obtain

$$\tan \theta = \frac{2a_{ij}}{(a_{ii} - a_{jj}) \pm \sqrt{(a_{ii} - a_{jj})^2 + 4a^2_{ij}}}, \tag{9.22}$$

or

$$\tan \theta = \frac{2a_{ij}}{|a_{ii} - a_{jj}| \pm \sqrt{(a_{ii} - a_{jj})^2 + 4a^2_{ij}}}, \tag{9.23a}$$

if $a_{ii} \geq a_{jj}$; and

$$\tan \theta = \frac{-2a_{ij}}{|a_{ii} - a_{jj}| \pm \sqrt{(a_{ii} - a_{jj})^2 + 4a^2_{ij}}}, \tag{9.23b}$$

if $a_{ii} < a_{jj}$. If we arbitrarily choose the plus sign in the denominators of both Eqs. (9.23a) and (9.23b), we have

$$\tan \theta = \frac{\pm 2a_{ij}}{|a_{ii} - a_{jj}| + \sqrt{(a_{ii} - a_{jj})^2 + 4a_{ij}^2}},\tag{9.24}$$

where the plus sign is used if $a_{ii} \geq a_{jj}$, the minus sign is used if $a_{ii} < a_{jj}$, and $\pi/2 \geq \theta \geq -\pi/2$. We also note that

$$\cos \theta = (1 + \tan^2 \theta)^{-1/2}\tag{9.25}$$

and

$$\sin \theta = \cos \theta \tan \theta.\tag{9.26}$$

Equations (9.24) through (9.26), rather than Eq. (9.13), will be used in discussing machine computation of the rotation of angles in Section 9.5.

9.4 EIGENVECTORS

Following our discussion on obtaining eigenvalues by Jacobi's method, we now turn to the numerical solution of eigenvectors. Eigenvectors are column matrices $[x]$ of the given matrix $[A]$, which are associated with the eigenvalue λ. Since there are n eigenvalues, $\lambda_1 \lambda_2, \ldots, \lambda_n$, the corresponding n column matrices for $[x]$ may be assembled together in a square matrix $[V]$. Let $[\lambda]$, a diagonal matrix, represent the known values of the eigenvalues. From Eq. (9.3), we have

$$[A][V] = [V][\lambda].\tag{9.27}$$

Premultiplying both sides by $[V]^{-1}$, the inverse of $[V]$, we obtain

$$[V]^{-1}[A][V] = [V]^{-1}[V][\lambda],\tag{9.28}$$

and since $[V]^{-1}[V] = [I]$, the unit matrix, Eq. (9.28) becomes

$$[V]^{-1}[A][V] = [\lambda].\tag{9.29}$$

Comparing Eq. (9.29) with Eqs. (9.10) and (9.16), we see that $[V]$, a square matrix representing eigenvectors, is equal to the successive multiplication of the matrices $[T]$ used in obtaining the eigenvalues. Thus we have

$$[V] = [T_1][T_2] \cdots [T_m].$$

As a first example, let us consider the following symmetric matrix of second order from Example 9.1:

$$[A] = \begin{bmatrix} 4 & 1 \\ 1 & 2 \end{bmatrix}.$$

Since only one rotation is needed to diagonalize the matrix (see Example 9.1), $[V]$ is therefore equal to $[T]$, or

$$[V] = \begin{bmatrix} \cos \theta & -\sin \theta \\ \sin \theta & \cos \theta \end{bmatrix},$$

where $\theta = 22.5°$, as obtained in Example 9.1. Hence

$$[V] = \begin{bmatrix} 0.9238 & -0.3826 \\ 0.3826 & 0.9238 \end{bmatrix}.$$

As a second numerical example, the $[V]$ for the given $[A_0]$ in Table 9.1 is simply a succession of multiplications,

$$[V] = [T_1][T_2][T_3][T_4][T_5],$$

where $[T_1]$, $[T_2]$, ... are given in Table 9.1. Thus we have

$$[V] = \begin{bmatrix} 0.5 & 0.707 & 0.5 \\ -0.707 & 0 & 0.707 \\ 0.5 & -0.707 & 0.5 \end{bmatrix}.$$

We note that the first column, or the first *modal column*, corresponds to the first eigenvalue 3.4142. Similarly, the second and third modal columns correspond to the eigenvalues 1.9998 and 0.5859, respectively. These three eigenvalues are shown in Table 9.1 ($m = 5$).

9.5 MACHINE COMPUTATION

Following our discussion on Jacobi's method for obtaining eigenvalues and eigenvectors, we now examine the details for the development of a computer program. A subroutine JACOBI corresponding with the procedures described in the previous sections is written in FORTRAN IV. In particular, Eqs. (9.24) through (9.26) are used to calculate the various trigonometric functions. The calculation stops when the following condition is met: the off-diagonal elements a_{ij} are smaller than a preassigned value $\epsilon = 10^{-8}$.

The arguments used in the subroutine JACOBI are defined as follows:

  N = order of the given real symmetric matrix [Q], N $\geq$ 2,

  Q = the matrix [Q] to be diagonalized (this input matrix is later destroyed),

  JVEC = a fixed point index,

    JVEC = 0 when eigenvalues alone are to be found,

    JVEC = $\pm 1, \pm 2, \ldots$ and up when eigenvalues and eigenvectors are to be found,

  M = the number of rotations performed,

  V = storage for eigenvectors. this storage is required even for the case JVEC = 0.

A program and a flow chart for the subprogram JACOBI are shown in Figs. 9.3 and 9.4, respectively.

```
      SUBROUTINE JACOBI(N,Q,JVEC,M,V)
C     SUBPROGRAM FOR DIAGONALIZATION OF MATRIX Q BY SUCCESSIVE
C     ROTATIONS
      DIMENSION Q(12,12),V(12,12),X(12),IH(12)
C
C     NEXT 8 STATEMENTS FOR SETTING INITIAL VALUES OF MATRIX V
C
      IF(JVEC.EQ.0) GO TO 15
      DO 14 I=1,N
      DO 14 J=1,N
   14 V(I,J)=(I/J)*(J/I)
C
   15 M=0
C
C     NEXT 8 STATEMENTS SCAN FOR LARGEST OFF DIAG. ELEM. IN EACH ROW
C     X(I) CONTAINS LARGEST ELEMENT IN ITH ROW
C     IH(I) HOLDS SECOND SUBSCRIPT DEFINING POSITION OF ELEMENT
C
   17 MI=N-1
      DO 30 I=1,MI
      X(I)=0.
      MJ=I+1
      DO 30 J=MJ,N
      IF(X(I).GT.ABS(Q(I,J))) GO TO 30
   20 X(I)=ABS(Q(I,J))
      IH(I)=J
   30 CONTINUE
C
C     NEXT 7 STATEMENTS FIND FOR MAXIMUM OF X(I)S FOR PIVOT ELEMENT
C
   40 DO 70 I=1,MI
      IF(I.LE.1) GO TO 60
   45 IF(XMAX.GT.X(I)) GO TO 70
   60 XMAX=X(I)
      IP=I
      JP=IH(I)
   70 CONTINUE
C
C     NEXT TWO STATEMENTS TEST FOR XMAX, IF LESSTHAN 10**-8,GO TO 1000
C
      EPSI=1.E-8
      IF(XMAX.LE.EPSI) GO TO 1000
C
  148 M=M+1
C
C     NEXT 11 STATEMENTS FOR COMPUTING TANG,SIN,COS,Q(I,I),Q(J,J)
C
      IF(Q(IP,IP).GT.Q(JP,JP)) GO TO 151
  150 TANG=-2.*Q(IP,JP)/(ABS(Q(IP,IP)-Q(JP,JP))+SQRT((Q(IP,IP)-Q(JP,JP))
     1**2+4.*Q(IP,JP)**2))
      GO TO 160
  151 TANG=+2.*Q(IP,JP)/(ABS(Q(IP,IP)-Q(JP,JP))+SQRT((Q(IP,IP)-Q(JP,JP))
     1**2+4.*Q(IP,JP)**2))
  160 COSN=1.0/SQRT(1.0+TANG**2)
      SINE=TANG*COSN
      QII=Q(IP,IP)
      Q(IP,IP)=COSN**2*(QII+TANG*(2.*Q(IP,JP)+TANG*Q(JP,JP)))
      Q(JP,JP)=COSN**2*(Q(JP,JP)-TANG*(2.*Q(IP,JP)-TANG*QII))
C
      Q(IP,JP)=0.
```

Fig. 9.3. Subroutine JACOBI for diagonalization.

```
C
C NEXT 4 STATEMENTS FOR PSEUDO RANK THE EIGENVALUES
C
      IF(Q(IP,IP).GE.Q(JP,JP)) GO TO 153
  152 TEMP=Q(IP,IP)
      Q(IP,IP)=Q(JP,JP)
      Q(JP,JP)=TEMP
C
C     NEXT 6 STATEMENTS ADJUST SIN, COS FOR COMPUTATION OF Q(I,K),V(I,K)
C
      IF(SINE.GE.0.) GO TO 155
  154 TEMP=+COSN
      GO TO 170
  155 TEMP=-COSN
  170 COSN=ABS(SINE)
      SINE=TEMP
C
C     NEXT 10 STATEMENTS FOR INSPECTING THE IHS BETWEEN I+1 AND N-1 TO
C     DETERMINE WHETHER A NEW MAXIMUM VALUE SHOULD BE COMPUTED SINCE
C     THE PRESENT MAXIMUM IS IN THE I OR J ROW
C
  153 DO 350 I=1,MI
      IF((((I.EQ.IP).OR.(I.EQ.JP)).OR.((IH(I).NE.IP).AND.(IH(I).NE.JP)))
     1 GO TO 350
  240 K=IH(I)
  250 TEMP=Q(I,K)
      Q(I,K)=0.
      MJ=I+1
      X(I)=0.
C
C     NEXT 5 STATEMENTS SEARCH IN DEPLETED ROW FOR NEW MAXIMUM
C
      DO 320 J=MJ,N
      IF(X(I).GT.ABS(Q(I,J))) GO TO 320
  300 X(I)=ABS(Q(I,J))
      IH(I)=J
  320 CONTINUE
      Q(I,K)=TEMP
  350 CONTINUE
C
      X(IP)=0.
      X(JP)=0.
C
C     NEXT 30 STATEMENTS FOR CHANGING THE OTHER ELEMENTS OF Q
C
      DO 530 I=1,N
C
      IF(I.EQ.IP) GO TO 530
      IF(I.GT.IP) GO TO 420
      TEMP=Q(I,IP)
      Q(I,IP)=COSN*TEMP+SINE*Q(I,JP)
      IF(X(I).GE.ABS(Q(I,IP))) GO TO 390
  380 X(I)=ABS(Q(I,IP))
      IH(I)=IP
  390 Q(I,JP)=-SINE*TEMP+COSN*Q(I,JP)
      IF(X(I).GE.ABS(Q(I,JP))) GO TO 530
  400 X(I)=ABS(Q(I,JP))
      IH(I)=JP
      GO TO 530
C
  420 IF(I-JP) 430,530,480
  430 TEMP=Q(IP,I)
```

(cont.)

Fig. 9.3 *(concl.)*

```
      Q(IP,I)=COSN*TEMP+SINE*Q(I,JP)
      IF(X(IP).GE.ABS(Q(IP,I))) GO TO 450
  440 X(IP)=ABS(Q(IP,I))
      IH(IP)=I
  450 Q(I,JP)=-SINE*TEMP+COSN*Q(I,JP)
      IF(X(IP).GE.ABS(Q(I,JP))) GO TO 530
C
  480 TEMP=Q(IP,I)
      Q(IP,I)=COSN*TEMP+SINE*Q(JP,I)
      IF(X(IP).GE.ABS(Q(IP,I))) GO TO 500
  490 X(IP)=ABS(Q(IP,I))
      IH(IP)=I
  500 Q(JP,I)=-SINE*TEMP+COSN*Q(JP,I)
      IF(X(JP).GE.ABS(Q(JP,I))) GO TO 530
  510 X(JP)=ABS(Q(JP,I))
      IH(JP)=I
  530 CONTINUE
C
C     NEXT 6 STATEMENTS TEST FOR COMPUTATION OF EIGENVECTORS
C
      IF(JVEC.EQ.0) GO TO 40
  540 DO 550 I=1,N
      TEMP=V(I,IP)
      V(I,IP)=COSN*TEMP+SINE*V(I,JP)
  550 V(I,JP)=-SINE*TEMP+COSN*V(I,JP)
      GO TO 40
 1000 RETURN
      END
```

Table 9.2 Answers for Example 9.2

```
      MATRIX A

          2.00000           -1.00000            0.00000
         -1.00000            2.00000           -1.00000
          0.00000           -1.00000            2.00000


      THE NUMBER OF ROTATION =  9


      EIGENVALUE ( 1 ) =           3.41421

      EIGENVECTORS
              .50000           -.70710             .50000

      EIGENVALUE ( 2 ) =           2.00000

      EIGENVECTORS
              .70710            0.00000            -.70710

      EIGENVALUE ( 3 ) =            .58578

      EIGENVECTORS
              .50000             .70710             .50000
```

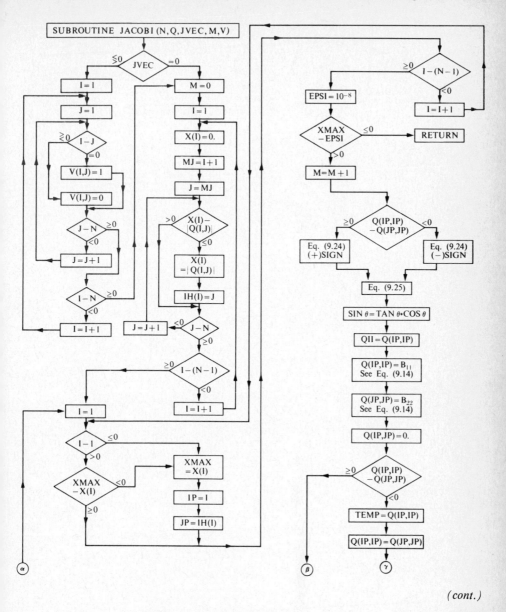

Fig. 9.4. Flow chart of subroutine JACOBI.

(cont.)

Fig. 9.4 (*cont.*)

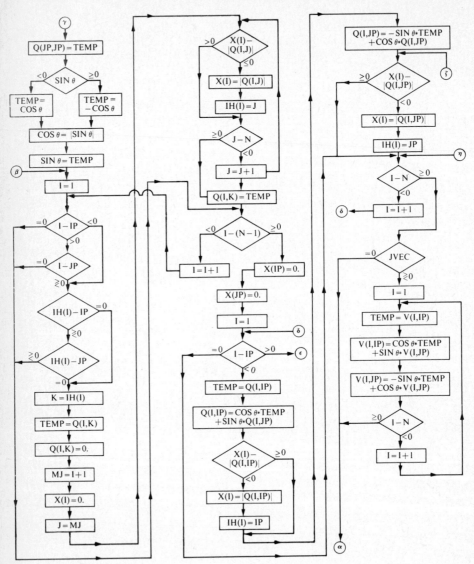

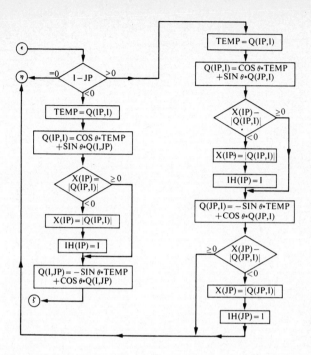

Example 9.2

Write a main program to call the subroutine JACOBI and to satisfy the input and output requirements. The primary objective of this example is to find the eigenvalues and eigenvectors of the real symmetric matrix $[A]$ shown in Eq. (9.17). The first data card is for N, with format I2, where N is the order of the given $[A]$. Following the first data cards, there will be the data cards for the elements a_{ij}. They are punched row by row on cards for all elements in the upper triangular matrix.

An echo print of $[A]$ is required before transferring the operation to the subroutine JACOBI. The required output is shown in Table 9.2, where the eigenvalue is printed first, followed by its associated eigenvectors. It is important that the size of each matrix in the main program agree with that in the subroutine. Figure 9.5 shows a possible main program. In the DIMENSION statement, both A and X are arbitrarily dimensioned as 12 × 12. Both the corresponding matrices Q and V in subroutine JACOBI must also be dimensioned as 12 × 12.

9.6 GENERAL EIGENVALUE PROBLEMS

In the previous sections we discussed the eigenvalue problem which can be expressed as

$$[A][x] = \lambda[B][x], \qquad (9.30)$$

where $[B]$ is a unit matrix. In this section we shall focus our attention on two cases: first, $[B]$ as a diagonal matrix and, second and this is the more general case, $[B]$ as a symmetric matrix.

```
C        THIS IS MAIN PROGRAM FOR EIGENVALUE PROBLEM IN THE FORM OF
C            AX=LX
C        A IS SYMMETRICAL MATRIX
C        L ARE EIGENVALUES, X ARE EIGENVECTORS
C
C
         DATA IREAD,IWRITE/5,6/
         DIMENSION A(12,12),X(12,12)
C
C        READ IN THE ORDER OF MATRIX A
C
         READ(IREAD,10) N
      10 FORMAT(I2)
C
C        READ IN UPPER TRIANGULAR MATRIX
C
         DO 11 I=1,N
      11 READ(IREAD,20)  (A(I,J),J=I,N)
      20 FORMAT(6F10.5)
         WRITE(IWRITE,30)
      30 FORMAT('1',' MATRIX A'/)
         DO 32 I=1,N
         DO 31 J=1,N
      31 A(J,I)=A(I,J)
      32 WRITE(IWRITE,40) (A(I,J),J=1,N)
      40 FORMAT(5F15.5)
         CALL JACOBI(N,A,1,NR,X)
C
C        NR=NUMBER OF ROTATIONS
C
         WRITE(IWRITE,80) NR
      80 FORMAT(//,' THE NUMBER OF ROTATION=',I3//)
C
C        WRITE OUT EIGENVALUES AND THE CORRESPONDING EIGENVECTORS
C
         DO 46 J=1,N
      44 WRITE(IWRITE,50) J,A(J,J)
      50 FORMAT(//' EIGENVALUE (',I2,')=',F15.5)
      45 WRITE(IWRITE,60)
      60 FORMAT(/,' EIGENVECTORS'/)
         DO 46 I=1,N
      46 WRITE(IWRITE,70) X(I,J)
      70 FORMAT(2X,F15.5)
         STOP
         END


03
2.        -1.        0.
2.        -1.
2.
/*
```

Fig. 9.5. Main program and data for $AX = LX$.

Let us consider the case where $[B]$ is a diagonal matrix. We assume that

$$[B] = [G]^T[G], \tag{9.31}$$

where $[G]$ is also a diagonal matrix. We have

$$[B] = [G][G]; \tag{9.32}$$

hence each element g_{ii} in $[G]$ is equal to the square root of the corresponding element b_{ii} in $[B]$, and this element is required to be positive:

$$g_{ii} = \sqrt{b_{ii}}.$$

Substituting Eq. (9.32) into

$$[A][x] = \lambda[B][x], \tag{9.33}$$

we obtain

$$[A][x] = \lambda[G][G][x]. \tag{9.34}$$

Premultiplying Eq. (9.34) by $[G]^{-1}$, we have

$$[G]^{-1}[A][x] = \lambda[G][x], \tag{9.35}$$

or

$$[G]^{-1}[A][G]^{-1}[G][x] = \lambda[G][x]. \tag{9.36}$$

If we define

$$[Y] = [G][x] \quad \text{and} \quad [Q] = [G]^{-1}[A][G]^{-1},$$

then Eq. (9.36) becomes

$$[Q][Y] = \lambda[Y]. \tag{9.37}$$

The above equation has precisely the same form as Eq. (9.3). Thus, the eigenvalues, obtained from diagonalization of $[Q]$, are the true eigenvalues. On the other hand, the true eigenvectors are obtained only after postmultiplying $[G]^{-1}$ by the resulting eigenvectors; that is,

$$[\text{true eigenvectors}] = [G]^{-1}[\text{resulting vectors}].$$

Each diagonal element in the diagonal matrix $[G]^{-1}$ is equal to the reciprocal of the square root of each diagonal element of $[B]$.

Example 9.3

We know from the study of the free vibrations of a system consisting of three masses (Fig. 9.6), that the equations of motion lead to the following eigenvalue formulation:

$$\begin{bmatrix} 2k - m\omega^2 & -k & 0 \\ -k & 2k - 2m\omega^2 & -k \\ 0 & -k & 2k - 3m\omega^2 \end{bmatrix} \begin{bmatrix} x_1 \\ x_2 \\ x_3 \end{bmatrix} = [0], \tag{9.38}$$

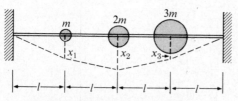

Fig. 9.6. The three-mass system.

where

$$\omega = \text{frequency,}$$
$$[x] = \text{displacement vector,}$$
$$k = s/l,$$
$$s = \text{tension in string,}$$
$$4l = \text{length of string.}$$

Letting $\lambda = m\omega^2/k$, we find that Eq. (9.38) becomes

$$\begin{bmatrix} 2 & -1 & 0 \\ -1 & 2 & -1 \\ 0 & -1 & 2 \end{bmatrix} \begin{bmatrix} x_1 \\ x_2 \\ x_3 \end{bmatrix} = \lambda \begin{bmatrix} 1 & 0 & 0 \\ 0 & 2 & 0 \\ 0 & 0 & 3 \end{bmatrix} \begin{bmatrix} x_1 \\ x_2 \\ x_3 \end{bmatrix},$$

or

$$[A][X] = \lambda[B][X], \tag{9.39}$$

which has precisely the same form as Eq. (9.33).

We wish to write a main program which will compute three λ's and associated model shapes (eigenvectors). The first data card is for N, the order of the given $[A]$. The data for elements A_{ij} are then read in row by row for all elements in the upper triangular matrix. Finally, the diagonal elements of $[B]$ are read in one per card.

An echo print of $[A]$ and $[B]$ is required. In the output, shown in Table 9.3, the rotation number is printed before each eigenvalue and its associated eigenvectors are printed. A possible main program is shown in Fig. 9.7. The program requires two subroutines: JACOBI and MATMPY. The subroutine MATMPY, which was discussed in Section 8.2, is used to multiply two given matrices.

We now turn to the general eigenvalue problem

$$[A][x] = \lambda[B][x], \tag{9.40}$$

where $[A]$ is real and symmetric, $[x]$ is a column matrix, λ is a scalar, and $[B]$ is real, symmetric, and positive definite.

At this point it is necessary to explain what is meant by *positive definite*. If we let $[B]$ be a real symmetric matrix, and $[x]$ be a real column matrix, then $[x]^T[B][x]$ is a quadratic in n variables $x_1, x_2, \ldots, x_n$; that is,

$$\begin{aligned} [x]^T[B][x] = {}& b_{11}x_1^2 + 2b_{12}x_1x_2 + 2b_{13}x_1x_3 + \cdots + 2b_{1n}x_1x_n \\ & + b_{22}x_2^2 + 2b_{23}x_2x_3 + \cdots + 2b_{2n}x_2x_n \\ & + b_{33}x_3^2 + \cdots + 2b_{3n}x_3x_n + \cdots + b_{nn}x_n^2. \end{aligned} \tag{9.41}$$

The quadratic form $[x]^T[B][x]$ is called positive definite if its values are always positive for $x_i \neq 0$ $(i = 1, 2, \ldots, n)$. Its value is zero for $x_i = 0$ $(i = 1, 2, \ldots, n)$. Also, if $[x]^T[B][x]$ is positive definite, then the real symmetric matrix $[B]$ is said to be *positive definite*. Of particular importance is the fact that if $[B]$ is positive definite, the eigenvalues are real and positive.

Table 9.3 Answers for Example 9.3

MATRIX A

2.00000	−1.00000	0.00000
−1.00000	2.00000	−1.00000
0.00000	−1.00000	2.00000

MATRIX B

1.00000	0.00000	0.00000
0.00000	2.00000	0.00000
0.00000	0.00000	3.00000

THE NUMBER OF ROTATION = 8

EIGENVALUE (1) =   2.38742

EIGENVECTORS
        .87134   −.33758   .06539

EIGENVALUE (2) =   1.00000

EIGENVECTORS
        .40824   .40824   −.40824

EIGENVALUE (3) =   .27924

EIGENVECTORS
        .27218   .46837   .40297

To find the eigenvalues and eigenvectors in Eq. (9.40), we use a procedure similar to that described previously in this section, but with one difference: $[G]$ cannot be calculated directly from $[B]$. However, $[B]$ can be diagonalized beforehand so that

$$[B] = [V][D][V]^T, \qquad (9.42)$$

where $[D]$ is a diagonal matrix and $[V]$ is a square matrix representing the eigenvectors (see Eqs. 9.27 and 9.29). Substituting Eq. (9.42) into (9.40), we obtain

$$[A][x] = \lambda[V][D][V]^T[x]. \qquad (9.43)$$

Premultiplying both sides by $[V]^T$, we get

$$[V]^T[A][x] = \lambda[D][V]^T[x], \qquad (9.44)$$

or

$$[V]^T[A][V][V]^T[x] = \lambda[D][V]^T[x]. \qquad (9.45)$$

```
C
C         THIS IS MAIN PROGRAM FOR EIGENVALUE PROBLEM IN THE FORM OF
C              AX=LBX
C         A IS SYMMETRIC MATRIX
C         B IS DIAGONAL MATRIX
C         L ARE EIGENVALUES, X ARE EIGENVECTORS
C
C
          DATA IREAD,IWRITE/5,6/
          DIMENSION A(12,12),B(12,12),C(12,12),AA(12,12),X(12,12),XX(12,12)
C
C         READ IN THE ORDER OF THE GIVEN MATRIX A
C
          READ(IREAD,10) N
   10 FORMAT(I2)
C
C         THE DATA FOR ELEMENTS A(I,J) ARE READ IN ROW BY ROW FOR ALL
C         ELEMENTS IN THE UPPER TRIANGULAR MATRIX.
C
          DO 11 I=1,N
   11 READ(IREAD,20) (A(I,J),J=I,N)
   20 FORMAT(7F10.5)
          DO 12 I=1,N
          DO 12 J=1,N
   12 B(I,J)=0.
C
C         READ IN THE DIAGONAL ELEMENTS OF MATRIX B
C
          DO 14 I=1,N
   14 READ(IREAD,18) B(I,I)
   18 FORMAT(F10.5)
C
C         ECHO PRINT OF MATRIX A AND MATRIX B
C
          WRITE(IWRITE,30)
   30 FORMAT('1',' MATRIX A'/)
          DO 32 I=1,N
          DO 31 J=1,N
   31 A(J,I)=A(I,J)
   32 WRITE(IWRITE,40) (A(I,K),K=1,N)
   40 FORMAT(5F15.5)
          WRITE(IWRITE,70)
   70 FORMAT(//,' MATRIX B'//)
          DO 75 I=1,N
   75 WRITE(IWRITE,40) (B(I,J),J=1,N)
          DO 85 I=1,N
   85 B(I,I)=1./SQRT(B(I,I))
C
C         CALL SUBROUTINE MATMPY TO MULTIPLY TWO GIVEN MATRICES
C         CALL SUBROUTINE JACOBI TO DIAGONALIZE MATRIX AA
C
          CALL MATMPY(B,N,N,A,N,C)
          CALL MATMPY(C,N,N,B,N,AA)
          CALL JACOBI(N,AA,1,NR,XX)
          CALL MATMPY(B,N,N,XX,N,X)
C
```

Fig. 9.7. Main program and data for Example 9.3.

```
C      NR=NUMBER OF ROTATIONS
C
       WRITE(IWRITE,80) NR
    80 FORMAT(//,' THE NUMBER OF ROTATION=',I3//)
C
C      WRITE OUT EACH EIGENVALUE AND ITS ASSOCIATED EIGENVECTORS
C
       DO 46 J=1,N
    44 WRITE(IWRITE,50) J,AA(J,J)
    50 FORMAT(/,' EIGENVALUE (',I2,')=',F15.5)
       WRITE(IWRITE,60)
    60 FORMAT(//' EIGENVECTORS'/)
       DO 46 I=1,N
    46 WRITE(IWRITE,90) X(I,J)
    90 FORMAT(2X,F15.5)
       STOP
       END
       SUBROUTINE MATMPY(A,N,M,B,L,C)
C      SUBROUTINE FOR TWO MATRICES MULTIPLICATION
C
C
       DIMENSION A(12,12),B(12,12),C(12,12)
       DO 5 I=1,N
       DO 5 J=1,L
       C(I,J)=0.
       DO 5 K=1,M
     5 C(I,J)=C(I,J)+A(I,K)*B(K,J)
       RETURN
       END


   03
   2.         -1.        0.
   2.         -1.
   2.
   1.
   2.
   3.
```

If we define $[H] = [V]^T[A][V]$ and $[Y] = [V]^T[x]$, then Eq. (9.45) takes the form

$$[H][Y] = \lambda[D][Y]. \tag{9.46}$$

So long as $[D]$ is positive definite, Eq. (9.46) can be solved readily by the method discussed in the first part of this section. In other words, we take

$$[D] = [G][G] \tag{9.47}$$

such that

$$[H][Y] = \lambda[G][G][Y] \tag{9.48}$$

or

$$[G]^{-1}[H][G]^{-1}[G][Y] = \lambda[G][Y]. \tag{9.49}$$

We now let

$$[G]^{-1}[H][G]^{-1} = [Q] \tag{9.50}$$

and

$$[G][Y] = [Z], \tag{9.51}$$

and we have the final form

$$[Q][Z] = \lambda[Z], \tag{9.52}$$

which has the precise form of Eq. (9.3).

The eigenvalues obtained from the diagonalization of $[Q]$ are the true eigenvalues. The true eigenvectors can be obtained by using the relation

$$[x] = [V][G]^T[Z]. \tag{9.53}$$

Example 9.4

Write a main program to call the subroutines JACOBI (Fig. 9.2) and MATMPY (see Section 8.2) to compute the eigenvalues and eigenvectors in the equation

$$[A][x] = \lambda[B][x],$$

where

$$A = \begin{bmatrix} 1 & 1 & 0.5 \\ 0 & 1 & 0.25 \\ 0.5 & 0.25 & 2 \end{bmatrix} \quad \text{and} \quad B = \begin{bmatrix} 2 & 2 & 2 \\ 2 & 5 & 5 \\ 2 & 5 & 11 \end{bmatrix}.$$

A sample main program is shown in Fig. 9.8, where the first data card is for N, the order of both the given $[A]$ and $[B]$. Elements a_{ij} and b_{ij} are then punched row by row on input cards for all elements in the upper triangular positions. They are echo-printed as shown in Table 9.4.

```
C       MAIN PROGRAM FOR GENERAL EIGENVALUE PROBLEM AX=LBX
C       A AND B ARE SYMMETRIC MATRICES. B IS POSITIVE DEFINITE
C       L=EIGENVALUES     X=EIGENVECTORS
C
C
        DATA IREAD,IWRITE/5,6/
        DIMENSION A(12,12),B(12,12),T(12,12),TP(12,12),S(12,12),X(12,12),
       1 BAB(12,12),W(12,12),AA(12,12),AB(12,12),XX(12,12)
C
C       READ IN THE ORDER OF BOTH THE GIVEN MATRICES A AND B
C
        READ(IREAD,10) N
     10 FORMAT(I2)
C
C       READ IN THE ELEMENTS IN THE UPPER TRIANGULAR POSITIONS OF MATRICES
C       A AND B
C
        DO 11 I=1,N
     11 READ(IREAD,20) (A(I,J),J=I,N)
     20 FORMAT(6F10.5)
        DO 12 I=1,N
     12 READ(IREAD,20) (B(I,J),J=I,N)
C
C       ECHO PRINT THE GIVEN MATRICES A AND B
```

Fig. 9.8. Main program for $Ax = LBx$, where A and B are symmetric.

```
C
      WRITE(IWRITE,30)
   30 FORMAT('1',' MATRIX A'/)
      DO 32 I=1,N
      DO 32 J=1,N
      A(J,I)=A(I,J)
   32 B(J,I)=B(I,J)
      DO 35 I=1,N
   35 WRITE(IWRITE,20) (A(I,J),J=1,N)
      WRITE(IWRITE,70)
   70 FORMAT(//,' MATRIX B'/)
      DO 75 I=1,N
   75 WRITE(IWRITE,20) (B(I,J),J=1,N)
C
C     CALL SUBROUTINE JACOBI TO DIAGONALIZE MATRIX B
C
      CALL JACOBI(N,B,1,NR,T)
      WRITE(IWRITE,25)
   25 FORMAT(//,' DIAGONALIZED MATRIX B'/)
C
C     COMPLETE THE ELEMENTS IN THE LOWER TRINGULAR POSITIONS OF
C     THE DIAGONALIZED MATRIX B
C
      DO 24 I=1,N
      DO 24 J=1,N
   24 B(J,I)=B(I,J)
      DO 26 I=1,N
   26 WRITE(IWRITE,20) (B(I,J),J=1,N)
C
C     MATRIX T IS THE MATRIX CONTAINING THE EIGENVECTORS OF MATRIX B
C
      WRITE(IWRITE,27)
   27 FORMAT(//,' EIGENVECTORS OF MATRIX B'/)
      DO 28 I=1,N
   28 WRITE(IWRITE,20) (T(I,J),J=1,N)
C
C     MATRIX TP IS THE TRANSPOSE OF THE MATRIX T
C
      DO 76 I=1,N
      DO 76 J=1,N
   76 TP(I,J)=T(J,I)
C
C     CALL SUBROUTINE MATMPY FOR MULTIPLICATION OF TWO GIVEN MATRICES.
C     MATRIX AA IS THE PRODUCTS OF MATRIX TP TIMES A TIMES T
C
      CALL MATMPY(TP,N,N,A,N,W)
      CALL MATMPY (W,N,N,T,N,AA)
C
C     CHECK DIAGONAL ELEMENTS OF MATRIX B TO SEE THEY ARE POSITIVE.
C     IF NOT, STOP THE PROGRAM
C
      DO 85 I=1,N
      IF(B(I,I).GE.0.) GO TO 85
   84 WRITE(IWRITE,83)
   83 FORMAT(//,' MATRIX B IS NOT POSITIVE DEFINITE')
      GO TO 100
C
C     B=G*G  G IS A DIAGONAL MATRIX. EACH DIAGONAL ELEMENT IN THE
C     INVERSE OF MATRIX G IS EQUAL TO THE RECIPROCAL OF THE SQUARE ROOT
C     OF EACH DIAGONAL ELEMENT OF MATRIX B
C     FOLLOWING STATEMENT IS TO CHANGE MATRIX B TO A NEW MATRIX WHICH
C     IS THE INVERSE OF MATRIX G
C
```

(cont.)

```
   85 B(I,I)=1./SQRT(B(I,I))
      CALL MATMPY(B,N,N,AA,N,AB)
      CALL MATMPY(AB,N,N,B,N,BAB)
C
C     CALL SUBROUTINE JACOBI TO DIAGONALIZE MATRIX BAB
C
      CALL JACOBI(N,BAB,1,NR,XX)
C
C     THE ELEMENTS OF DIAGONALIZED MATRIX BAB ARE THE TRUE EIGENVALUES
C
C     MATRIX B=INVERSE OF MATRIX G
C     THE ELEMENTS OF THE MATRIX OBTAINED FROM THE MULTIPLICATION OF
C     MATRICES A,B,AND XX ARE THE TRUE EIGENVECTORS
C
      CALL MATMPY(T,N,N,B,N,S)
      CALL MATMPY(S,N,N,XX,N,X)
C
C     NR=NUMBER OF ROTATIONS
C
      WRITE(IWRITE,80) NR
   80 FORMAT(//,' THE NUMBER OF ROTATION=',I3//)
C
C     WRITE OUT EACH EIGENVALUE AND ITS ASSOCIATED EIGENVECTORS
C
      DO 46 J=1,N
   44 WRITE(IWRITE,50) J,BAB(J,J)
   50 FORMAT(//' EIGENVALUE (',I2,')=',F15.5)
      WRITE(IWRITE,60)
   60 FORMAT(/,' EIGENVECTORS'/)
C
      DO 46 I=1,N
   46 WRITE(IWRITE,90) X(I,J)
   90 FORMAT(2X,F15.5)
  100 STOP
      END
```

Fig. 9.8   *(concl.)*

Table 9.4   Output for $Ax = LBx$

MATRIX A

1.00000	1.00000	.50000
1.00000	1.00000	.25000
.50000	.25000	2.00000

MATRIX B

2.00000	2.00000	2.00000
2.00000	5.00000	5.00000
2.00000	5.00000	11.00000

DIAGONALIZED MATRIX B

14.43089	0.00000	0.00000
0.00000	2.61519	0.00000
0.00000	0.00000	.95390

```
EIGENVECTORS OF MATRIX B

    .21493     -.50489      .83599
    .49265     -.68305     -.53919
    .84326      .52774      .10192


THE NUMBER OF ROTATION =   7



EIGENVALUE ( 1 ) =              .61064

EIGENVECTORS
             .52639          .28178         -.24794

EIGENVALUE ( 2 ) =              .31504

EIGENVECTORS
             .51459         -.40261          .31839

EIGENVALUE ( 3 ) =             -.00902

EIGENVECTORS
             .53984         -.50842         -.06174
```

BIBLIOGRAPHY

The Symmetric Matrices

Bowdler, H., R. S. Martin, C. Reinsch, and J. H. Wilkinson, "The QR and QL algorithms for symmetric matrices," *Num. Math.*, **11**, 293–306 (1968).

Causey, R. L., and P. Henrici, "Convergence of approximate eigenvectors in Jacobi methods," *Num. Math.*, **2**, 67–78 (1960).

Givens, W., "A method of computing eigenvalues and eigenvectors suggested by classical results on symmetric matrices," *Natn. Bur. Stand. Appl. Math. Ser.*, **29**, 117–122 (1953).

Givens, W., "Numerical computation of the characteristic values of a real symmetric matrix," *ORNL*-1574, Oak Ridge National Laboratory (1954).

Goldstine, H. H., F. J. Murray, and J. von Neumann, "The Jacobi method for real symmetric matrices," *J. Assoc. Comput. Mach.*, **6**, 59–96 (1959).

Gregory, R. T., "Computing eigenvalues and eigenvectors of a symmetric matrix on the ILLIAC," *Math. Tables Aids Comput.*, **7**, 215–220 (1953).

Henrici, P., "On the speed of convergence of cyclic and quasicyclic Jacobi methods for computing eigenvalues of Hermitian matrices," *J. Soc. Ind. Appl. Math.*, **6**, 144–162 (1958).

Householder, A. S., and F. L. Bauer, "On certain methods for expanding the characteristic polynomial," *Num. Math.*, **1**, 29–37 (1959).

Kuo, S. S., "A note on Jacobi's method for real symmetric matrices," *J. Aerospace Sci.*, **28**, no. 3 (1961).

Peters, G., and J. H. Wilkinson, "Eigenvalues of $Ax = \lambda Bx$ with band symmetric A and B," *Comput. J.*, **12**, no. 4, 398–404 (1969).

Pope, D. A., and C. B. Tompkins, "Maximizing functions of rotations. Experiments concerning the speed of diagonalization of symmetric matrices using Jacobi's method," *J. Assoc. Comput. Mach.*, **4**, 459–466 (1957).

Reinsch, C., and F. L. Bauer, "Rational QR transformations with Newton shift, for symmetrical tridiagonal matrices," *Handbook Series Linear Algebra, Num. Math.*, **11**, 264–272 (1968).

Stewart, G. W., "Incorporating origin shifts into the QR algorithm for symmetric tridiagonal matrices," *Commun. Assoc. Comput. Mach.*, **13**, no. 6 (1970).

Wilkinson, J. H., "Householder's method for the solution of the algebraic eigenproblem," *Comput. J.*, **3**, 23–27 (1960).

Arbitrary Matrices

Eberlein, P. J., "A Jacobi-like method for the automatic computation of eigenvalues and eigenvectors of an arbitrary matrix," *J. Soc. Ind. Appl. Math.*, **10**, 74–88 (1962).

Francis, J. G. F., "The QR transformation," Parts I and II, *Comput. J.*, **4**, 265–271, 332–345 (1961, 1962).

Martin, R. S., and J. H. Wilkinson, "The implicit QL algorithm," *Num. Math.*, **12**, 377–383 (1968).

Martin, R. S., and J. H. Wilkinson, "Solution of symmetric and unsymmetric band equations on the calculation of eigenvectors of band matrices," *Num. Math.*, **9**, 279–301 (1967).

Ostrowski, A. M., "On the convergence of the Rayleigh quotient iteration for the computation of the characteristic roots and vectors," *I–VI*, *Arch. Rational Mech. Ann.*, **1**, 233–241 (1958); **2**, 423–428 (1959); **3**, 325–340, 341–367, 472–481 (1959); **4**, 153–165 (1960).

Parlett, B., "Laguerre's method applied to the matrix eigenvalue problem," *Maths. Computat.*, **18** (1964).

Parlett, B., "The development and use of methods of LR type," *SIAM Rev.*, **6**, 275–310 (1964).

Parodi, M., *La localisation des valeurs caractéristiques des matrices et ses applications*. Gauthier-Villars, Paris, 1959.

Peters, G., and J. H. Wilkinson, "$Ax = \lambda Bx$ and the generalized eigenproblem," *SIAM J. Num. Anal.*, **7**, 479–492 (1970).

Rutishauser, H., "Solution of eigenvalue problems with the LR-transformation," *Natn. Bur. Stand. Appl. Math. Ser.*, **49**, 47–81 (1958).

Rutishauser, H., "Computational aspects of F. L. Bauer's simultaneous iteration method," *Num. Math.*, **13**, 4–13 (1969).

White, P. A., "The computation of eigenvalues and eigenvectors of a matrix," *J. Soc. Ind. Appl. Math.*, **6**, 393–437 (1958).

Wilkinson, J. H., "Global convergence of tridiagonal QR algorithm with origin shifts," *Lin. Alg. and Applics.*, **1**, 409–420 (1968).

Wilkinson, J. H., *The Algebraic Eigenvalue Problem*. Clarendon Press, Oxford, 1965.

Wilkinson, J. H., "The calculation of eigenvectors by the method of Lanczos," *Comput. J.*, **1**, 148–152 (1958).

Wilkinson, J. H., "The calculation of the eigenvectors of codiagonal matrices," *Comput. J.*, **1**, 90–96 (1958).

PROBLEMS

1. Find the eigenvalues and eigenvectors of the system

$$x_1 + x_2 + x_3 - \lambda x_1 = 0,$$
$$x_1 + 2x_2 + 2x_3 - \lambda x_2 = 0,$$
$$x_1 + 2x_2 + 3x_3 - \lambda x_3 = 0.$$

2. Find the principal moments of inertia and principal axes of rotation of a three-dimensional rigid body when the moment-of-inertia matrix is

$$I = mr^2 \begin{bmatrix} 10 & 0.134 & -0.866 \\ 0.134 & 6.5 & -1.0 \\ -0.866 & -1.0 & 7.5 \end{bmatrix}.$$

Note: The principal moments of inertia are the eigenvalues, the principal axes of rotation are the associated eigenvectors, and m and r are constants.

3. In a three-dimensional stress problem, we have the symmetric matrix

$$\begin{bmatrix} \sigma_x & \tau_{xy} & \tau_{zx} \\ \tau_{xy} & \sigma_y & \tau_{yz} \\ \tau_{zx} & \tau_{yz} & \sigma_z \end{bmatrix}.$$

Find the principal stresses (eigenvalues) and the associated direction cosines (eigenvectors). We are given that $\sigma_x = 120$, $\sigma_y = 200$, $\sigma_z = 150$, $\tau_{xy} = 65$, $\tau_{zx} = 180$, and $\tau_{yz} = 75$.

4. It is known that, if a symmetric matrix $[A]$ is positive definite, the Gauss-Seidel method discussed in Section 8.10 converges independently of the initial values assumed. Determine whether $[A]$ in Eq. (9.17) is positive definite.

5. Using the finite-difference approach, we can treat the buckling of a column (Fig. 9.9) as an eigenvalue-eigenvector problem.

$$\begin{bmatrix} 7 & -4 & 1 & 0 \\ -4 & 6 & -4 & 1 \\ 1 & -4 & 5 & -2 \\ 0 & 1 & -2 & 1 \end{bmatrix} \begin{bmatrix} w_b \\ w_c \\ w_d \\ w_e \end{bmatrix} = \begin{bmatrix} 2k & -k & 0 & 0 \\ -k & 2k & -k & 0 \\ 0 & -k & 2k & -k \\ 0 & 0 & -k & k \end{bmatrix} \begin{bmatrix} w_b \\ w_c \\ w_d \\ w_e \end{bmatrix}.$$

Find the buckling load $P = 16kEI/L^2$, where the k's are the eigenvalues, and relative values of w_b, w_c, w_d, and w_e are represented by the eigenvectors.

Fig. 9.9

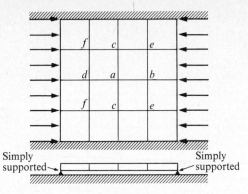

Fig. 9.10

6. The buckling of a plate (see Fig. 9.10) can be treated as an eigenvalue problem:

$$
\begin{bmatrix}
20 & -8 & -16 & -8 & 4 & 4 \\
-8 & 19 & 4 & 1 & -16 & 0 \\
-16 & 4 & 44 & 4 & -16 & -16 \\
-8 & 1 & 4 & 19 & 0 & -16 \\
4 & -16 & -16 & 0 & 42 & 2 \\
4 & 0 & -16 & -16 & 2 & 42
\end{bmatrix}
\begin{bmatrix}
w_a \\ w_b \\ w_c \\ w_d \\ w_e \\ w_f
\end{bmatrix}
= k
\begin{bmatrix}
-2 & 1 & 0 & 1 & 0 & 0 \\
1 & -2 & 0 & 0 & 0 & 0 \\
0 & 0 & -4 & 0 & 2 & 2 \\
1 & 0 & 0 & -2 & 0 & 0 \\
0 & 0 & 2 & 0 & -2 & 0 \\
0 & 0 & 2 & 0 & 0 & -2
\end{bmatrix}
\begin{bmatrix}
w_a \\ w_b \\ w_c \\ w_d \\ w_e \\ w_f
\end{bmatrix}.
$$

Find the eigenvalues in terms of k and the eigenvectors.

7. Determine the frequency ω and the modes of the system shown in Fig. 9.11. The equation of motion is

$$
\begin{bmatrix} y_1 \\ y_2 \\ y_3 \end{bmatrix}
= \frac{\omega^2 \delta_{11}}{27}
\begin{bmatrix}
27 & 14 & 4 \\
14 & 8 & 2.5 \\
4 & 2.5 & 1
\end{bmatrix}
\begin{bmatrix}
M_1 & 0 & 0 \\
0 & M_2 & 0 \\
0 & 0 & M_3
\end{bmatrix}
\begin{bmatrix} y_1 \\ y_2 \\ y_3 \end{bmatrix},
$$

where $M_1 = M_2 = 2$, $M_3 = 3$ and δ_{11} is a constant.

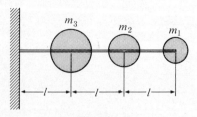

Fig. 9.11

8. a) Write the transformation matrix for the following problem $[A][X] = \lambda[I][X]$ where:

$$
[A] = \begin{bmatrix} 1 & 2 \\ 2 & 1 \end{bmatrix}.
$$

b) Write the eigenvalues for Part (a).

c) What is the eigenvector matrix for Part (a)?

d) Find the true eigenvalues in the problem $[C][X] = \lambda[A][X]$, where $[A]$ is shown in Part (a) and

$$[C] = \begin{bmatrix} 2 & 0 \\ 0 & 3 \end{bmatrix}.$$

e) What are the true eigenvectors for Part (d)?

9. The following matrix can be diagonalized using only two rotations.

$$[A] = \begin{bmatrix} 1 & 2 & 1 \\ 2 & 1 & 1 \\ 1 & 1 & 1 \end{bmatrix}.$$

Write out the eigenvalues in fractional form.

CHAPTER 10

Polynomial Interpolation

10.1 INTERPOLATION

Table 10.1 shows the values of a set of equally spaced data. It is sometimes necessary to determine the y-value at a given x between two known data points, for example, the y-value at $x = 6.3$. This type of problem is usually referred to as an *interpolation problem* for equally spaced data. As a *crude* estimate, it is possible to obtain the value y at $x = 6.3$ by linear interpolation.

Table 10.1 Given data points

k	0	1	2	3	4
x_k	0	2	4	6	8
y_k	-1	1	6	9	11

As shown in Fig. 10.1, we replace the true curve (which is actually unknown) between $x = 6$ and $x = 8$ by its chord, and then simply use $\overline{AB}$ as the required value:

$$\overline{AB} = y_{(at\ x=6)} + \tfrac{1}{2}(0.3)(y_{(at\ x=8)} - y_{(at\ x=6)}). \tag{10.1}$$

In general, $\overline{AB}$ represents the ordinate at

$$x = x_k + r\,\Delta x, \tag{10.2}$$

where Δx is the given interval, r represents the ratio $\overline{A'A}/\Delta x$, and k is the station number where point A' is located (see Fig. 10.1). We note that $0 \le r \le 1$. However, the true answer is $\overline{AC}$; and the difference $\overline{BC}$ between the two values $\overline{AB}$ and $\overline{AC}$ is the error produced by the linear interpolation.

242

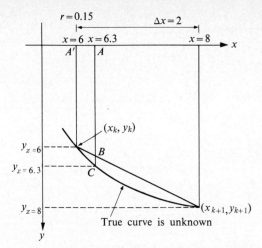

Fig. 10.1. Linear interpolation; BC is ignored.

As an improvement over the linear approximation, we now employ a second-order interpolation, as shown in Fig. 10.2. Here, three data points are given, namely, (x_k, y_k), $(x_k + \Delta x, y_{k+1})$, and $(x_k + 2\,\Delta x, y_{k+2})$; and a parabola of the form

$$y = c_0 + c_1 x + c_2 x^2 \tag{10.3}$$

can be found to pass through the three given points.

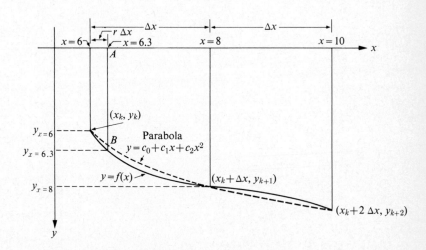

Fig. 10.2. Second-order interpolation.

To determine the coefficients c_0, c_1, and c_2, one substitutes the three given points into Eq. (10.3). The three resulting simultaneous equations yield the values

$$c_0 = \frac{y_{k+2}x_k}{2(\Delta x)^2}(x_k + \Delta x) - \frac{y_{k+1}x_k}{(\Delta x)^2}(x_k + 2\,\Delta x) + \frac{2y_k[x_k^2 + 2(\Delta x)x_k + 2(\Delta x)^2]}{2(\Delta x)^2},$$

$$c_1 = \frac{-y_{k+2}(2x_k + \Delta x) + 4y_{k+1}(x_k + \Delta x) - (2x_k + 3\,\Delta x)y_k}{2(\Delta x)^2}, \qquad (10.4)$$

$$c_2 \doteq \frac{y_{k+2} - 2y_{k+1} + y_k}{2(\Delta x)^2}.$$

Hence we can readily compute the interpolated value by substituting the appropriate x-value into Eq. (10.3).

It is not always necessary to use this cumbersome line of reasoning in order to obtain the coefficients for polynomials of higher order. One can arrive at these polynomials with the aid of the Lagrange polynomials. Before we present this procedure, which basically involves unequally spaced data, we will first introduce the Lagrange polynomials.

10.2 THE LAGRANGE POLYNOMIALS

Suppose that we are given $n + 1$ data points,

$$(x_0, y_0), (x_1, y_1), \ldots, (x_n, y_n),$$

and we wish to find the coefficient $c_0, c_1, \ldots, c_n$ of the polynomial

$$P_n(x) = c_0 + c_1 x + \cdots + c_n x^n, \qquad (10.5)$$

such that the curve represented by Eq. (10.5) will pass through all $n + 1$ distinct points, that is,

$$P_n(x_k) = y_k \qquad (k = 0, 1, \ldots, n). \qquad (10.6)$$

We now define the Lagrange polynomial $L_k(x)$ of degree n as

$$L_k(x_i) = 0 \qquad \text{if} \quad i \neq k, \qquad (10.7a)$$

$$= 1 \qquad \text{if} \quad i = k, \qquad (10.7b)$$

where x_i ($i = 0, 1, \ldots, n$) are the given $n + 1$ distinct arguments. We can now write $P_n(x)$ in the form

$$P_n(x) = L_0(x)y_0 + L_1(x)y_1 + \cdots + L_n(x)y_n$$

$$= \sum_{k=0}^{n} L_k(x)y_k, \qquad (10.8)$$

as is readily seen since we can obtain Eq. (10.6) when each x_k-value ($k = 0, 1, \ldots, n$ is substituted into Eq. (10.8) and when Eqs. (10.7) are taken into consideration. On the other hand, Eq. (10.7a) requires that each factor $x - x_i$, where $i \neq k$, must divide $L_k(x)$. We have

$$L_k(x) = \alpha(x - x_0)(x - x_1) \cdots (x - x_{k-1})(x - x_{k+1}) \cdots (x - x_n), \qquad (10.9)$$

where the factor $(x - x_k)$ is *not* present on the right-hand side.

To determine α, we note that $L_k(x_k) = 1$, and so

$$\alpha = \frac{1}{(x_k - x_0)(x_k - x_1) \cdots (x_k - x_{k-1})(x_k - x_{k+1}) \cdots (x_k - x_n)}. \qquad (10.10)$$

Substituting Eq. (10.10) into Eq. (10.9), we obtain

$$L_k(x) = \frac{(x - x_0)(x - x_1) \cdots (x - x_{k-1})(x - x_{k+1}) \cdots (x - x_n)}{(x_k - x_0)(x_k - x_1) \cdots (x_k - x_{k-1})(x_k - x_{k+1}) \cdots (x_k - x_n)}. \qquad (10.11)$$

As a practical procedure, we note from Eq. (10.9) that

$$L_k(x) = \frac{\alpha(x - x_0)(x - x_1) \cdots (x - x_n)}{x - x_k}, \qquad (10.12)$$

and, using L'Hôpital's rule, we have

$$L_k(x_k) = \lim_{x \to x_k} \frac{(d/dx)[\alpha F_{n+1}(x)]}{(d/dx)(x - x_k)} = \frac{\alpha F'_{n+1}(x_k)}{1}, \qquad (10.13)$$

where

$$F_{n+1}(x) = (x - x_0)(x - x_1) \cdots (x - x_n). \qquad (10.14)$$

Since $L_k(x_k) = 1$, we have

$$\alpha = 1/F'_{n+1}(x_k). \qquad (10.15)$$

Substituting Eq. (10.15) into Eq. (10.12), we have

$$L_k(x) = \frac{1}{F'_{n+1}(x_k)} \frac{F_{n+1}(x)}{x - x_k} = \frac{G_k(x)}{F'_{n+1}(x_k)}, \qquad (10.16)$$

where

$$G_k(x) = \frac{F_{n+1}(x)}{x - x_k}. \qquad (10.17)$$

Hence, a Lagrange polynomial of degree n can be constructed by the following steps:

1) Set up $F_{n+1}(x) = (x - x_0)(x - x_1) \cdots (x - x_n)$.
2) Find $F'_{n+1}(x)$ and then compute $F'_{n+1}(x_k)$; $k = 0, 1, \ldots, n$.
3) Using Eq. (10.17), find $G_k(x)$ for $x = x_0, x_1, \ldots, x_n$.
4) Finally, find $L_k(x)$ by use of Eq. (10.16).

Example 10.1

Find the Lagrange polynomials $L_0(x)$, $L_1(x)$, $L_2(x)$, and $L_3(x)$ from the unequally spaced data given below:

k	0	1	2	3
x_k	2	3	6	7

We have four data points; thus $n = 4 - 1 = 3$. We first set up $F_4(x)$ and $F_4'(x)$:

$$F_4(x) = (x - x_0)(x - x_1)(x - x_2)(x - x_3),$$
$$F_4'(x) = (x - x_1)(x - x_2)(x - x_3) + (x - x_0)(x - x_2)(x - x_3)$$
$$+ (x - x_0)(x - x_1)(x - x_3) + (x - x_0)(x - x_1)(x - x_2).$$

We then formulate $G_0(x)$ through $G_3(x)$:

$$G_0(x) = (x - x_1)(x - x_2)(x - x_3),$$
$$G_1(x) = (x - x_0)(x - x_2)(x - x_3),$$
$$G_2(x) = (x - x_0)(x - x_1)(x - x_3),$$
$$G_3(x) = (x - x_0)(x - x_1)(x - x_2).$$

It follows that

$$L_0(x) = \frac{(x - x_1)(x - x_2)(x - x_3)}{(x_0 - x_1)(x_0 - x_2)(x_0 - x_3)} = \frac{(x - 3)(x - 6)(x - 7)}{-20},$$

$$L_1(x) = \frac{(x - x_0)(x - x_2)(x - x_3)}{(x_1 - x_0)(x_1 - x_2)(x_1 - x_3)} = \frac{(x - 2)(x - 6)(x - 7)}{12},$$

$$L_2(x) = \frac{(x - x_0)(x - x_1)(x - x_3)}{(x_2 - x_0)(x_2 - x_1)(x_2 - x_3)} = \frac{(x - 2)(x - 3)(x - 7)}{-12},$$

$$L_3(x) = \frac{(x - x_0)(x - x_1)(x - x_2)}{(x_3 - x_0)(x_3 - x_1)(x_3 - x_2)} = \frac{(x - 2)(x - 3)(x - 6)}{20}.$$

10.3 LAGRANGE INTERPOLATION FORMULA FOR UNEQUALLY SPACED DATA

Let us examine Eq. (10.11) carefully:

$$L_k(x) = \frac{(x - x_0)(x - x_1) \cdots (x - x_{k-1})(x - x_{k+1}) \cdots (x - x_n)}{(x_k - x_0)(x_k - x_1) \cdots (x_k - x_{k-1})(x_k - x_{k+1}) \cdots (x_k - x_n)}. \tag{10.11}$$

At first glance, this expression appears to be very complicated. However, it is not difficult to see that Eq. (10.11) is equivalent to the following shorthand version:

$$L_k(x) = \prod_{\substack{i=0 \\ i \neq k}}^{n} \frac{x - x_i}{x_k - x_i}, \tag{10.18}$$

where the pi notation has the meaning

$$\prod_{k=1}^{n} x_k = x_1 \cdot x_2 \cdot x_3 \cdots x_n. \tag{10.19}$$

Substituting Eq. (10.18) into Eq. (10.8), we obtain

$$P_n(x) = \sum_{k=0}^{n} \left[\prod_{\substack{i=0 \\ i \neq k}}^{n} \left(\frac{x - x_i}{x_k - x_i} \right) \right] y_k. \tag{10.20}$$

This equation is known as the Lagrange interpolation formula for unequally spaced data.

Example 10.2

The following ten tabulated values are given:

x	y	x	y
0.1	0.9003	1.5	0.0158
0.3	0.7077	1.6	−0.0059
0.7	0.3798	1.8	−0.0376
1.0	0.1988	2.0	−0.0563
1.2	0.1091	2.4	−0.0669

We are required (1) to compute the values of y corresponding to a given argument, and (2) to find the error resulting from the interpolation by subtracting the answer based on (1) from the answer using

$$y = e^{-x} \cos x.$$

The following six arguments are to be read in, one per data card.

$$0.4, \ 0.6, \ 0.9, \ 1.4, \ 1.7, \ 1.9.$$

Write a FORTRAN program to perform the Lagrangian interpolation. The program should be written in a flexible way so that a maximum of 12 given points can be considered. These points should also be read in via data cards. A possible FORTRAN program and its partial output are shown in Fig. 10.3 and Table 10.2, respectively.

```
C         MAIN PROGRAM OF LAGRANGIAN INTERPOLATION
C         L=NUMBER OF TABULATED POINTS CONSIDERED
C         LL=NUMBER OF GIVEN ARGUMENTS
C         X(K)=A GIVEN X ARGUMENT, ZX( )=A TABULATED VALUE OF X
C         Y( )=A TABULATED VALUE OF Y
C         E=DIFFERENCE BETWEEN EXACT VALUE AND LAGRANGIAN VALUE
C
C
          DATA IREAD,IWRITE/5,6/
          DIMENSION X(12), ZX(12), Y(12)
C         READ THE GIVEN POINTS AND ARGUMENTS
          READ(IREAD,10) L,LL
       10 FORMAT(I3,I3)
          WRITE(IWRITE,11) L,LL
       11 FORMAT('1',5X,'L=',I3,5X,'LL=',I3///)
          DO 15 I=1,L
       15 READ(IREAD,16) ZX(I),Y(I)
       16 FORMAT(F5.1,F10.5)
          WRITE(IWRITE,17)
       17 FORMAT(5X,'GIVEN POINTS',//,5X,'ZX',12X,'Y')
          DO 18 I=1,L
       18 WRITE(IWRITE,19) ZX(I),Y(I)
       19 FORMAT(5X,F5.1,5X,F10.5)
          WRITE(IWRITE,20)
       20 FORMAT(///,5X,'A GIVEN X ARGUMINT'/)
          DO 25 K=1,LL
       25 READ(IREAD,26) X(K)
       26 FORMAT(F5.1)
          DO 27 K=1,LL
       27 WRITE(IWRITE,28) X(K)
       28 FORMAT(5X,F5.1)
          WRITE(IWRITE,29)
       29 FORMAT(///)
          WRITE(IWRITE,1)
        1 FORMAT(9X,'(1)',15X,'(2)',16X,'(3)',15X,'(4)')
          WRITE(IWRITE,5)
        5 FORMAT(7X,'ARGUMENT',9X,'LAGRANGIAN',11X,'EXACT',12X,'(3)-(2)'///)
          DO 95 K=1,LL
          XK=X(K)
          C=0.
          DO 55 I=2,L
          ZXI=ZX(I)
          P=1.
C
C         DO LOOP 40 IS CARRYING OUT LAGRANGE POLYNOMIAL(EXPRESSED BY P)
C         A IS ONE COMMON TERM OF THE LAGRANGE POLYNOMIAL
C
          DO 40 J=1,L
          IF(I.EQ.J) GO TO 40
       35 ZXJ=ZX(J)
          A=(XK-ZXJ)/(ZXI-ZXJ)
          P=P*A
       40 CONTINUE
          B=P*Y(I)
          C=C+B
       55 CONTINUE
```

Fig. 10.3. FORTRAN program and data for Lagrangian interpolation.

```
C
C       YY IS THE EXACT ANSWER CALCULATED BY THE KNOWN CURVE
C
        YY=FXP(-XK)*COS(XK)
        E=YY-C
        WRITE(IWRITE,90) XK,C,YY,E
     90 FORMAT(7X,F4.1,12X,F9.5,10X,F9.5,9X,F9.5/)
     95 CONTINUE
        STOP
        END
```

```
 10  6
    .1      .9003
    .3      .7077
    .7      .3798
   1.       .1988
   1.2      .1091
   1.5      .0158
   1.6    -0.0059
   1.8    -0.0376
   2.0    -0.0563
   2.4    -0.0669
    .4
    .6
    .9
   1.4
   1.7
   1.9
```

Table 10.2 Partial output for Lagrangian interpolation

(1) ARGUMENT	(2) LAGRANGIAN	(3) EXACT	(4) (3)-(2)
.4	.61698	.61740	.00042
.6	.45273	.45295	.00022
.9	.25282	.25272	-.00009
1.4	.04192	.04191	0.00000
1.7	-.02357	-.02353	.00003
1.9	-.04836	-.04835	.00001

BIBLIOGRAPHY

Interpolation

Gershinsky, M., and D. A. Levine, "Aitken-Hermite interpolation," *J. Assoc. Comput. Mach.*, **11**, no. 3, 352–356 (1964).

Greville, T. N. E., "Numerical procedures for interpolation by spline functions," *J. Soc. Ind. Math., Numer. Anal. Ser. B.*, **1**, 53–68 (1964).

Interpolation and Allied Tables. H.M. Stationery Office, London, 1956.

Kuntzmann, J., *Méthodes numériques, interpolation, dérivées.* Dunod, Paris, 1959.

Lanczos, C., "Trigonometric interpolation of empirical and analytic functions," *J. Math. Phys.*, **17**, 123–199 (1938).

Salzer, H. E., "A new formula for inverse interpolation," *Bull. Am. Math. Soc.*, **50**, 513–516 (1946).

Southard, T. H., "Everett's formula for bivariate interpolation and throw-back of fourth differences," *Math. Tables Aids Comput.*, **10**, 216–223 (1956).

Steffensen, J. F., *Interpolation.* Chelsea, New York, 1927.

Thacher, H. C., "Derivation of interpolation formulas in several independent variables," *Ann. N.Y. Acad. Sci.*, **86**, 758–775 (1960).

Traub, J. F., "On Lagrange-Hermite interpolation," *J. Soc. Ind. Math.*, **12**, 886–891 (1964).

Walsh, J. L., *Interpolation and Approximation in the Complex Domain.* Am. Math. Soc., Providence, R.I., 1955.

Iterated Interpolation

Aitken, A. C., "On interpolation by iteration of proportional parts, without the use of differences," *Proc. Edinburgh Math. Soc.*, **3**, 56–76 (1932).

Neville, E. H., "Iterative interpolation," *J. Indian Math. Soc.*, **20**, 87–120 (1934).

Finite Differences

Abramowitz, M., "Note on modified second differences for use with Everett's interpolation formula," in *Tables of Bessel Functions of Fractional Order*, Natn. Bur. Stand. Columbia Univ. Press Ser., **10**, pp. XXXIII–XXXVI (1948).

Bickley, W. G., "Differences and associated operators, with some applications," *J. Math. Phys.*, **27**, 182–192 (1948).

Freeman, H., *Finite Differences for Actuarial Students*, 2nd ed. (1st ed. published in 1939 as *Mathematics for Actuarial Students, Part II*). Cambridge University Press, London, 1960.

Jordan, C., *Calculus of Finite Differences*, 2nd ed. Chelsea, New York, 1950.

Michel, J. G. L., "Central-difference formulae obtained by means of operator expansions," *J. Inst. Actu.*, **72**, 470–480 (1946).

Milne-Thomson, L. M., *The Calculus of Finite Differences*, reprint (1st ed. 1933). Macmillan, London, 1951.

PROBLEMS

1 Find the Lagrange polynomials $L_0(x)$, $L_1(x)$, and $L_2(x)$ from the data given below:

i	0	1	2
x_i	2	4	6
y_i	3	4	5

Also find $P_2(x)$, using Eqs. (10.5) and (10.8) independently.

2. Ten tabulated values are given below:

x	y	x	y
0.1	0.9907	1.5	0.2384
0.3	0.9267	1.7	0.1576
0.7	0.6997	2.0	0.0667
0.9	0.5712	2.1	0.0439
1.2	0.3899	2.3	0.0080

Write a program in FORTRAN to compute the values of y at each of the eight x's, 0.2, 0.5, 0.6, 0.8, 1.0, 1.4, 1.8, 2.2. The values are to be computed by each of the following two methods: 1. Lagrangian interpolation. 2. Closed form solution: $y = e^{-x}(\cos x + \sin x)$. Compute the error resulting from the first method by comparing your results with those obtained by using the closed form.

Notes. (a) The ten tabulated values should be read in via data cards (one x and one y per card). (b) Each of the eight x's for which y must be computed should be read in via data cards (one x per card). (c) The program should be written to handle 12 given points.

3. If $v_t = (1/\sqrt{2\pi})\int_{-\infty}^{t} e^{-t^2/2}\, dt$, then we can construct the following table:

t	v_t	t	v_t
0	0.50000000	2.0	0.97724987
.5	0.69146246	2.5	0.99379033
1.0	0.84134475	3.0	0.99865010
1.5	0.93319280		

Find $v_{0.75}$, $v_{1.75}$, $v_{2.75}$, and $v_{2.9}$ using the Lagrangian interpolation equation.

4. Given the table below, find e^π.

x	e^x	x	e^x
3.10	22.197951	3.13	22.873980
3.11	22.421044	3.14	23.103867
3.12	22.646380	3.15	23.336065

CHAPTER 11

Least-Squares Curve Fitting

11.1 INTRODUCTION

In the previous chapter we demonstrated how to obtain the equation of a curve which passes exactly through all given points. This is accomplished by means of interpolation formulas, such as Eq. (10.20). We now ask for the equation of a smooth curve (Fig. 11.1) which does not pass through each of a number of given points, but which passes near each of them in a plane. The "nearness" is usually obtained by imposing the least-squares criterion, and the application of this criterion is the basis of the method of least squares, which we shall develop in Sections 11.2 and 11.3.

11.2 NORMAL EQUATIONS FOR CURVE FITTING

Let us begin by considering the problem of fitting a given number of function values (see Table 11.1) by a straight line in the form

$$Y = k_0 + k_1 x. \tag{11.1}$$

Table 11.1 Seven function values ($n = 7$)

x	y	x	y
0	2	4	8
1	3	5	9
2	5	6	10
3	5		

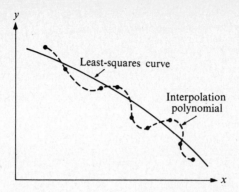

Fig. 11.1. Interpolation curve versus least-squares curve.

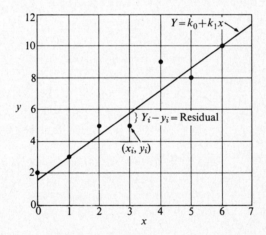

Fig. 11.2. Straight line representation for data in Table 11.1.

We first plot the values in Table 11.1 on a graph paper to see whether or not it is reasonable to approximate the given values by a straight line (Fig. 11.2). We next determine the coefficients k_0 and k_1 using the least-squares criterion which requires that $S = \sum(Y_i - y_i)^2$ be a minimum, where Y_i is evaluated from Eq. (11.1). The values $Y_i - y_i$ are called *residuals*.

To obtain the minimum value for S, which is a function of two variables k_0 and k_1, we set the following two first partial derivatives to zero:

$$\frac{\partial S}{\partial k_0} = 0 \quad \text{and} \quad \frac{\partial S}{\partial k_1} = 0.$$

This yields the two simultaneous linear equations

$$\frac{\partial S}{\partial k_0} = \sum_{i=1}^{n} \frac{\partial}{\partial k_0} (k_0 + k_1 x_i - y_i)^2 = \sum_{i=1}^{n} 2(k_0 + k_1 x_i - y_i) = 0 \qquad (11.2a)$$

and

$$\frac{\partial S}{\partial k_1} = \sum_{i=1}^{n} \frac{\partial}{\partial k_1} (k_0 + k_1 x_i - y_i)^2 = \sum_{i=1}^{n} 2x_i(k_0 + k_1 x_i - y_i) = 0. \qquad (11.2b)$$

From Eqs. (11.2a) and (11.2b) one has

$$\sum_{i=1}^{n} y_i = nk_0 + k_1 \sum_{i=1}^{n} x_i \qquad (11.3a)$$

and

$$\sum_{i=1}^{n} x_i y_i = k_0 \sum_{i=1}^{n} x_i + k_1 \sum_{i=1}^{n} x_i^2. \qquad (11.3b)$$

Thus Eq. (11.3a) and (11.3b) represent two conditions which must be met in order to obtain the best fit of the straight line, based on the least-squares criterion.

Example 11.1

To illustrate the application of Eqs. (11.3a) and (11.3b), we shall now find k_0 and k_1, using the values in Table 11.1. We first compute $x_i y_i$, then x_i^2 for $i = 1, 2, \ldots, 7$, as shown in Table 11.2. Substituting the values of $\sum y_i$, $\sum x_i$, $\sum x_i y_i$, and $\sum x_i^2$ into Eqs. (11.3a) and (11.3b), we find that k_0 and k_1 satisfy the following two conditions, or *normal equations*:

$$42 = 7k_0 + 21k_1, \qquad 164 = 21k_0 + 91k_1.$$

Table 11.2 Computations for $\sum x_i y_i$ and $\sum x_i^2$

i	x_i	y_i	$x_i y_i$	$(x_i)^2$
1	0	2	0	0
2	1	3	3	1
3	2	5	10	4
4	3	5	15	9
5	4	9	36	16
6	5	8	40	25
7	6	10	60	36
Sum	21	42	164	91

The solution of these equations yields

$$k_0 = 1.928571 \qquad \text{and} \qquad k_1 = 1.357143.$$

The equation for the required straight line is therefore

$$Y = 1.928571 + 1.357143x. \qquad (11.4)$$

It is now possible to compute Y_i by means of Eq. (11.4). This is shown in column 3 of Table 11.3. Columns 4 and 5 are the values of $(Y_i - y_i)$ and $x_i(Y_i - y_i)$, respectively. As a check, note that the values of both $\sum (Y_i - y_i)$ and $\sum x_i(Y_i - y_i)$ should be equal to zero, as indicated in Eqs. (11.2a) and (11.2b).

Table 11.3 Residuals

(1)	(2)	(3)	(4)	(5)
x_i	y_i	Y_i	$Y_i - y_i$	$x_i(Y_i - y_i)$
0	2	1.928571	-0.071429	0.000000
1	3	3.285714	$+0.285714$	$+0.285714$
2	5	4.642857	-0.357143	-0.714286
3	5	6.000000	$+1.000000$	$+3.000000$
4	9	7.357143	-1.642857	-6.571428
5	8	8.714286	$+0.714286$	$+3.571430$
6	10	10.071429	$+0.071429$	$+0.428574$
Sum			0.000000	$+0.000004$

After our discussion on fitting given data with a straight line, we turn now to fitting a set of data with a polynomial of mth degree. The procedure is quite similar to the straight-line case. Basically, we wish to find a minimum value of S, where

$$S = \sum_{i=1}^{n} (Y_i - y_i)^2 = \sum_{i=1}^{n} (k_0 + k_1 x_i + k_2 x_i^2 + \cdots + k_m x_i^m - y_i)^2.$$

To obtain the minimum value of S, which is now a function of $m + 1$ variables $k_0, k_1, \ldots, k_m$, we set the following $m + 1$ first partial derivatives to zero:

$$\frac{\partial S}{\partial k_0} = \sum_{i=1}^{n} 2(k_0 + k_1 x_i + k_2 x_i^2 + \cdots + k_m x_i^m - y_i) = 0,$$

$$\frac{\partial S}{\partial k_1} = \sum_{i=1}^{n} 2x_i(k_0 + k_1 x_i + k_2 x_i^2 + \cdots + k_m x_i^m - y_i) = 0, \quad\quad (11.5)$$

$$\vdots$$

$$\frac{\partial S}{\partial k_m} = \sum_{i=1}^{n} 2x_i^m(k_0 + k_1 x_i + k_2 x_i^2 + \cdots + k_m x_i^m - y_i) = 0.$$

We then obtain the $m + 1$ simultaneous linear, or normal, equations

$$k_0 n \quad + k_1 \sum x_i \quad ` + k_2 \sum x_i^2 \quad + \cdots + k_m \sum x_i^m \quad - \sum y_i \quad = 0,$$
$$k_0 \sum x_i + k_1 \sum x_i^2 \quad + k_2 \sum x_i^3 \quad + \cdots + k_m \sum x_i^{m+1} - \sum x_i y_i = 0,$$
$$k_0 \sum x_i^2 + k_1 \sum x_i^3 \quad + k_2 \sum x_i^4 \quad + \cdots + k_m \sum x_i^{m+2} - \sum x_i^2 y_i = 0, \qquad (11.6)$$
$$\vdots$$
$$k_0 \sum x_i^m + k_1 \sum x_i^{m+1} + k_2 \sum x_i^{m+2} + \cdots + k_m \sum x_i^{m+m} - \sum x_i^m y_i = 0,$$

where the symbol $\sum$ implies summation for i from 1 to n.

It is convenient to express Eq. (11.6) in the following matrix notation,

$$[A][k] = [B], \qquad (11.7)$$

where

$$[A] = \begin{bmatrix} n & \sum x_i & \sum x_i^2 & \cdots & \sum x_i^m \\ \sum x_i & \sum x_i^2 & \sum x_i^3 & \cdots & \sum x_i^{m+1} \\ \vdots & & & & \\ \sum x_i^m & \sum x_i^{m+1} & \sum x_i^{m+2} & \cdots & \sum x_i^{2m} \end{bmatrix},$$

is a symmetric matrix, and

$$[k] = \begin{bmatrix} k_0 \\ k_1 \\ \vdots \\ k_m \end{bmatrix}, \qquad [B] = \begin{bmatrix} \sum y_i \\ \sum x_i y_i \\ \vdots \\ \sum x_i^m y_i \end{bmatrix}.$$

Example 11.2

We are to fit a parabola to a set of points (x_i, y_i) shown in columns 2 and 3 of Table 11.4. It should be noted that the given data are not equally spaced along the x-direction.

Normal equations can be obtained only after the values of the sum of the following expressions are known:

$$x_i, \ y_i, \ x_i y_i, \ x_i^2, \ x_i^3, \ x_i^4, \qquad \text{and} \qquad x_i^2 y_i.$$

These are tabulated in Table 11.4. From the table and Eq. (11.6), we see that three normal equations are

$$9k_0 + \quad 53k_1 + \quad 381k_2 = 76,$$
$$53k_0 + \quad 381k_1 + \quad 3017k_2 = 489,$$
$$381k_0 + 3017k_1 + 25317k_2 = 3547.$$

Table 11.4 Fitting a parabola to a set of points

(1)	(2)	(3)	(4)	(5)	(6)	(7)	(8)
i	x_i	y_i	$x_i y_i$	x_i^2	$x_i^2 y_i$	x_i^3	x_i^4
1	1	2	2	1	2	1	1
2	3	7	21	9	63	27	81
3	4	8	32	16	128	64	256
4	5	10	50	25	250	125	625
5	6	11	66	36	396	216	1296
6	7	11	77	49	539	343	2401
7	8	10	80	64	640	512	4096
8	9	9	81	81	729	729	6561
9	10	8	80	100	800	1000	10000
Σ	53	76	489	381	3547	3017	25317

Solving the equations, we have

$$k_0 = -1.4597,$$

$$k_2 = -0.2676,$$

$$k_1 = 3.6053,$$

and the equation of the desired parabola is

$$y = -1.4597 + 3.6053x - 0.2676x^2.$$

The resulting parabola is plotted in Fig. 11.3.

So far in this section we have considered a given set of data points having equal importance. Very often, due to measuring instruments of differing precision, some of the y_i values are more accurate and are therefore considered more important than others and these points are necessarily assigned more "weight." In other words, we should assign a weight, or weighting coefficient w_i, to each point (x_i, y_i).

By means of the least-squares criterion discussed in Eq. (11.5), the expression

$$S = \sum_{i=1}^{n} w_i^2 \left(y_i - \sum_{j=0}^{m} k_j x_i^j \right)^2$$

is minimized. The system of normal equations

$$\frac{\partial S}{\partial k_j} = 0 \qquad (j = 0, 1, \ldots, m)$$

is represented in matrix notation as

$$[A][k] = [B], \tag{11.8}$$

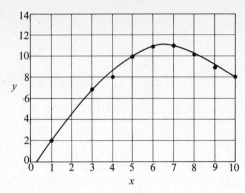

Fig. 11.3. Fitting a parabola to nine points.

where

$$[A] = \begin{bmatrix} \sum w_i & \sum w_i x_i & \sum w_i x_i^2 & \cdots & \sum w_i x_i^m \\ \sum w_i x_i & \sum w_i x_i^2 & \sum w_i x_i^3 & \cdots & \sum w_i x_i^{m+1} \\ \vdots & & & & \\ \sum w_i x_i^m & \sum w_i x_i^{m+1} & \sum w_i x_i^{m+2} & \cdots & \sum w_i x_i^{2m} \end{bmatrix},$$

$$[k] = \begin{bmatrix} k_0 \\ k_1 \\ \vdots \\ k_m \end{bmatrix}, \qquad [B] = \begin{bmatrix} \sum w_i y_i \\ \sum w_i x_i y_i \\ \vdots \\ \sum w_i x_i^m y_i \end{bmatrix}.$$

When $w_i = 1$ $(i = 1, \ldots, n)$, Eq. (11.8) becomes identical to Eq. (11.7). It is noted that the column matrix $[k]$, also referred to as the regression coefficient vector, can be readily obtained by the Gauss-Jordan Method, as discussed in Chapter 8.

11.3 FORTRAN PROGRAM FOR LEAST-SQUARES POLYNOMIAL FIT

The main program shown in Fig. 11.4 may be used to fit polynomials of several different degrees to a given set of N data points (x_i, y_i). Each data point has a weighting coefficient w_i. This program may determine the coefficients of polynomials of degree $0, 1, 2, 3, \ldots, N - 1$. In other words, up to N different sets of polynomial coefficients can be found for the same set of data points. The variables used in this main program are:

    N = the number of data points.

    M = the degree of the polynomial to be fitted.

    LL = the number of polynomials of different degrees which are to be fitted.

```
C       MAIN PROGRAM FOR LEAST SQUARE POLYNOMIAL FIT    SECTION 11.3
        DIMENSION A(15,15),B(15,15),X(15),Y(15,4),W(15),C(15,15)
        DATA NREAD,NWRITE/5,6/
        MM=0
        READ(NREAD,100) LL
    5   IF(MM.GE.LL) GO TO 1000
   10   READ (NREAD,200) N,M,L
        IF(MM.NE.0) GO TO 25
   15   WRITE(NWRITE,400)
        DO 20 I=1,N
        READ(NREAD,300) X(I),W(I),(Y(I,J),J=1,L)
        DO 20 J=1,L
   20   WRITE(NWRITE,500) X(I),W(I),Y(I,J)
C
C       FOLLOWING 11 STATEMENTS TO FORM THE COEFFICIENT MATRIX OF THE NORMAL
C       EQUATIONS
C
   25   DO 30 I=1,N
   30   C(I,1)=1.0
        MP1=M+1
        DO 35 J=2,MP1
        DO 35 I=1,N
   35   C(I,J)=C(I,J-1)*X(I)
        DO 40 I=1,MP1
        DO 40 J=1,MP1
        A(I,J)=0.0
        DO 40 K=1,N
   40   A(I,J)=A(I,J)+C(K,I)*C(K,J)*W(K)
C
C       FOLLOWING 5 STATEMENTS TO FORM THE CONSTANT MATRIX OF THE NORMAL EQUATIONS
C
        DO 45 J=1,L
        DO 45 I=1,MP1
        B(I,J)=0.0
        DO 45 K=1,N
   45   B(I,J)=B(I,J)+C(K,I)*Y(K,J)*W(K)
C
C       CALL SUBROUTINE CHAP8 TO SOLVE THE NORMAL EQUATIONS
C
        CALL CHAP8(A,MP1,B,L,DET)
        WRITE(NWRITE,600) N
        WRITE(NWRITE,700) M
        DO 50 J=1,L
        DO 50 I=1,MP1
        II=I-1
   50   WRITE(NWRITE,800) II,B(I,J)
        WRITE(NWRITE,900)
        MM=MM+1
        GO TO 5
  100   FORMAT(I4)
  200   FORMAT(3I4)
  300   FORMAT(6F10.5)
  400   FORMAT('1',14X,'X',20X,'WEIGHT',22X,'Y'//)
  500   FORMAT(8X,E12.6,12X,E12.6,13X,E12.6/)
  600   FORMAT(///,' NUMBER OF GIVEN DATA POINTS = ',I2)
  700   FORMAT(/7X,'DEGREE OF POLYNOMIAL = ',I2,///)
  800   FORMAT(5X,I2,' DEGREE COEFFICIENT = ',E14.8,/)
  900   FORMAT(/,10H - - - - -,///)
 1000   STOP
        END
```

Fig. 11.4 Main program for the normal-equation method.

Table 11.5 Normal-equation solutions

X	WEIGHT	Y
.100000E+01	.100000E+01	.200000E+01
.300000E+01	.100000E+01	.700000E+01
.400000E+01	.100000E+01	.800000E+01
.500000E+01	.100000E+01	.100000E+02
.600000E+01	.100000E+01	.110000E+02
.700000E+01	.100000E+01	.110000E+02
.800000E+01	.100000E+01	.100000E+02
.900000E+01	.100000E+01	.900000E+01
.100000E+02	.100000E+01	.800000E+01

NUMBER OF GIVEN DATA POINTS = 9

DEGREE OF POLYNOMIAL = 2

0 DEGREE COEFFICIENT = -.14596409E+01

1 DEGREE COEFFICIENT = .36052985E+01

2 DEGREE COEFFICIENT = -.26756969E+00

- - - - - -

NUMBER OF GIVEN DATA POINTS = 9

DEGREE OF POLYNOMIAL = 4

0 DEGREE COEFFICIENT = .44753310E+00

1 DEGREE COEFFICIENT = .11041027E+01

2 DEGREE COEFFICIENT = .61011515E+00

3 DEGREE COEFFICIENT = -.11443138E+00

4 DEGREE COEFFICIENT = .49863874E-02

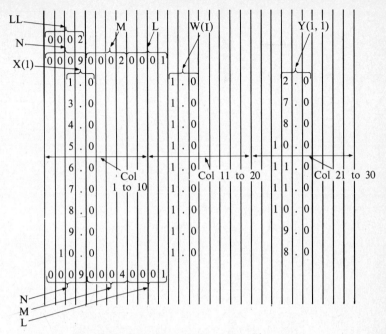

Fig. 11.5. Input format for the main program.

MM = an index to read data points for the first polynomial being fitted, to avoid rereading data points for the next polynomial, and to stop the program after fitting the last polynomial.

X(I) = the value of x at the ith data point.

W(I) = the weight of the ith data point. If all points are treated equally, W(I) = 1.

L = the number of y-values which correspond to each X(I) (in most cases, L = 1).

Y(I,J) = the jth value of y at the ith data point.

In Example 11.2, we were fitting a second-degree polynomial to nine points. If we wish to use this program to fit a fourth-degree polynomial to this set of points in the same run, the data would be read in the format shown in Fig. 11.5. This main program calls for the subprogram CHAP8, the subroutine for matrix inversion and simultaneous equations (see Section 8.9). The answers corresponding to the input (see Fig. 11.5) are listed in Table 11.5. We can see that the answers, obtained by machine computation for the polynomial of second degree, check closely with the known answers in Example 11.2.

11.4 ORTHOGONAL POLYNOMIALS

The normal-equation method discussed in the previous sections can be used to fit a number of function values by a polynomial in the form

$$Y = k_0 + k_1 x + k_2 x^2 + \cdots + k_m x^m.$$

To determine k_0 through k_m, we must solve a set of simultaneous linear equations. Unfortunately, this method fails when the resulting m normal equations are ill-conditioned, which is often the case when m becomes large. Another disadvantage of the normal-equation method lies in the fact that whenever the value of m is changed, a set of new values of k_0 through k_m must be computed, as can be seen in Examples 11.1 and 11.2.

In the remainder of this chapter, we shall discuss a method which makes it possible to overcome the two difficulties mentioned above. This technique is designed for digital computers using the Chebyshev polynomials $T_m(x)$ and fits the given data in the form

$$Y_m(x) = c_0 T_0(x) + c_1 T_1(x) + \cdots + c_m T_m(x). \tag{11.9}$$

Our immediate problem is to derive an expression for each of the coefficients c_j $(j = 0, 1, \ldots, m)$. This derivation requires the knowledge of certain properties of orthogonal polynomials discussed below.

The following two sets of definite integrals are known as orthogonal conditions:

$$\int_0^\pi \sin mx \sin nx \, dx = 0, \qquad m \neq n,$$
$$= \pi/2, \qquad m = n; \tag{11.10}$$

$$\int_0^\pi \cos mx \cos nx \, dx = 0, \qquad m \neq n,$$
$$= \pi/2, \qquad m = n. \tag{11.11}$$

They are valid provided that m and n are nonnegative integers and are not both zero.

In general, a set of functions $\phi_0(x), \phi_1(x), \ldots, \phi_m(x)$ is known as orthogonal in an interval $a \leq x \leq b$ if

$$\int_a^b w(x)\phi_m(x)\phi_n(x) \, dx = 0, \qquad m \neq n, \tag{11.12}$$

where the weighting function $w(x)$ is nonnegative in the given interval (a, b). When all the *orthogonal* functions $\phi_m(x)$ are polynomials, they are known as *orthogonal polynomials*.

A particular example of such orthogonal polynomials is the set of Chebyshev polynomials.

11.5 CHEBYSHEV POLYNOMIALS

Chebyshev polynomials of degree r in x are defined by

$$T_r(x) = \cos (r \cos^{-1} x). \qquad (11.13)$$

The first through sixth Chebyshev polynomials are

$$
\begin{aligned}
T_0(x) &= 1, \\
T_1(x) &= x, \\
T_2(x) &= 2x^2 - 1, \\
T_3(x) &= 4x^3 - 3x, \\
T_4(x) &= 8x^4 - 8x^2 + 1, \\
T_5(x) &= 16x^5 - 20x^3 + 5x, \\
T_6(x) &= 32x^6 - 48x^4 + 18x^2 - 1.
\end{aligned}
\qquad (11.14)
$$

Other Chebyshev polynomials can be readily obtained by using the recurrence relation

$$T_{r+1}(x) - 2xT_r(x) + T_{r-1}(x) = 0. \qquad (11.15)$$

The set of Chebyshev polynomials, one of several classical orthogonal polynomials,† has the following important property:

$$
\sum_{i=1}^{m+1} T_k(\bar{x}_i)T_l(\bar{x}_i) =
\begin{cases}
0, & \text{for } k \neq l, & (11.16) \\
(m + 1)/2, & \text{for } k = l \neq 0, & (11.17) \\
m + 1, & \text{for } k = l = 0, & (11.18)
\end{cases}
$$

where

$$\bar{x}_i = \cos \frac{(2i - 1)\pi}{2(m + 1)}, \qquad i = 1, 2, \ldots, m + 1. \qquad (11.19)$$

For example, if $m = 3$, the values of $\bar{x}_i$, as determined by Eq. 11.19 and listed in Table 11.6, would satisfy Eqs. (11.16), (11.17), and (11.18).

Table 11.6 Values of $\bar{x}_i$

i	1	2	3	4
$\bar{x}_i$	0.92388	0.38268	-0.38268	-0.92388

---

† Others are Laguerre, Legendre, Gegenbauer, Jacobi, Hermite, etc. The Legendre polynomials will be discussed in Section 12.8.

11.6 CHEBYSHEV POLYNOMIALS AND CURVE FITTING

We are now in a position to derive an expression for each of the coefficients c_j $(j = 0, 1, \ldots, m)$ in the equation

$$Y_m(x) = c_0 T_0(x) + c_1 T_1(x) + \cdots + c_m T_m(x). \tag{11.20}$$

We shall assume that the given n data points $(x_1, y_1), (x_2, y_2), \ldots, (x_n, y_n)$ are *equally spaced*.

As with the normal-equation solution, we apply the least-squares criterion discussed in Eq. (11.5) to minimize the expression

$$S = \sum_{i=1}^{n} \left[y_i - \sum_{j=0}^{m} c_j T_j(x_i) \right]^2. \tag{11.21}$$

The system of simultaneous equations

$$\frac{\partial S}{\partial c_j} = 0, \qquad j = 0, 1, \ldots, m,$$

can then be written in the following matrix form:

$$
\begin{bmatrix}
\sum T_0^2(x_i) & \sum T_0(x_i)T_1(x_i) & \cdots & \sum T_0(x_i)T_m(x_i) \\
\sum T_1(x_i)T_0(x_i) & \sum T_1^2(x_i) & \cdots & \sum T_1(x_i)T_m(x_i) \\
\vdots & & & \vdots \\
\sum T_m(x_i)T_0(x_i) & \sum T_m(x_i)T_1(x_i) & \cdots & \sum T_m^2(x_i)
\end{bmatrix}
\begin{bmatrix}
c_0 \\ c_1 \\ \vdots \\ c_m
\end{bmatrix}
=
\begin{bmatrix}
\sum y_i T_0(x_i) \\
\sum y_i T_1(x_i) \\
\vdots \\
\sum y_i T_m(x_i)
\end{bmatrix},
$$

$$\tag{11.22}$$

where $\sum$ implies $\sum_{i=1}^{n}$. This matrix equation can be expressed more simply as

$$[T][C] = [E]. \tag{11.23}$$

We would like to reduce $[T]$ to a diagonal matrix since there would then be no need to solve simultaneous equations, and the coefficients c_j could be obtained very easily. This reduction is indeed possible, and we shall now discuss the procedure used to perform it.

We observe that each off-diagonal element in $[T]$ is of the form

$$\sum_{i=1}^{n} T_k(x_i) T_l(x_i),$$

where $k \neq l$. We ask, Do the x_i's in each element meet the following two conditions, that is,

1) do they lie in the interval $(-1, 1)$?

2) do they satisfy Eq. (11.19)?

The answer is clearly negative. In order to meet these two conditions, the given data points $(x_1, y_1), (x_2, y_2), \ldots, (x_n, y_n)$ which lie in the interval (a, b) are first changed to a set of points (x_1', y_1), (x_2', y_2), ..., (x_n', y_n), which lie in the interval $(-1, 1)$, by the following linear transformation:

$$x_i' = \frac{x_i - (b + a)/2}{(b - a)/2} = \frac{2x_i - (b + a)}{b - a}, \qquad i = 1, 2, \ldots, n. \qquad (11.24)$$

We next employ Eq. (10.20), the Lagrange interpolation formula, to obtain $m + 1$ interpolated values for $\bar{y}_i$ which correspond to the values of $\bar{x}_i$ $(i = 1, \ldots, m + 1)$ generated by Eq. (11.19).

For example, if we are to fit the equation $(m = 2)$

$$Y_m(x) = c_0 T_0(x) + c_1 T_1(x) + c_2 T_2(x)$$

to a set of seven $(n = 7)$ data points then the necessary transformation would be as indicated in Table 11.7.

Table 11.7 Relation between x_i, x_i', $\bar{x}_i$, y_i, and $\bar{y}_i$

x_i	x_i'	$\bar{x}_i$	y_i	$\bar{y}_i$
Given	Using Eq. (11.24)	Using Eq. (11.19)	Given	Using Eq. (10.20)
1	-1		2	
		-0.866		3.1
2	-0.667		5	
3	-0.333		7	
4	0	0	8	8.0
5	$+0.333$		10	
6	$+0.667$		9	
		$+0.866$		7.7
7	$+1$		7	

Since the final set of modified $m + 1$ data points $(\bar{x}_1, \bar{y}_1)$, $(\bar{x}_2, \bar{y}_2)$, ..., $(\bar{x}_{m+1}, \bar{y}_{m+1})$ satisfies Eq. (11.16), the orthogonal relation of the Chebyshev polynomials, then all off-diagonal elements of $[T]$ will be equal to zero. In other words,

$$\begin{bmatrix} \sum T_0^2(\bar{x}_i) & 0 & \cdots & 0 \\ 0 & \sum T_1^2(\bar{x}_i) & \cdots & 0 \\ \vdots & & & \\ 0 & 0 & \cdots & \sum T_m^2(\bar{x}_i) \end{bmatrix} \begin{bmatrix} c_0 \\ c_1 \\ \vdots \\ c_m \end{bmatrix} = \begin{bmatrix} \sum \bar{y}_i T_0(\bar{x}_i) \\ \sum \bar{y}_i T_1(\bar{x}_i) \\ \vdots \\ \sum \bar{y}_i T_m(\bar{x}_i) \end{bmatrix}, \qquad (11.25)$$

where Σ implies $\sum_{i=1}^{m+1}$. There is, consequently, no need to solve simultaneous equations, and we can compute by simple division a value for each coefficient c_j,

$$c_j = \frac{\sum_{i=1}^{m+1} \bar{y}_i T_j(\bar{x}_i)}{\sum_{i=1}^{m+1} T_j^2(\bar{x}_i)}, \qquad j = 0, 1, \ldots, m. \tag{11.26}$$

Using the relation (11.17) and (11.18), we obtain

$$c_j = \frac{2}{m+1} \sum_{i=1}^{m+1} \bar{y}_i T_j(\bar{x}_i), \qquad j \neq 0, \tag{11.27}$$

and

$$c_j = \frac{1}{m+1} \sum_{i=1}^{m+1} \bar{y}_i T_j(\bar{x}_i), \qquad j = 0. \tag{11.28}$$

Once we know the coefficients c_j, the Chebyshev expansion of degree m for $Y_m(\bar{x})$ has been completely determined:

$$Y_m(\bar{x}) = \sum_{j=0}^{m} c_j T_j(\bar{x}).$$

The final answer may then be expressed as a Chebyshev series in the interval $(-1, 1)$ or, with the proper transformations, as a power series in the original interval (a, b).

For example, if $m = 2$, the Chebyshev series expansion for Y_m in the interval $(-1, 1)$ is

$$Y_2(\bar{x}) = c_0 T_0(\bar{x}) + c_1 T_1(\bar{x}) + c_2 T_2(\bar{x})$$

$$= (c_0 - c_2) + c_1 \bar{x} + 2c_2 \bar{x}^2.$$

In order to convert the Chebyshev series expansion for Y_2 in the interval $(-1, 1)$ to the interval (a, b), we use the transformation

$$x = \frac{(b-a)\bar{x}}{2} + \frac{(b+a)}{2} \tag{11.29}$$

and obtain $Y_2(x) = A_0 + A_1 x + A_2 x^2$, where

$$A_0 = c_0 - c_2 + c_1 \left(\frac{b+a}{b-a}\right) + 2c_2 \left(\frac{b+a}{b-a}\right)^2,$$

$$A_1 = \frac{2c_1}{b-a} - \frac{8(b+a)c_2}{(b-a)^2},$$

$$A_2 = \frac{8c_2}{(b-a)^2}.$$

At this point, $Y_2(x)$ has been expressed as a power series in the interval (a, b).

Before we turn, in the next section, to the FORTRAN program for curve fitting using Chebyshev polynomials, it is worth while to mention that Chebyshev polynomials are often used to approximate elementary functions,† such as e^x. The approximations using Chebyshev polynomials converge more rapidly than, say, Taylor series, and thus have two practical advantages: (1) the computer time is reduced and (2) error estimation can be made more accurately. This *economization* is the main reason why the Chebyshev-series approach is used in most computer subroutines for the evaluation of elementary functions.

11.7 CHEBYSHEV-POLYNOMIAL CURVE FITTING WITH A COMPUTER

The procedure presented in the previous section is well suited to high-speed digital computation. The following seven steps are involved:

1) Compute the $m + 1$ values of $\bar{x}_i$, where m is the degree of the polynomial Y_m.
2) Normalize the initial values of x_i to the interval $(-1, 1)$.
3) Perform the Lagrangian interpolation to obtain $m + 1$ values of $\bar{y}_i$ which correspond to the $m + 1$ values of $\bar{x}_i$.
4) Compute the coefficients c_j.
5) Convert the Chebyshev series for Y_m to its equivalent power series.
6) Convert the power series from the interval $(-1, 1)$ to the interval (a, b).
7) Print the coefficients of the final series expansion.

A FORTRAN program written to carry out the above seven steps is shown in Fig. 11.7. The input variables for the FORTRAN program are defined as follows:

$\qquad$ M = degree of the polynomial Y_m desired.

$\qquad$ N = number of original data points.

$\qquad$ XMIN = first value of x (smallest value of original x-coordinates).

$\qquad$ DELTX = increment between values of x, that is, $(x_i - x_{i-1})$.

$\qquad$ Y(J) = value of the original y corresponding to the jth value of x.

In the dimension statement we have

$\qquad$ R(I) = the ith root, or $\bar{x}_i$.

$\qquad$ V(I) = the ith value of x_i', or normalized x_i.

$\qquad$ C(I) = the ith coefficient of the Chebyshev series in $(-1, 1)$.

$\qquad$ F(I) = the intermediate storage used in computing interpolated $\bar{y}_i$, in computing C(I)'s ,and in converting C(I)'s to final power-series coefficients in (a, b). The final coefficients are stored in Y(J).

---

† See, for example, Fox, L., and I. B. Parker, *Chebyshev Polynomials in Numerical Analysis.* Oxford University Press, New York, 1968, p. 70.

Example 11.3

Using Chebyshev polynomials, we are to fit a parabola to the following ten data points:

x	y	x	y
1	2	6	11
2	5	7	11
3	7	8	10
4	8	9	9
5	10	10	8

This is the same set of data as that used in Example 11.2 and Table 11.4, except that a tenth point, (2, 5), has been added to make the x-coordinates equispaced. The results of this example may then be compared with those obtained by the normal-equation solution.

Applying the Chebyshev program to this example, the data would be read in the format shown in Fig. 11.6. A possible FORTRAN program for Chebyshev-polynomial curve fitting is shown in Fig. 11.7.

The output for the Chebyshev program gives the coefficients A_j for the polynomial Y_m in terms of the power series:

$$Y_m = \sum_{j=0}^{m} A_j x^j.$$

For our specific example, the series is expressed in the interval (1, 10), and the resulting coefficients are

$$A(0) = -1.5601329, \qquad A(1) = 3.8158582, \qquad A(2) = -0.2909781.$$

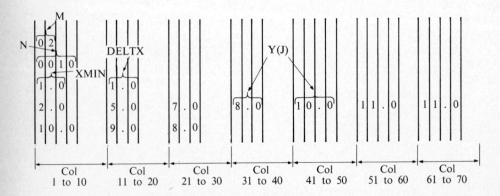

Fig. 11.6 Data cards.

```
C          CHEBYSHEV POLYNOMIAL APPROXIMATION-EQUIDISTANT DATA
           DATA IREAD,IWRITE/5,6/
           DIMENSION R(25), V(100), Y(100), C(25), F(25)
           READ(IREAD,5) M,N,XMIN,DELTX
         5 FORMAT(I2/I4/F10.5,F10.5)
C
C          COMPUTE ROOTS XBAR
C
           M= M+1
           X2=M
           DO 1000 I=1,M
           X1=I
           ARG=1.5707963268*(2.0*X1-1.0)/X2
           MSUB=M+1-I
           R(MSUB)=COS(ARG)
      1000 CONTINUE
C
C          NORMALIZE VECTORS
C
           X1=N-1
           DV=2.0/X1
           V(1)=-1.0
           L=N-1
           DO 1003 I=1,L
           V(I+1)=V(I)+DV
      1003 CONTINUE
           READ(IREAD,10) (Y(J),J=1,N)
        10 FORMAT(7F10.5)
C
C          PERFORM LAGRANGIAN INTERPOLATION
C
           I=1
           DO 150 L=1,N
           IF(R(I).GT.V(L)) GO TO 150
       151 U=(R(I)-V(L-1))/(V(L)-V(L-1))
           IF(L.GT.2) GO TO 155
       154 F(I)=U*(Y(L)-Y(L-1))+Y(L-1)
           GO TO 157
       155 IF(L.GE.N) GO TO 154
       156 ZIP=-U*(U-1.0)*(U-2.)*Y(L-2)/6. + (U*U-1.)*(U-2.)*Y(L-1)/2.
           F(I)=ZIP-(U+1.)*(U-2.)*U*Y(L)/2. +U*(U*U-1.)*Y(L+1)/6.
       157 I=I+1
           IF((M+1).LE.I) GO TO 153
           IF(R(I).LE.V(L)) GO TO 151
       150 CONTINUE
C
C          COMPUTE COEFFICIENTS
C
       153 ZM=M
           DO 280 I=1,M
           SUM=0.0
           IF(I-2) 260,265,270
       260 DO 261 J=1,M
       261 SUM=SUM+F(J)
           GO TO 275
       265 DO 266 J=1,M
       266 SUM=SUM+R(J)*F(J)
           GO TO 275
       270 V(1)=1.0
           DO 272 J=1,M
           V(2)=R(J)
           DO 271 K=3,I
       271 V(K)=2.*R(J)*V(K-1)-V(K-2)
       272 SUM=SUM+F(J)*V(I)
       275 C(I)=2.0*SUM/ZM
       280 CONTINUE
           C(1)=C(1)/2.0
```

(cont.)

Fig. 11.7 FORTRAN program for Chebyshev polynomial solution.

Fig. 11.7 *(concl.)*

```
          ZIP=XMIN
          NN=N-1
          DO 593 J=1,NN
      593 XMIN=XMIN+DELTX
          SUM1=(XMIN+ZIP)/2.
          SUM2=(XMIN-ZIP)/2.
          NN=M-1
          WRITE(IWRITE,305) NN
      305 FORMAT('1',' POLYNOMIAL COEFFICIENTS - DEGREE ',I2//)
    C
    C     CONVERT CHEBYSHEV SERIES TO ITS EQUIVALENT POWER SERIES
    C
          F(1)=C(1)
          F(2)=C(2)
          IF(M.LE.2) GO TO 597
      596 DO 598 K=1,M
          V(K)=0.0
          Y(K)=0.0
      598 F(K+2)=0.0
          V(2)=1.0
          DO 599 K=3,M
          ZIP=K-1
          P=3.14159265*ZIP/2.0
          Y(1)=COS(P)
          DO 594 J=2,K
      594 Y(J)=2.*V(J-1)-Y(J)
          DO 595 J=1,K
          F(J)=F(J)+C(K)*Y(J)
          ZIP=V(J)
          V(J)=Y(J)
      595 Y(J)=ZIP
      599 CONTINUE
    C
    C     GO BACK TO ORIGINAL INTERVAL
    C
      597 Y(1)=F(1)
          DO 580 K=2,M
      580 Y(K)=0.0
          DO 583 K=2,M
          L=K-1
          Y(K)=Y(K)+F(K)/SUM2**L
          SUM3=1.0
          ZIP=1.0
          X2=K
          DO 582 J=1,L
          X1=J
          SUM3=SUM3*X1
          DV=SUM3*SUM2**L
          ZIP=ZIP*(X2-X1)
      582 Y(K-J)=Y(K-J)+(ZIP*SUM1**J*(-1.)**J*F(K))/DV
      583 CONTINUE
    C
    C     COEFFICIENTS STORED IN Y(K)
    C
          DO 584 K=1,M
          I=K-1
      584 WRITE(IWRITE,585) I,Y(K)
      585 FORMAT(30X,'A(',I2,') =',E16.8)
          STOP
          END


    2
     10
      1.0         1.0
      2.0         5.0         7.0         8.0        10.0        11.0        11.0
     10.0         9.0         8.0
```

Using the same set of data points with the normal-equation approach, we obtain the coefficients

$$A(0) = -1.2999683, \qquad A(1) = 3.5651364, \qquad A(2) = -0.2651502.$$

Although the two sets of coefficients appear to differ considerably, an evaluation of the polynomial Y_m at each of the x-coordinates with both sets of coefficients indicates that the curve fits are remarkably close. This similarity can be seen in Fig. 11.8.

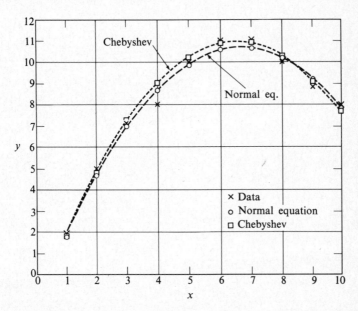

Fig. 11.8 Normal equation solution vs. Chebyshev polynomial solution.

BIBLIOGRAPHY

General References

Andersen, R. S., and M. R. Osborne (eds.), *Least Square Methods in Data Analysis.* The Australian National University, Canberra, Australia, Computer Centre Publication, 1969.

Ascher, M., and G. A. Forsythe, "SWAC experiments on the use of orthogonal polynomials for data fitting," *J. Assoc. Comput. Mach.*, **5** (1958).

Berztiss, A. T., "Least squares fitting of polynomials to irregularly spaced data," *SIAM Rev.*, **6**, no. 3, 203–227 (1964).

Birkhoff, G., and C. de Boor, "Piecewise polynomial surface fitting," *Proceedings, General Motors Research Laboratories Symposium: "Approximation of Functions."* Elsevier, Amsterdam, 1965.

Blum, E. K., "Polynomial approximation," *U.S. Naval Ordinance Laboratory Report 3740* (1956).

Braess, D., Über Daempfung bei Minimalisierungsverfahren," *Comput.*, **1**, no. 3, 264–272 (1966).

Clenshaw, C. W., "Curve fitting with a digital computer," *Comput. J.*, **2**, 170–173 (1960).

Clenshaw, C. W., "Polynomial approximations to elementary functions," *Math. Tables Aids Comput.*, **8**, 143–147 (1954).

Forsythe, G. E., "Generation and use of orthogonal polynomials for data fitting with a digital computer," *J. Soc. Ind. Appl. Math.*, **5**, 74–88 (1957).

Golub, G., "Numerical method for solving linear least squares problems," *Num. Math.*, **7**, 206–216 (1965).

Greville, T. N. E., "Numerical procedures for interpolation by spline functions," *J. SIAM, Num. Anal. Ser. B.*, **1**, 53–68 (1964).

Handscomb, D. C. (ed.), *Methods of Numerical Approximation.* Pergamon Press, New York, 1966.

Hayes, J. G., and T. Vickers, "The fitting of polynomials to unequally spaced data," *Phil. Mag.*, **42**, 1387–1400 (1951).

Lanczos, C., *Approximations by Orthogonal Polynomials.* University of California, Los Angeles, 1952.

Marquardt, D. W., "An algorithm for least square estimation of non-linear parameters," *J. SIAM*, **2**, 431–441 (1963).

von Holdt, R. E., and R. J. Brousseau, *Weighted Least-Squares Polynomial Approximation to a Continuous Function of a Single Value.* University of California Radiation Laboratory, Livermore, 1956.

Chebyshev Polynomials

Clenshaw, C. W., "A note on the summation of Chebyshev series," *Math. Tab., Wash.*, **9**, 118–120 (1955).

Lanczos, C., Introduction to "Tables of Chebyshev polynomials," *Natn. Bur. Stand. Appl. Math. Ser.*, **9** (1952); *Applied Analysis.* Prentice-Hall, Englewood Cliffs, N.J., 1956.

Murnaghan, F. D., and J. W. Wrench, "The determination of the Chebyshev approximating polynomial for a differentiable function," *Math. Tab., Wash.*, **13**, 185–193 (1959).

PROBLEMS

1. The number of graduate students enrolled in the College of Engineering of a major eastern university for a period of 11 years is tabulated below:

t	N	t	N
1952	920	1958	1260
1953	970	1959	1260
1954	940	1960	1325
1955	1000	1961	1350
1956	1100	1962	1360
1957	1180		

Fit these points to the polynomials

$$N = a_0 + a_1 t + a_2 t^2 + a_3 t^3,$$

by (a) normal equation method and (b) Chebyshev polynomials.

2. An oxyacetylene torch is used to cut a one-inch piece of metal of varying thickness. The following data are given.

Metal thickness, in. (t)	Time to cut 1 in., min (T)	Metal thickness, in. (t)	Time to cut 1 in., min (T)
$\frac{1}{4}$	0.036	$3\frac{1}{2}$	0.077
$\frac{3}{8}$	0.037	4	0.084
$\frac{1}{2}$	0.039	$4\frac{1}{2}$	0.091
$\frac{3}{4}$	0.042	5	0.100
1	0.046	$5\frac{1}{2}$	0.106
$1\frac{1}{4}$	0.050	6	0.111
$1\frac{1}{2}$	0.053	$6\frac{1}{2}$	0.118
2	0.059	7	0.125
$2\frac{1}{2}$	0.065	$7\frac{1}{2}$	0.134
3	0.072	8	0.143

Find coefficients for a least-squares curve of the form

(a) $T = c_0 + c_1 t$,

(b) $T = c_0 + c_1 t + c_2 t^2$,

(c) $T = c_0 + c_1 t + c_2 t^2 + c_3 t^3$.

3. A cooling fin is placed in an air stream and the center of the fin is heated by a heater to simulate an air-cooled engine cylinder. Twelve thermocouples are placed at equal distances radially on the fin for temperature measurements. During one test run, the following data were taken:

Radius, in. (R)	Temperature, °F (T)	Radius, in. (R)	Temperature, °F (T)
1	138.0	7	148.1
2	138.8	8	150.2
3	139.7	9	152.3
4	141.5	10	152.9
5	143.4	11	153.5
6	145.6	12	153.8

Fit the data into the following curve:

$$T = c_0 + c_1 R + c_2 R^2 + c_3 R^3.$$

Use normal equations and Chebyshev polynomials.

CHAPTER 12

Numerical Integration

12.1 PRELIMINARY REMARKS

Two facts which relate to the evaluation of a definite integral are not commonly recognized:

1) An integral does not always have its closed-form expression.
2) Even when this closed-form expression is available, it is sometimes preferable to compute the integral by numerical methods.

For example, consider the integral

$$\int_0^x \sqrt{ax^2 + b}\ x^2\ dx = \frac{x}{4a}(ax^2 + b)^{3/2} - \frac{bx}{8a}\sqrt{ax^2 + b}$$

$$- \frac{b^2}{8a\sqrt{a}}\ln(x\sqrt{a} + \sqrt{ax^2 + b}), \qquad \text{when } a \geq 0;$$

$$= \frac{x}{4a}(ax^2 + b)^{3/2} - \frac{bx}{8a}\sqrt{ax^2 + b}$$

$$- \frac{b^2}{8a\sqrt{-a}}\sin^{-1}\left(x\sqrt{\frac{-a}{b}}\right), \qquad \text{when } a < 0.$$

When we are to compute the values corresponding to a large number of values of x, it is often advantageous to make a numerical evaluation of the integrals. Numerical integration is often known as *quadrature*.

In this chapter, we shall discuss two widely used procedures: the Romberg method and the Gaussian quadrature. The algorithm used in the Romberg method is simple and efficient; it calculates the approximate values of increasing order *recursively*. For the sake of discussion, we shall start this topic with the well-known trapezoidal and Simpson's rules. As will be seen later, these two rules actually are two special cases in a class of quadrature formulas known as the Newton-Cotes closed-end formulas. However, the Romberg method is not a reformulation of the Newton-Cotes formulas.

We shall review the trapezoidal and Simpson's rules in the next two sections. For easy adaptation to modern computers, the particular emphasis is on the repeated halving of a given original width. Thus the integration interval is divided into 1, 2, 4, 8, 16, ... equal parts.

12.2 TRAPEZOIDAL RULE

Figure 12.1 shows the area bounded by the curve $y = f(x)$, the x-axis and the ordinates erected at $x = a$ and $x = b$. If we divide the interval from $x = a$ to $x = b$ into any number of equal subintervals and inscribe the same number of trapezoids within the area, the sum of trapezoidal areas will yield a good approximation to the area under the curve. As mentioned before, we shall be concerned with 1, 2, 4, 8, 16, ..., (not 1, 2, 3, 4, 5, ...) equal subdivisions. For example, Fig. 12.2(a) shows that only one trapezoid is used; its area (shaded) is

$$T^{(0)} = \frac{(f_0 + f_1)l}{2}, \tag{12.1}$$

where f_0 and f_1 are values of the function $f(x)$ evaluated at $x = a$ and $x = b$ respectively.

Better approximations to the area under the curve can be obtained if the interval (a, b) is divided into 2, 4, or 8 equal subintervals, erect ordinates at these points of

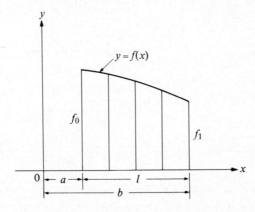

Fig. 12.1. A number of trapezoids are used to approximate the area under curve $y = f(x)$.

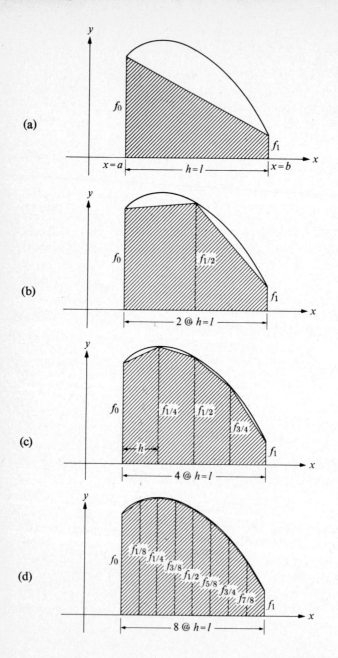

Fig. 12.2. Approximate areas $T^{(0)}$, $T^{(1)}$, $T^{(2)}$, and $T^{(3)}$.

division, and complete trapezoids as shown in Figs. 12.2(b), 12.2(c) and 12.2(d). The shaded areas $T^{(1)}$, $T^{(2)}$ and $T^{(3)}$ are respectively

$$T^{(1)} = \frac{(f_0 + 2f_{1/2} + f_1)l}{4}, \tag{12.2}$$

$$T^{(2)} = \frac{(f_0 + 2f_{1/4} + 2f_{1/2} + 2f_{3/4} + f_1)l}{8}, \tag{12.3}$$

$$T^{(3)} = \frac{(f_0 + 2f_{1/8} + 2f_{1/4} + 2f_{3/8} + 2f_{1/2} + 2f_{5/8} + 2f_{3/4} + 2f_{7/8} + f_1)l}{16}. \tag{12.4}$$

The superscript associated with each T indicates the number of subdivisions used. For example, the superscript 0 indicates that the full interval is used; 1, that two equal subdivisions are used; 2, that the full interval is subdivided into 4 equal parts; and so on. In general

$$h = \frac{l}{2^k}, \tag{12.5}$$

where

$\quad\quad h =$ width of each subdivision,

$\quad\quad k =$ superscript,

$\quad\quad l =$ interval $= b - a$. (see Figs. 12.1 and 12.2),

and

$$T^{(k)} = \frac{h}{2} \left\{ f(a) + \left[2 \sum_{j=1}^{2^k - 1} f(a + jh) \right] + f(b) \right\}. \tag{12.6}$$

As a computational detail, Eqs. (12.2) through (12.4) can be rewritten as follows:

$$T^{(1)} = \tfrac{1}{2}[T^{(0)} + f_{1/2}l], \tag{12.7}$$

$$T^{(2)} = \frac{1}{2} \left[T^{(1)} + \frac{(f_{1/4} + f_{3/4})l}{2} \right], \tag{12.8}$$

$$T^{(3)} = \frac{1}{2} \left[T^{(2)} + \frac{(f_{1/8} + f_{3/8} + f_{5/8} + f_{7/8})l}{4} \right]. \tag{12.9}$$

In general,

$$T^{(k)} = \frac{1}{2} \left[T^{(k-1)} + \frac{\sum\limits_{j=1,3,5,\ldots}^{2^k} \left(a + \frac{j}{2^k} l \right)}{2^{k-1}} l \right] \tag{12.10}$$

$$= \tfrac{1}{2} T^{(k-1)} + \frac{l}{2^k} \sum_{j=1}^{2^{(k-1)}} f \left[a + \frac{(2j-1)l}{2^k} \right]. \tag{12.10a}$$

Example

Approximate the following integral, using the trapezoidal rule with 1, 2, 4, 8, and 16 equal subdivisions (see Fig. 12.3):

$$\int_1^2 \frac{1}{x}\, dx$$

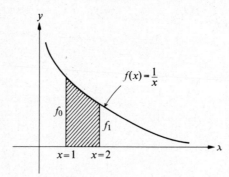

Fig. 12.3. $f(x) = 1/x$.

Solution

Here $f(x) = 1/x$. Considering the full interval as a single subdivision, or $k = 0$,

$$f_0 = f(1) = 1,$$
$$f_1 = f(2) = \tfrac{1}{2}.$$

From Eq. (12.1), we have

$$T^{(0)} = \frac{1 + 0.5}{2} \times (2 - 1) = 0.75.$$

For two subdivisions, $(k = 1)$

$$f_{1/2} = f(1.5) = \frac{1}{1.5}.$$

From Eq. (12.7), we obtain

$$T^{(1)} = \frac{1}{2}\left[0.75 + \frac{1}{1.5}\right] = 0.70833.$$

If we divide the original interval into four equal parts, or $k = 2$, we first calculate

$$f_{1/4} = f(1.25) = \frac{1}{1.25}$$

and

$$f_{3/4} = f(1.75) = \frac{1}{1.75},$$

substituting in Eq. (12.8)

$$T^{(2)} = \frac{1}{2} \left[0.7083 + \frac{(1/1.25) + (1/1.75)}{2} \right] = 0.69702.$$

Similarly, if eight subdivisions ($k = 3$) are used, we need the following values:

$$f_{1/8} = f(1.125) = \frac{1}{1.125}$$

$$f_{3/8} = f(1.375) = \frac{1}{1.375}$$

$$f_{5/8} = f(1.625) = \frac{1}{1.625}$$

$$f_{7/8} = f(1.875) = \frac{1}{1.875}$$

Substituting in Eq. (12.9),

$$T^{(3)} = \frac{1}{2} \left[0.6970 + \frac{\dfrac{1}{1.125} + \dfrac{1}{1.375} + \dfrac{1}{1.625} + \dfrac{1}{1.875}}{4} \right] = 0.69412.$$

Continuing in this manner, it can be shown that $T^{(4)} = 0.69339$. We summarize values of $T^{(0)}$ through $T^{(4)}$ in Table 12.1.

Table 12.1 $\int_1^2 (1/x)\, dx$ by trapezoidal rule using various numbers of equal subdivisions

Number of equal subdivisions	k	$T^{(k)}$ (Trapezoidal rule)	Closed-form solution
1	0	$T^{(0)} = 0.75000$	
2	1	$T^{(1)} = 0.70833$	
4	2	$T^{(2)} = 0.69702$	0.69314
8	3	$T^{(3)} = 0.69412$	
16	4	$T^{(4)} = 0.69339$	

From the closed form solution, it is known that the correct answer is equal to

$$\int_1^2 \frac{1}{x}\, dx = \log_e 2 = 0.6931.$$

Before we turn to Simpson's rule in the next section, it is worth while to note that the shaded areas $T^{(0)}$, $T^{(1)}$, $T^{(2)}$ and $T^{(3)}$ in Fig. 12.2 are each equal to the area under the curve $y = f(x)$, if the given intergrand $f(x)$ represents a straight line, or $f(x) = c_0 + c_1 x$ where c_0 and c_1 are coefficients.

12.3 SIMPSON'S RULE

Instead of plotting straight lines, or chords, between the points of a curve and forming trapezoids, one can get a much closer approximation to the area by connecting the points with segments of parabolas. As shown in Fig. 12.4(a) a parabola with a vertical axis may be passed through three points P_0, P_1 and P_2 on a curve $y = f(x)$; this para-

(a)

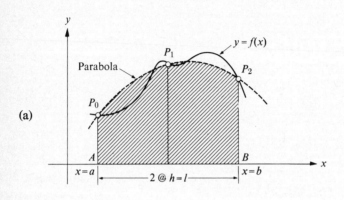

(b)

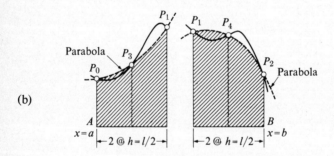

(c)

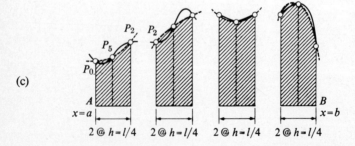

Fig. 12.4. Simpson's rule (parabolic rule).

bola segment will fit the curve more closely than the straight line. According to the well-known Simpson's rule,† the shaded area of the parabolic strip $AP_0P_1P_2B$ is

$$S^{(0)} = \frac{h}{3} [f_0 + 4f_{1/2} + f_1],$$ (12.11)

where $h = l/2$.

Similarly, two parabola segments will fit the given curve more closely as shown in Fig. 12.4(b). The area of the left parabolic strip

$$S^{(0)L} = \frac{l/4}{3} (f_0 + 4f_{1/4} + f_{1/2}),$$

and that of the right strip

$$S^{(0)R} = \frac{l/4}{3} (f_{1/2} + 4f_{3/4} + f_1).$$

Adding, we get

$$S^{(1)} = S^{(0)R} + S^{(0)L} = \frac{l/4}{3} (f_0 + 4f_{1/4} + 2f_{1/2} + 4f_{3/4} + f_1).$$ (12.12)

If we halve the interval once more, we would have four equal parts, each $l/4$ wide, as shown in Fig. 12.4(c). Each dotted curve is the graph of a parabola that agrees with the given function $f(x)$ at three points, such as P_0, P_5 and P_2 in the interval $(a, a + l/4)$. The total shaded area is

$$S^{(2)} = \frac{l/8}{3} (f_0 + 4f_{1/8} + 2f_{1/4} + 4f_{3/8} + \cdots + 4f_{7/8} + f_1).$$ (12.13)

The general rule for the approximate area is called the *extended Simpson's rule*

$$S^{(k)} = \frac{h}{3} \left\{ [f(a) + f(b)] + 4 \sum_{j=1}^{2^k} f[a + (2j - 1)h] + 2 \sum_{j=1}^{2^k-1} f(a + 2jh) \right\},$$ (12.14)

where $h = l/2^{k+1}$.

As with the $T^{(k)}$ based on the trapezoidal rule, the superscript k in each S shows how many interpolating parabola segments are used. When $k = 0$, one parabola is used; $k = 1$, two parabolas are used. In general, the number of the interpolating parabola segments used is 2^k.

Before we turn to the next section, let us examine the relationship between the approximate areas using Trapezoidal rule $T^{(k)}$ and those from Simpson's rules $S^{(k)}$. From Eqs. (12.1), (12.2) and (12.11), we have

$$S^{(0)} = \frac{4T^{(1)} - T^{(0)}}{3}.$$ (12.15)

---

† For example, See Thomas, G. B. *Calculus and Analytic Geometry*, 3rd ed. Addison-Wesley, Reading, Mass., 1960, p. 422.

Likewise, from Eqs. (12.2), (12.3) and 12.12), we have

$$S^{(1)} = \frac{4T^{(2)} - T^{(1)}}{3}. \tag{12.16}$$

Similarly, the relationship between $S^{(2)}$, $T^{(2)}$, and $T^{(3)}$ can be obtained from Eqs. (12.3), (12.4), and (12.13) as follows:

$$S^{(2)} = \frac{4T^{(3)} - T^{(2)}}{3}. \tag{12.17}$$

In fact, we have the general relation

$$S^{(k)} = \frac{4T^{(k+1)} - T^{(k)}}{3}. \tag{12.18}$$

Example

Approximate the following integral using Simpson's rule:

$$\int_1^2 \frac{1}{x} \, dx$$

Compute $S^{(0)}$, $S^{(1)}$, $S^{(2)}$, and $S^{(3)}$.

Solution: The approximate values of this integral using the trapezoidal rules were calculated in the previous section, they are $T^{(0)}$, $T^{(1)}$, $T^{(2)}$, ... in Table 12.1. Using Eqs. (12.15) through (12.18), we get

$$S^{(0)} = \frac{4 \times 0.70833 - 0.75}{3} = 0.69444,$$

$$S^{(1)} = \frac{4 \times 0.69702 - 0.70833}{3} = 0.69325,$$

$$S^{(2)} = \frac{4 \times 0.69412 - 0.69702}{3} = 0.69315,$$

$$S^{(3)} = \frac{4 \times 0.69339 - 0.69412}{3} = 0.693148.$$

12.4 ROMBERG'S TABLE

In the previous section, we pointed out the connection between the approximate areas based on Simpson's rule and those based on Trapezoidal rule. Equation (12.15) is an example. When $T^{(0)}$ and $T^{(1)}$ are given, $S^{(0)}$ can be readily computed; this relationship is conveniently represented by the three elements shown in Fig. 12.5. The basic relationship can also be extended to include $T^{(0)}$, $T^{(1)}$, $T^{(2)}$, $T^{(3)}$ ... and $S^{(0)}$, $S^{(1)}$, $S^{(2)}$ as illustrated in Fig. 12.6. For example, when $T^{(1)}$ and $T^{(2)}$ are known, $S^{(1)}$ can be calculated from Eq. (12.16).

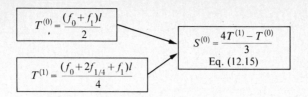

Fig. 12.5. $S^{(0)}$ is computed from $T^{(0)}$ and $T^{(1)}$.

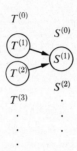

Fig. 12.6. $S^{(1)}$ can be computed from the value of $T^{(1)}$ and $T^{(2)}$.

It is important to note that Fig. 12.6 actually represents the leftmost two columns of the so-called Romberg table, which is designed to evaluate the definite integrals. In this section, we shall be primarily concerned with the procedures and some typical examples illustrating how the Romberg table is constructed. The detailed derivations are given in Section 12.7.

In Table 12.2 is shown a Romberg table. The elements $T^{(0)}$, $T^{(1)}$, $T^{(2)}$, ... in column 1 may be computed from the trapezoidal rule. The elements in the second column, i.e., $S^{(0)}$, $S^{(1)}$, $S^{(2)}$, ... can then be calculated directly using Eqs. (12.15) through (12.18). It is not necessary to use Simpson's rule, i.e., Eqs. (12.11) through (12.14). As soon as the values of $S^{(0)}$, $S^{(1)}$, $S^{(2)}$ ... are known, it is possible to compute $C^{(0)}$, $C^{(1)}$ ... in column 3, and so on. In this step-by-step manner, construction of the table is continued until a cluster of elements in the lower right has practically the same value.

We shall illustrate the construction of the Romberg table for the approximate evaluation of the following definite integral:

$$I = \int_1^2 \frac{1}{x}\, dx$$

The values based on both the trapezoidal and Simpson's rules have already been calculated in the previous two sections. Thus the values $T^{(0)}$, $T^{(1)}$, ..., as shown in column 1 of Table 12.3 are taken directly from Table 12.1 except that they are carried out to more digits. Likewise, the values of $S^{(0)}$, $S^{(1)}$, ... shown in column 2 in Table 12.3 are taken from the example in Section 12.3; the values in columns 3, 4, and 5 are

Table 12.2 Romberg's table

Row	(1)	(2)	(3)	(4)	(5)
1	$T^{(0)} = \dfrac{f_0 + f_1}{2}\,l$ Eq. (12.1)				
2		$S^{(0)} = \dfrac{4T^{(1)} - T^{(0)}}{4 - 1}$			
3	$T^{(1)}$ Use Eq. (12.7)		$C^{(0)} = \dfrac{4^2 S^{(1)} - S^{(0)}}{4^2 - 1}$		
4		$S^{(1)} = \dfrac{4T^{(2)} - T^{(1)}}{4 - 1}$		$D^{(0)} = \dfrac{4^3 C^{(1)} - C^{(0)}}{4^3 - 1}$	
5	$T^{(2)}$ Use Eq. (12.8)		$C^{(1)} = \dfrac{4^2 S^{(2)} - S^{(1)}}{4^2 - 1}$		$E^{(0)} = \dfrac{4^4 D^{(1)} - D^{(0)}}{4^4 - 1}$
6		$S^{(2)} = \dfrac{4T^{(3)} - T^{(2)}}{4 - 1}$		$D^{(1)} = \dfrac{4^3 C^{(2)} - C^{(1)}}{4^3 - 1}$	
7	$T^{(3)}$ Use Eq. (12.9)		$C^{(2)} = \dfrac{4^2 S^{(3)} - S^{(2)}}{4^2 - 1}$		
8		$S^{(3)} = \dfrac{4T^{(4)} - T^{(3)}}{4 - 1}$			
9	$T^{(4)}$ Use Eq. (12.10)				

Table 12.3   Romberg's table for $\int_1^2 dx/x$

Column \ Row	(1)	(2)	(3)	(4)	(5)
1	$T^{(0)} = 0.75$				
2		$S^{(0)} = 0.69444444$			
3	$T^{(1)} = 0.708333333$		$C^{(0)} = \frac{1}{15}(16 \times 0.693253967 - 0.6944444) = 0.693174603$		
4		$S^{(1)} = 0.693253967$		$D^{(0)} = \frac{1}{63}(64 \times 0.693147901 - 0.693174603) = 0.693147479$	
5	$T^{(2)} = 0.697023809$		$C^{(1)} = \frac{1}{15}(16 \times 0.693154532 - 0.693253967) = 0.693147901$		$E^{(0)} = \frac{1}{255}(256 \times 0.693147182 - 0.693147479) = 0.693147181$
6		$S^{(2)} = 0.693154532$		$D^{(1)} = \frac{1}{63}(64 \times 0.693147193 - 0.693147901) = 0.693147182$	
7	$T^{(3)} = 0.694121851$		$C^{(2)} = \frac{1}{15}(16 \times 0.693147652 - 0.693154543) = 0.693147193$		
8		$S^{(3)} = 0.693147652$			
9	$T^{(4)} = 0.693391202$				

calculated using the equations shown in Table 12.2. It should be noted that, in Table 12.3, the value of $E^{(0)}$, or 0.693147181, already agrees well with the exact solution $\log_e 2 = 0.693147180$.

12.5 FORTRAN PROGRAM FOR ROMBERG'S QUADRATURE

In this section we shall be concerned with a FORTRAN program which generates the elements in a Romberg table. Toward this end, it is convenient to rename each element in Table 12.2 as $A_m^{(n)}$, where A_1's represent the elements in column 1 which are the values $T^{(0)}, T^{(1)}, \ldots$ computed from the trapezoidal rule; A_2's represent the elements $S^{(0)}, S^{(1)}, \ldots$ in column 2; and so on. These elements are now rearranged in Table 12.4. Each element $A_m^{(n)}$ is obtained by using the following two important formulas:

$$A_m^{(n)} = \frac{4^{m-1} A_{m-1}^{(n+1)} - A_{m-1}^{(n)}}{4^{m-1} - 1} \qquad (m \neq 1), \qquad (12.19\text{a})$$

and, from Eq. (12.10a),

$$A_1^{(n)} = \tfrac{1}{2} A_1^{(n-1)} + \frac{l}{2^{n-1}} \sum_{j=1}^{2^{n-2}} f\left[a + \frac{(2j-1)l}{2^{n-1}}\right], \qquad (12.19\text{b})$$

where $n = 1, 2, 3, \ldots$.

As shown in Fig. 12.7, the element $A_m^{(n)}$ can be readily computed, if the values of elements $A_{m-1}^{(n)}$ and $A_{m-1}^{(n+1)}$ are known.

Table 12.4 Romberg's table

$A_1^{(1)} = T^{(0)}$				
	$A_2^{(1)} = S^{(0)}$			
$A_1^{(2)} = T^{(1)}$		$A_3^{(1)} = C^{(0)}$		
	$A_2^{(2)} = S^{(1)}$		$A_4^{(1)} = D^{(0)}$	
$A_1^{(3)} = T^{(2)}$		$A_3^{(2)} = C^{(1)}$		$A_5^{(1)} = E^{(0)}$
	$A_2^{(3)} = S^{(2)}$		$A_4^{(2)} = D^{(1)}$	
$A_1^{(4)} = T^{(3)}$		$A_3^{(3)} = C^{(2)}$		
	$A_2^{(4)} = S^{(3)}$			
$A_1^{(5)} = T^{(4)}$				

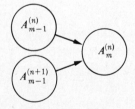

Fig. 12.7. Three-element graph to show the value of $A_m^{(n)}$ from $A_{m-1}^{(n)}$ and $A_{m-1}^{(n+1)}$.

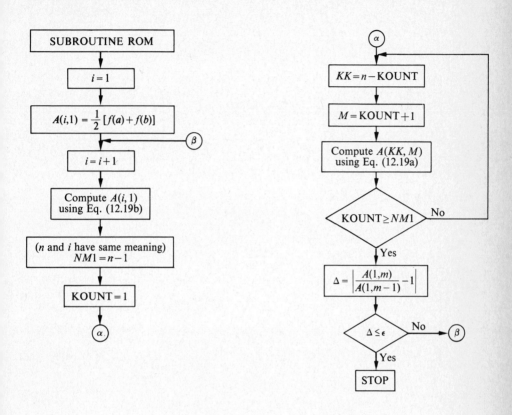

Fig. 12.8. Flow chart for subroutine ROM.

A flowchart and a subroutine subprogram ROM to generate the elements $A_m^{(n)}$ for $n = 1, 2, 3, \ldots$ and $m = 1, 2, 3, \ldots$ are shown in Figs. 12.8 and 12.9 respectively. The arguments used in the subroutine ROM are defined as follows:

ALIMIT = lower limit of integration

BLIMIT = upper limit of integration

          (L = BLIMIT − ALIMIT)

   EPS = ϵ, a preassigned small quantity

   ANS = the required answer

FUNCT = function describing the integrand $f(x)$.

```
         SUBROUTINE ROM (ALIMIT, BLIMIT, EPS, ANS, FUNCT, A, ATEMP)
C        USER MUST CODE FUNCTION SUBPROGRAM FUNCT(X)
         COMMON N
         EQUIVALANCE (I,N)
         I=1
C        THE UPWARD DIAGONAL ELEMENTS ARE COMPUTED AND STORED IN ATEMP.
         DIMENSION ATEMP(12)
C        ARRAY A IS USED ONLY FOR PRINTING A COMPLETE TABLE.  NEXT NO. IS
C 510
         DIMENSION A(12,12)
         ATEMP(1)=0.5*(FUNCT(BLIMIT)+FUNCT(ALIMIT))
   520 A(I,1)=ATEMP(1)
C        INTERVAL (WIDTH) MUST NOT BE ZERO.
         WIDTH=BLIMIT-ALIMIT
         ZL=WIDTH
C        POWER USED TO COMPUTE 2**(I-1)
C        JJ USED TO CONTROL DO LOOP 12 BELOW.
         POWER=1.
         JJ=1
C        BEGIN THE OUTERMOST LOOP
   711 I=I+1
         ANS=ATEMP(1)
C        FOLLOWING 10 STATEMENTS COMPUTE A NEW A(I,1) BY TRAPEZOIDAL RULE.
C          EQ.(12.19B).
         TEMPL=ZL
         ZL=0.5*ZL
         POWER=0.5*POWER
         X=ALIMIT+ZL
         SUM=0.
         DO 12 JCOUNT=1,JJ
         SUM=SUM+FUNCT(X)
    12 X=X+TEMPL
         ATEMP(I)=0.5*ATEMP(I-1)+SUM*POWER
   530 A(I,1)=ATEMP(I)
C        NEXT 9 STATEMENTS USED TO CALCULATE A(N,M) BY EQ.(12.19A).
C          M IS NOT EQUAL TO 1.
         R=1.
         NM1=N-1
         DO 16 KOUNT=1,NM1
         KK=N-KOUNT
         R=R+R
         R=R+R
         ATEMP(KK)=ATEMP(KK+1)+(ATEMP(KK+1)-ATEMP(KK))/(R-1.)
         M=KOUNT+1
   540 A(KK,M)=ATEMP(KK)
    16 CONTINUE
         DELTA=ABS((ATEMP(1)/ANS)-1.)
         IF(DELTA-EPS)40,40,30
    30 JJ=JJ+JJ
         GO TO 711
    40 ANS=ATEMP(1)*WIDTH
   550 DO 580 IROW=1,N
   560   NN=N+1-IROW
   570 DO 580 ICOL=1,NN
   580 A(IROW,ICOL)=A(IROW,ICOL)*WIDTH
         RETURN
         END
```

Fig. 12.9. Subroutine ROM.

The elements $A_m^{(n)}$ are generated in the upward diagonal direction in the following manner:

Step 1. Compute $A_1^{(1)}$;
Step 2. Compute $A_1^{(2)}$, $A_2^{(1)}$;
Step 3. Compute $A_1^{(3)}$, $A_2^{(2)}$, $A_3^{(1)}$;
Step 4. Compute $A_1^{(4)}$, $A_2^{(3)}$, $A_3^{(2)}$, $A_4^{(1)}$; and so on.

The calculation stops when the following condition is met:

$$\left| \frac{A_m^{(1)}}{A_{m-1}^{(1)}} - 1 \right| < \epsilon,$$

where ϵ is a preassigned number such as 10^{-6}; $A_m^{(1)}$ is then taken as the approximate value of a given definite integral.

In Subroutine ROM, the values of $A_m^{(n)}$ in each step (along the upward diagonal direction), such as $A_1^{(4)}$, $A_2^{(3)}$, $A_3^{(2)}$, and $A_4^{(1)}$, are stored in a vector (one-dimensional array) ATEMP:

$$\begin{array}{ll} A_1^{(n)} & \text{is stored in ATEMP(N)} \\ A_1^{(n-1)} & \text{is stored in ATEMP(N-1),} \\ & \vdots \\ A_n^{(1)} & \text{is stored in ATEMP(1).} \end{array}$$

It should be noted that the array A(N,M) specified in Subroutine ROM is used to store the values of the vector ATEMP right after each element of this vector is computed. The purpose of the array A(N,M) is twofold: forming the Romberg table, and later printing out the table. If this table is not required, the eight statements with statement numbers 510, 520, 530, ..., 580 can be removed from the subroutine ROM. The mathematical symbol $A_m^{(n)}$ is written in the subroutine as A(N,M), not as A(M,N).

There are a number of practical programming details worth mentioning in connection with subroutine ROM. First, the variable names I and N both represent the same quantity: the row number in the Romberg table. This is due to, say, lack of communication between two programmers working on the same program. Since the names I and N have each been used many times, it would be error-prone and expensive to change them. A single statement:

```
EQUIVALENCE (I,N)
```

will resolve this difficulty.

Second, the quantity $R = 4^k$ ($k = 1, 2, \ldots$) is not coded as 4**K. Rather, it is computed as follows:

```
      R = 1.
      DO 16 K=1,LAST
      R = R + R
      R = R + R
   16 CONTINUE
```

Example 12.1

Write a main program to call the subroutine ROM for the following definite integral

$$I = \int_1^2 \frac{1}{x}\, dx.$$

The input data card is for ALIMIT, BLIMIT, and EPS. A sample output is shown in Table 12.5, where the Romberg table is printed first, followed by the approximate value of the definite integral. It is essential that the size of the matrix $A(n, m)$ agrees

Table 12.5.

```
          LOWER LIMIT =    1.00      UPPER LIMIT =    2.00

                         ROMBERG TABLE ROW   1

     .7500000   .6944444   .6931744   .6931471   .6931468

                         ROMBERG TABLE ROW   2

     .7083333   .6932539   .6931476   .6931468

                         ROMBERG TABLE ROW   3

     .6970237   .6931543   .6931468

                         ROMBERG TABLE ROW   4

     .6941217   .6931473

                         ROMBERG TABLE ROW   5

     .6933907


               ANSWER =   .6931468
```

```
C       MAIN PROGRAM.   THE CALL STATEMENT MUST BE PRECEDED BY AN EXTERNAL
C       STATEMENT LISTING THE NAME OF THE FUNCTION SUBPROGRAM.
        COMMON N
C        READ LOWER AND UPPER LIMITS, AND EPSILON, A SMALL NUMBER.
         READ (5,10) XL,XU,EPS
   10 FORMAT(2F10.4,F10.6)
         DIMENSION A(12,12),ATEMP(12)
         EXTERNAL ONEX
         CALL ROM(XL,XU,EPS,ANS,ONEX,A,ATEMP)
         WRITE(6,20) XL,XU
   20 FORMAT('1',' LOWER LIMIT = ',F7.2,5X,' UPPER LIMIT = ',F7.2/)
         DO 21 IROW=1,N
         NN=N+1-IROW
         WRITE(6,22) IROW
   21 WRITE(6,24) (A(IROW,ICOL),ICOL=1,NN)
   22 FORMAT(16X,'ROMBERG TABLE ROW ',I2/)
   24 FORMAT(2X,7F10.7/)
         WRITE(6,25) ANS
   25 FORMAT(3X,'ANSWER = ',F10.7)
         STOP
         END
```

Fig. 12.10. Main Program.

with that in the subroutine.   Fig. 12.10 shows a possible main program.   In the DIMENSION statement, the matrix is arbitrarily dimensioned as 12×12.

The subroutine ROM "calls in" the integrand by a "function name call."   An external function defining the integrand must be supplied by the user.   Fig. 12.11 shows a possible function ONEX.

```
FUNCTION ONEX(X)
ONEX=1./X
RETURN
END
```

Fig. 12.11.   Function ONEX.

12.6 GEOMETRIC INTERPRETATION OF ELEMENTS IN ROMBERG'S TABLE

In Section 12.2, we pointed out that $T^{(0)}$, $T^{(1)}$, $T^{(2)}$, ... (or $A_1^{(n)}$) are the approximate areas based on the trapezoidal rule.   Likewise, in Section 12.3, it was shown that $S^{(0)}$, $S^{(1)}$, $S^{(2)}$, ... (or $A_2^{(n)}$) are the approximate values, which are calculated using Simpson's rule.   In deriving each rule, an interpolating polynomial is substituted for the original integrand $f(x)$.   The interpolating polynomial used in connection with the trapezoidal rule is a two-term polynomial $y = c_0 + c_1 x$ (a straight line), while that used for Simpson's rule is a quadratic polynomial $y = c_0 + c_1 x + c_2 x^2$ (See Figs. 10.2 and 12.4).

In this section we shall summarize some formulas obtained by interpolating with polynomials of high order.   In addition, we shall examine whether the approximate areas using these formulas bear any relationship with the C's, D's and E's listed in Table 12.1 (or $A_3^{(n)}$, $A_4^{(n)}$ and $A_5^{(n)}$ in Table 12.4).

Let us begin now with a cubic interpolating polynomial $y = c_0 + c_1 x + c_2 x^2 + c_3 x^3$.   If only one cubic parabola is used in the interval (a, b) as shown in Fig. 12.12, the approximate area can be computed using the following Cote's rule:

$$A = \frac{3h}{8} (f_0 + 3f_{1/3} + 3f_{2/3} + f_1) \tag{12.20}$$

where $h = l/3$.

If the original integrand $f(x)$ represents a parabola, the area based on Simpson's rule will be exact, with no error involved.   It is interesting to note that, for the integrand $f(x) = c_0 + c_1 x + c_2 x^2$, Cote's rule will also yield the same exact area.†

The principles used to derive the element $T^{(0)}$ in Eq. (12.1) and the element $S^{(0)}$ in Eq. (12.10) can be extended to cover the case in which a fourth-order interpolating polynomial $y = c_0 + c_1 x + c_2 x^2 + c_3 x^3 + c_4 x^4$ is used.   The approximate area $AP_0P_1P_2P_3P_4B$ shaded in Fig. 12.13 is

$$A = \frac{2h}{45} (7f_0 + 32f_{1/4} + 12f_{1/2} + 32f_{3/4} + 7f_1) \tag{12.21}$$

---

† A proof is given in: Apostol, T. M., *Calculus*, Vol. 2, p. 439.   Blaisdell, Waltham, Mass., 1962.

Table 12.6 Newton-Cotes closed-end formulas compared with the elements in Romberg's table

Row	Newton-Cotes closed-end formulas		Corresponding element in Romberg's table	
	Rules	Interpolating polynomial used	Table 12.1	Table 12.4
1	Trapezoidal rule $\dfrac{h}{2}(f_0 + f_1)$	Straight line $y = c_0 + c_1 x$	$T^{(0)}$	$A_1^{(1)}$
2	Simpson's rule $\dfrac{h}{3}(f_0 + 4f_{1/2} + f_1)$	Parabola $y = c_0 + c_1 x + c_2 x^2$	$S^{(0)}$	$A_2^{(1)}$
3	Cotes' rule $\dfrac{3h}{8}(f_0 + 3f_{1/3} + 3f_{2/3} + f_1)$	Cubic parabola $y = c_0 + c_1 x + c_2 x^2 + c_3 x^3$		
4	5-abscissa rule $\dfrac{2h}{45}(7f_0 + 32f_{1/4} + 12f_{1/2} + 32f_{3/4} + 7f_1)$	$y = \displaystyle\sum_{j=0}^{4} c_j x^j$	$C^{(0)}$	$A_3^{(1)}$
5	6-abscissa rule $\dfrac{5h}{288}(19f_0 + 75f_{1/5} + 50f_{2/5} + 50f_{3/5} + 75f_{4/5} + 19f_1)$	$y = \displaystyle\sum_{j=0}^{5} c_j x^j$		
6	7-abscissa rule $\dfrac{h}{140}(41f_0 + 216f_{1/6} + 27f_{1/3} + 272f_{1/2} + 27f_{2/3} + 216f_{5/6} + 41f_1)$	$y = \displaystyle\sum_{j=0}^{6} c_j x^j$	$D^{(0)}$	$A_4^{(1)}$
7	8-abscissa rule $\dfrac{7h}{17280}(751f_0 + 3577f_{1/7} + 1323f_{2/7} + 2989f_{3/7} + 2989f_{4/7} + \cdots + 751f_1)$	$y = \displaystyle\sum_{j=0}^{7} c_j x^j$		No connections between Newton-Cotes formulas and elements in Romberg's table.
8	9-abscissa rule $\dfrac{4h}{14175}(989f_0 + 5888f_{1/8} - 928f_{1/4} + 10946f_{3/8} - 4540f_{1/2} + \cdots + 989f_1)$	$y = \displaystyle\sum_{j=0}^{8} c_j x^j$		
...	...	...		

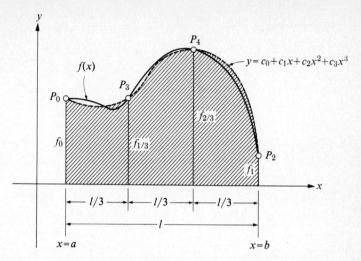

Fig. 12.12. Cote's rule. Area under a cubic interpolating polynomial.

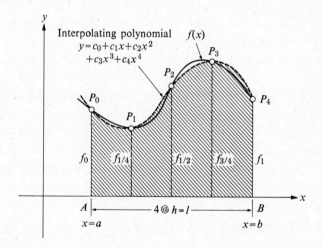

Fig. 12.13. Area under a fourth-order interpolating polynomial.

Equations (12.1), (12.10), (12.20) and (12.21), together with the approximate areas based on higher-order interpolating polynomials, are known as "Newton-Cotes closed-end formulas". The term *closed-end* serves to indicate that both f_0 and f_1 are included in each formula. Each rule and its corresponding element in Romberg's Table are summarized in Table 12.6. It is significant to note that no elements in Romberg's table correspond to any Newton-Cotes formula beginning with row 8 in Table 12.6. As a result, no geometrical interpretation seems to be possible for the elements $A_5^{(1)}$, $A_6^{(1)}$, ..., $A_5^{(2)}$, $A_6^{(2)}$, ..., $A_5^{(3)}$, $A_6^{(3)}$, ...

12.7 DERIVATION OF ROMBERG'S FORMULA

In Section 12.4, the Romberg's procedure was described and illustrated with an example. Then in Section 12.5, the general formula for Romberg's table was stated as Eq. (12.19). We turn now to the derivation of this important formula in three steps.

Step 1. In this step, we shall establish that the difference between the true value of the integral

$$I = \int_a^b f(x)\, dx$$

and the approximate areas $T^{(k)} = A_1^{(k+1)}$ (using the trapezoidal rule) is

$$A_1^{(k+1)} - I = C_1 4^{-k} + C_2 4^{-2k} + C_3 4^{-3k} + \cdots, \qquad (12.22)$$

where $C_1, C_2, C_3, \ldots$ are constants so long as the limits of integration a and b, as well as the integrand $f(x)$, are fixed. Consider a basic strip (shaded in Fig. 12.14), which can be represented by the integral

$$I_1 = \int_x^{x+h} f(\xi)\, d\xi.$$

If we define

$$I(x) = \int_a^x f(\xi)\, d\xi,$$

then

$$I(x + h) = \int_a^{x+h} f(\xi)\, d\xi$$

and

$$I_1 = I(x + h) - I(x).$$

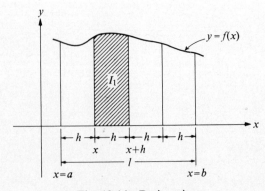

Fig. 12.14. Basic strip.

Using the Taylor-series expansion in Eq. (6.7) for $I(x)$ at $x + h$, we have

$$I(x + h) = I(x) + hI'(x) + \cdots + h^k \frac{I^{(k)}(x)}{k!} + \cdots \tag{12.23}$$

The derivatives of I with respect to x can be given in terms of $f(x)$ and its derivatives:

$$I'(x) = f(x),$$
$$I''(x) = f'(x),$$
$$\vdots$$
$$I^{(k)}(x) = f^{(k-1)}(x). \tag{12.24}$$

Substituting into Eq. (12.23), we obtain

$$I_1 = (x + h) - I(x)$$

$$= hf(x) + \frac{h^2}{2} f'(x) + \frac{h^3}{6} f''(x) + \frac{h^4}{24} f'''(x) + \frac{h^5}{120} f^{IV}(x) + \cdots \tag{12.25}$$

The Taylor-series expansions for $f(x + h)$ and $f'(x + h)$ about x are respectively

$$f(x + h) = f(x) + hf'(x) + \frac{h^2}{2} f''(x) + \frac{h^3}{6} f'''(x) + \frac{h^4}{24} f^{IV}(x) + \cdots \tag{12.26}$$

and

$$f'(x + h) = f'(x) + hf''(x) + \frac{h^2}{2} f'''(x) + \frac{h^3}{6} f^{IV}(x) + \frac{h^4}{24} f^{V}(x) + \cdots \tag{12.27}$$

Solving $f'(x)$ from Eq. (12.26) and substituting into Eq. (12.25), we have

$$I_1 = hf(x) + \frac{h^2}{2} \left\{ \frac{1}{h} [f(x + h) - f(x)] - \frac{h}{2} f''(x) - \frac{h^2}{6} f'''(x) - \frac{h^3}{24} f^{IV}(x) - \cdots \right\}$$

$$+ \frac{h^3}{6} f''(x) + \frac{h^4}{24} f'''(x) + \frac{h^5}{120} f^{IV}(x) + \cdots,$$

$$= \frac{h}{2} [f(x + h) + f(x)] - \frac{h^3}{12} f''(x) - \frac{h^4}{24} f'''(x) - \frac{h^5}{80} f^{IV}(x) - \cdots \tag{12.28}$$

Solving $f''(x)$ from Eq. (12.27) and substituting into Eq. (12.28), we have

$$I_1 = \frac{h}{2} [f(x + h) + f(x)] - \frac{h^2}{12} \left\{ [f'(x + h) - f'(x)] \right.$$

$$\left. - \frac{h^2}{2} f'''(x) - \frac{h^3}{6} f^{IV}(x) - \frac{h^4}{24} f^{V}(x) - \cdots \right\}$$

$$- \frac{h^4}{24} f'''(x) - \frac{h^5}{80} f^{IV}(x) - \cdots$$

$$= \frac{h}{2} [f(x + h) + f(x)] - \frac{h^2}{12} [f'(x + h) - f'(x)] + \frac{h^5}{720} f^{IV}(x) + \cdots$$

Continuing in this manner, we have

$$I_1 = \int_x^{x+h} f(\xi)\, d\xi$$

$$= \frac{h}{2}\left[f(x+h) + f(x)\right] - \frac{h^2}{12}\left[f'(x+h) - f'(x)\right] + \frac{h^4}{720}\left[f'''(x+h) - f'''(x)\right]$$

$$- \frac{h^6}{30{,}240}\left[f^V(x+h) - f^V(x)\right] + \cdots \tag{12.29}$$

If we now apply Eq. (12.29) to each of the basic strips of width h between $x = a$ and $x = b$, we obtain

$$I = \int_a^b f(x)\, dx = \frac{h}{2}\left[f(a) + 2\sum_{j-1}^{k-1} f(a + jh) + f(b)\right] - \frac{h^2}{12}\left[f'(b) - f'(a)\right]$$

$$+ \frac{h^4}{720}\left[f'''(b) - f'''(a)\right] - \frac{h^6}{30{,}240}\left[f^V(b) - f^V(a)\right] + \cdots \tag{12.30}$$

The first term in the right-hand side of Eq. (12.30) is the approximate area $T^{(k)}$ or $A_1^{(k+1)}$ based on the trapezoidal rule. [See Eq. (12.6)]. Eq. (12.30) becomes

$$I - A_1^{(k+1)} = \frac{-h^2}{12}\left[f'(b) - f'(a)\right] + \frac{h^4}{720}\left[f'''(b) - f'''(a)\right]$$

$$- \frac{h^6}{30{,}240}\left[f^V(b) - f^V(a)\right] + \cdots \tag{12.31}$$

If $f(x)$, as well as a and b, is fixed, the expression in each square bracket on the right-hand side of Eq. (12.31) yields a constant. If we further substitute $h = l/2^k$ into Eq. (12.31), we obtain

$$A_1^{(k+1)} - I = C_1 4^{-k} + C_2 4^{-2k} + C_3 4^{-3k} + \cdots, \tag{12.22}$$

where $C_1, C_2, C_3, \ldots$ are constants.

Step 2. We shall now show that the difference between I and the areas based on Simpson's rules, i.e., $A_2^{(k+1)}$, can be expressed as

$$A_2^{(k+1)} - I = -\tfrac{1}{4}C_2 4^{-2k} - \tfrac{5}{16}C_3 4^{-3k} - \tfrac{21}{64}C_4 4^{-4k} - \cdots \tag{12.32}$$

Let us begin this step by substituting $k + 1$ for k in Eq. (12.22). We have

$$A_1^{(k+2)} - I = C_1 4^{-(k+1)} + C_2 4^{-2(k+1)} + C_3 4^{-3(k+1)} + \cdots \tag{12.33}$$

Multiply both sides of Eq. (12.22) by 4^{-1}, getting

$$\tfrac{1}{4}(A_1^{(k+1)} - I) = C_1 4^{-(k+1)} + C_2 4^{-2k-1} + C_3 4^{-3k-1} + \cdots \tag{12.34}$$

Now, subtracting Eq. (12.34) from Eq. (12.33) to eliminate C_1, we have

$$\tfrac{1}{4}A_1^{(k+1)} - A_1^{(k+2)} + \tfrac{3}{4}I = C_2(4^{-2k-1} - 4^{-2k-2}) + C_3(4^{-3k-1} - 4^{-3k-3}) + \cdots,$$

or,

$$A_1^{(k+2)} + \tfrac{1}{3}(A_1^{(k+2)} + A_1^{(k+1)}) = I - \tfrac{3}{4}C_2 4^{-2k} - \tfrac{1}{3}\cdot\tfrac{15}{16}C_3 4^{-3k} - \tfrac{1}{3}\cdot\tfrac{63}{64}C_4 4^{-4k} - \cdots \tag{12.35}$$

If we define the left-hand expression in (12.35) as $A_2^{(k+1)}$, we obtain

$$A_2^{(k+1)} - I = -\frac{4}{4-1}\left[\frac{4-1}{4^2}C_2 4^{-2k} + \frac{4^2-1}{4^3}C_3 4^{-3k} + \frac{4^3-1}{4^4}C_4 4^{-4k} + \cdots\right], \tag{12.36}$$

which yields (12.32).

Step 3. Let us continue to derive expression for $A_3^{(k+1)} - I$. This step is essentially a repetition of Step 2.

Substitute $k+1$ for k in Eq. (12.36), we obtain

$$A_2^{(k+2)} - I$$
$$= \frac{-4}{4-1}\left[\frac{4-1}{4^2}C_2 4^{-2k-2} + \frac{4^2-1}{4^3}C_3 4^{-3k-3} + \frac{4^3-1}{4^4}C_4 4^{-4k-4} + \cdots\right]. \tag{12.37}$$

Multiplying both sides of Eq. (12.36) by 4^{-2}, we have

$$4^{-2}A_2^{(k+1)} - 4^{-2}I$$
$$= \frac{-4}{4-1}\left[\frac{4-1}{4^2}C_2 4^{-2k-2} + \frac{4^2-1}{4^3}C_3 4^{-3k-2} + \frac{4^3-1}{4^4}C_4 4^{-4k-2} + \cdots\right]. \tag{12.38}$$

Subtracting Eq. (12.38) from Eq. (12.37) and simplifying,

$$\frac{4^2 A_2^{(k+2)} - A_2^{(k+1)}}{4^2-1} = I - \frac{4}{4-1}\left(\frac{4^2}{4^2-1}\right)\left[\frac{4^2-1}{4^3}C_3 4^{-3k}(4^{-3}-4^{-2})\right.$$
$$\left. + \frac{4^3-1}{4^4}C_4 4^{-4k}(4^{-4}-4^{-2}) + \cdots\right]. \tag{12.39}$$

We define the left-hand side in the above equation as $A_3^{(k+1)}$ to find

$$A_3^{(k+1)} - I = \tfrac{1}{64}C_3 4^{-3k} - \tfrac{21}{256}C_4 4^{-4k} - \cdots \tag{12.40}$$

Continuing in the fashion of Steps 2 and 3, we find the general element in the Romberg table

$$A_m^{(n)} = \frac{4^{m-1}A_{m-1}^{(n+1)} - A_{m-1}^{(n)}}{4^{m-1}-1}, \tag{12.19a}$$

where $n = 1, 2, 3, \ldots$ and $m = 2, 3, 4, \ldots (m \neq 1)$. Also,

$$A_m^{(k+1)} - I = \alpha_{m+1}4^{-mk} + \alpha_{m+2}4^{-(m+1)k} + \cdots, \tag{12.41}$$

where $k = 0, 1, 2, \ldots$ as defined in Eq. (12.5) and $m = 2, 3, 4, \ldots (m \neq 1)$.

Before we conclude our discussion on Romberg's quadrature, let us inspect the leading error terms from several equations. In Eq. (12.22), the leading error term for trapezoidal rule is a constant times 4^{-k}, or times 2^{-2k}. It is also equal to a constant C_1/l^2 times h^2, since $h = l/2^k$. Likewise, from Eq. (12.32), the leading error term for Simpson's rule is a constant $C_2/4$ times 4^{-2k}, or times 2^{-4k}. This term is equal to C_2/l^4 times h^4. Finally, as shown in Eq. (12.41), the leading error term in general for $A_m^{(k+1)}$ may be expressed in the standard form of a constant times 4^{-mk}, or times 2^{-2mk}. In terms of h, the leading error term is a constant times h^{2m}.

12.8 LEGENDRE POLYNOMIALS

In the previous sections, we have studied Romberg's method for numerical evaluation of an integral. It is in many ways better than other methods; it is accurate, simple, and computer-oriented. However, the method has its disadvantages. In particular, it seems to be not very well suited for integrals of the type

$$\int_0^1 \frac{f(x)}{1 - x^2}\, dx,$$

which may be readily evaluated by the Gaussian quadrature procedure. A study of this procedure demands a complete understanding of the two types of special polynomials, namely, Lagrange polynomials and Legendre polynomials. The Lagrange polynomials and their associated interpolation formula were discussed in Section 10.3, while a brief discussion was presented in Section 11.4 on general orthogonal polynomials, of which the Legendre polynomials are special cases.

To help build up background, we shall summarize the principal properties of the Legendre polynomials in the remainder of this section. Once they are fully understood, it is a relatively simple matter to study the Gaussian quadrature procedure.

The first five orthogonal polynomials of Legendre are

$$
\begin{aligned}
P_0(x) &= 1, \\
P_1(x) &= x, \\
P_2(x) &= \tfrac{1}{2}(3x^2 - 1), \\
P_3(x) &= \tfrac{1}{2}(5x^3 - 3x), \\
P_4(x) &= \tfrac{1}{8}(35x^4 - 30x^2 + 3).
\end{aligned}
\tag{12.42}
$$

The Legendre polynomial of degree n can be obtained† from Rodrigues' formula

$$P_n(x) = \frac{1}{2^n n!} \frac{d^n}{dx^n} (x^2 - 1)^n \tag{12.43}$$

or from the recurrence relation

$$(n + 1)P_{n+1}(x) - (2n + 1)xP_n(x) + nP_{n-1}(x) = 0. \tag{12.44}$$

---

† For example, see Hildebrand, F. B., *Advanced Calculus for Applications*. Prentice-Hall, Englewood Cliffs, N.J., p. 164, 1962.

The orthogonality and normalization relations, with the weighting function equal to unity, are

$$\int_{-1}^{1} P_n(x)P_m(x)\,dx = \begin{cases} 0 & \text{if } n \neq m, & (12.45a) \\ \dfrac{2}{2n+1} & \text{if } n = m. & (12.45b) \end{cases}$$

It is worth while to note that all the roots of each $P_n(x) = 0$ are real and distinct, and are between -1 and $+1$.

12.9 GAUSSIAN QUADRATURE

The purpose of this section is to discuss the Gaussian integration formula which approximates the definite integral

$$\int_{-1}^{1} f(x)\,dx,$$

and to show that by a simple change of variables, the procedure can be extended to limits of integration other than $(-1, 1)$.

This method serves to approximate the definite integral

$$\int_{-1}^{1} f(x)\,dx$$

by the expression

$$\int_{-1}^{1} f(x)\,dx = w_0 f(x_0) + w_1 f(x_1) + \cdots + w_n f(x_n) = \sum_{k=0}^{n} w_k f(x_k), \qquad (12.46)$$

where $w_0, w_1, \ldots, w_n$ are the weighting coefficients and $x_0, x_1, \ldots, x_n$, are the associated points. The basic problem is to determine these $2n + 2$ constants and our basic assumption is that Eq. (12.46) should involve no approximations if the integrand $f(x)$ is a polynomial of degree $2n + 1$ or less.

We shall first show that the associated points x_k $(k = 0, 1, \ldots, n)$ are equal to the values of the roots of a Legendre polynomial $P_{n+1}(x)$. These polynomials were discussed in Section 12.8.

Let us arbitrarily take a polynomial $g_n(x)$ of degree n. This polynomial can be written in terms of the Legendre polynomials as

$$g_n(x) = \beta_0 P_0(x) + \beta_1 P_1(x) + \cdots + \beta_n P_n(x). \qquad (12.47)$$

As an illustration, let us suppose that

$$g_3(x) = 1 + 3x + 4x^2 - 7x^3. \qquad (12.48)$$

From Eq. (12.47) we have

$$g_3(x) = \beta_0 + \beta_1 x + \frac{\beta_2}{2}(3x^2 - 1) + \frac{\beta_3}{2}(5x^3 - 3x). \qquad (12.49)$$

Expanding the right-hand side of Eq. (12.49) and comparing the coefficients with Eq. (12.48), we find that

$$g_3(x) = \tfrac{7}{3}P_0(x) - \tfrac{6}{5}P_1(x) + \tfrac{8}{3}P_2(x) - \tfrac{14}{5}P_3(x).$$

This simple example serves to show that any polynomial $g_n(x)$ can be written in terms of the Legendre polynomials.

From Eq. (12.45a), the orthogonality relation, we have

$$\int_{-1}^{1} g_n(x)P_{n+1}(x)\,dx = \int_{-1}^{1} \beta_0 P_0(x)P_{n+1}(x)\,dx + \int_{-1}^{1} \beta_1 P_1(x)P_{n+1}(x)\,dx + \cdots$$

$$+ \int_{-1}^{1} \beta_n P_n(x)P_{n+1}(x)\,dx = 0. \tag{12.50}$$

It is worth while to note that $g_n(x)P_{n+1}(x)$ is a polynomial of degree $2n + 1$. Hence it satisfies the basic requirement in the selection of the weighting coefficients w_k and associated points x_k $(k = 0, 1, \ldots, n)$. Comparing Eq. (12.50) with Eq. (12.46) and noting that $g_n(x)P_{n+1}(x)$ is the integrand, we have

$$w_0 g_n(x_0)P_{n+1}(x_0) + w_1 g_n(x_1)P_{n+1}(x_1) + \cdots + w_n g_n(x_n)P_{n+1}(x_n) = 0. \tag{12.51}$$

In the above equation, $g_n(x)$ is an *arbitrarily* chosen polynomial. For each value of x_k $(k = 0, 1, \ldots, n)$, we find that $g_n(x_k)$ has a corresponding value. Not all of these $n + 1$ corresponding values can be equal to zero, and not all of the $n + 1$ weighting coefficients w_k $(k = 0, 1, \ldots, n)$ can be equal to zero. Should this be so, then Eq. (12.46) would become identically zero, which is a trivial case. Therefore, the only condition which must be satisfied for Eq. (12.51) is that

$$P_{n+1}(x_0) = 0,$$
$$P_{n+1}(x_1) = 0,$$
$$\vdots$$
$$P_{n+1}(x_n) = 0.$$

In other words, the associated points $x_0, x_1, \ldots, x_n$ are the roots of the Legendre polynomial $P_{n+1}(x) = 0$. There are $n + 1$ distinct real roots in the interval $(-1, 1)$. For example, for $n = 2$, the roots of $P_3(x) = \tfrac{1}{2}(5x^3 - 3x) = 0$ are $-\sqrt{3/5}$, 0, and $\sqrt{3/5}$, respectively. In the remainder of this chapter we shall use the subscripted x, such as $x_0, x_1, \ldots, x_k, \ldots, x_n$ to indicate the roots of the Legendre polynomial $P_{n+1}(x)$.

Having selected the values of x_k, we now turn to the determination of the values of the weighting coefficients w_k $(k = 0, 1, \ldots, n)$. We recall that in accordance with our basic requirement, Eq. (12.46) must involve no approximation if the integrand $f(x)$ is a polynomial of degree $2n + 1$ or less. By the definition of the Lagrange polynomial, any polynomial $h_n(x)$ of degree n passing through x_k $(k = 0, 1, \ldots, n)$ points may be expressed in the form

$$h_n(x) = \sum_{k=0}^{n} h_n(x_k)L_k(x). \tag{12.52}$$

Hence

$$\int_{-1}^{1} h_n(x)\, dx = \int_{-1}^{1} \sum_{k=0}^{n} h_n(x_k) L_k(x), \tag{12.53}$$

and since $h_n(x_k)$ is a constant, we have

$$\int_{-1}^{1} h_n(x)\, dx = \sum_{k=0}^{n} h_n(x_k) \int_{-1}^{1} L_k(x)\, dx. \tag{12.54}$$

Comparing Eq. (12.46) with Eq. (12.54), we obtain

$$w_k = \int_{-1}^{1} L_k(x)\, dx, \qquad k = 0, 1, \ldots, n. \tag{12.55}$$

As a practical computational detail, one frequently calculates the weighting coefficients w_k in terms of the Legendre polynomials $P_n(x)$:

$$w_k = \frac{1}{P'_{n+1}(x_k)} \int_{-1}^{1} \frac{P_{n+1}(x)\, dx}{x - x_k}. \tag{12.56}$$

To see why Eq. (12.56) is valid, we first note that the polynomial

$$\frac{P_{n+1}(x)}{x - x_k}$$

has zero value for $x = x_j$ ($j \neq k$, $j = 0, 1, \ldots, n$). Then, by L'Hôpital's rule, we have

$$\lim_{x \to x_k} \frac{P_{n+1}(x)}{x - x_k} = \left[\frac{dP_{n+1}(x)/dx}{d(x - x_k)/dx} \right]_{x = x_k} = P'_{n+1}(x_k), \tag{12.57}$$

where the x_k is one of the $n + 1$ roots of the Legendre polynomial $P_{n+1}(x) = 0$. Hence the Lagrange polynomial can be written as

$$L_k(x) = \frac{1}{P'_{n+1}(x_k)} \frac{P_{n+1}(x)}{x - x_k}, \tag{12.58}$$

since it assumes the value of 0 at $x = x_j$ ($j \neq k$), and the value of 1 at $x = x_k$, where the derivative of $P_{n+1}(x_k)$ has been established in Eq. (12.57). Substituting Eq. (12.58) into Eq. (12.55), one readily obtains Eq. (12.56).

For purposes of illustration, we now use Eq. (12.56) to compute w_0, w_1, and w_2 ($n = 2$). Since $n = 2$, we have

$$P_{n+1}(x) = P_3(x) = \tfrac{1}{2}(5x^3 - 3x),$$

whose roots are $x_0 = -\sqrt{\tfrac{3}{5}}$, $x_1 = 0$, $x_2 = \sqrt{\tfrac{3}{5}}$, and whose derivative is

$$P'_3(x) = \tfrac{3}{2}(5x^2 - 1).$$

Thus

$$w_0 = \frac{1}{\frac{3}{2}(5 \cdot \frac{3}{5} - 1)} \int_{-1}^{1} \frac{\frac{1}{2}(5x^3 - 3x)\, dx}{x + \sqrt{\frac{3}{5}}} = \frac{2}{3(3-1)} \int_{-1}^{1} \frac{1}{2}(x^2 - \sqrt{\tfrac{3}{5}}x)\, dx = \tfrac{5}{9},$$

$$w_1 = \frac{1}{\frac{3}{2}(5 \cdot 0 - 1)} \int_{-1}^{1} \frac{\frac{1}{2}(5x^3 - 3x)}{x - 0}\, dx = \tfrac{8}{9},$$

$$w_2 = \frac{1}{\frac{3}{2}(5 \cdot \frac{3}{5} - 1)} \int_{-1}^{1} \frac{\frac{1}{2}(5x^3 - 3x)}{x - \sqrt{\frac{3}{5}}}\, dx = \frac{2}{3(3-1)} \int_{-1}^{1} \frac{1}{2}(x^2 + \sqrt{\tfrac{3}{5}}x)\, dx = \tfrac{5}{9}.$$

Table 12.7 lists the weighting coefficients and associated points for a number of values of n (from $n = 2$ to $n = 5$).

Table 12.7 Weighting coefficients w_k and associated points x_k for the Gaussian quadrature formula

n	Weighting coefficients w_k	Associated points x_k
2	$\frac{8}{9}$	0
	$\frac{5}{9}$	$\pm 0.774\ 596\ 669$
3	0.652 145 154 9	$\pm 0.339\ 981\ 043\ 6$
	0.347 854 845 1	$\pm 0.861\ 136\ 311\ 6$
4	0.568 888 888 8	0
	0.478 628 670 5	$\pm 0.538\ 469\ 310\ 1$
	0.236 926 885 1	$\pm 0.906\ 179\ 845\ 9$
5	0.467 913 934 6	$\pm 0.238\ 619\ 186\ 1$
	0.360 761 573 0	$\pm 0.661\ 209\ 386\ 5$
	0.171 324 492 4	$\pm 0.932\ 469\ 514\ 2$

Example

Approximate the following definite integral by using the Gaussian quadrature with $n = 2$:

$$I = \int_{-1}^{1} x^2 \cos x\, dx.$$

Solution: Let

$$f(x) = x^2 \cos x \quad \text{and} \quad I \doteq w_0 f(x_0) + w_1 f(x_1) + w_2 f(x_2).$$

We then obtain the following:

Weighting coefficients	Associated points
$w_0 = \frac{5}{9}$	$x_0 = -\sqrt{\frac{3}{5}}$
$w_1 = \frac{8}{9}$	$x_1 = 0$
$w_2 = \frac{5}{9}$	$x_2 = \sqrt{\frac{3}{5}}$

Hence

$$I \doteq \tfrac{5}{9}(x_0^2 \cos x_0) + \tfrac{8}{9}(x_1^2 \cos x_1) + \tfrac{5}{9}(x_2^2 \cos x_2)$$
$$= 0.47650.$$

It may be noted that this integral has the following closed-form solution:

$$\int_{-1}^{1} x^2 \cos x \, dx = [2x \cos x + (x^2 - 2) \sin x]_{-1}^{1}$$
$$= 0.47830.$$

To extend our discussion to the case where the lower and upper limits of integration are a and b, respectively, we must first obtain a relation that holds for the limits of integration, -1 to 1. We do this by simply using

$$x = \frac{(b - a)t + (b + a)}{2}, \tag{12.59}$$

$$dx = \frac{b - a}{2} \, dt. \tag{12.60}$$

For example, suppose that we wish to approximate the following integral, using the three-point ($n = 2$) Gaussian quadrature:

$$I_1 = \int_{0}^{3} x^2 \cos x \, dx.$$

We have $a = 0$ and $b = 3$; hence

$$x = 3(t + 1)/2, \quad \text{and} \quad dx = \tfrac{3}{2} \, dt.$$

Therefore

$$I_1 = \int_{0}^{3} x^2 \cos x \, dx$$

$$= \frac{3}{2} \int_{-1}^{1} \left[\frac{3(t + 1)}{2}\right]^2 \cos \left[\frac{3(t + 1)}{2}\right] dt.$$

Using the three-point Gaussian quadrature, we find that the associated points in terms of the new variable t are $t_0 = -\sqrt{\frac{3}{5}}$, $t_1 = 0$, and $t_2 = \sqrt{\frac{3}{5}}$. Therefore we have

$$I_1 = \frac{3}{2} \left(\frac{5}{9} \left\{ \left[\frac{3}{2}(t_0 + 1) \right]^2 \cos \frac{3(t_0 + 1)}{2} \right\} + \frac{8}{9} \left\{ \left[\frac{3}{2}(t_1 + 1) \right]^2 \cos \frac{3(t_1 + 1)}{2} \right\} \right.$$

$$\left. + \frac{5}{9} \left\{ \left[\frac{3}{2}(t_2 + 1) \right]^2 \cos \frac{3(t_2 + 1)}{2} \right\} \right) = -4.936.$$

For comparison, the closed-form solution is equal to

$$[2x \cos x + (x^2 - 2) \sin x]_0^3 = -4.952.$$

12.10 GAUSSIAN QUADRATURE IN FORTRAN IV LANGUAGE

We shall now develop a FORTRAN IV subprogram to evaluate the integration of $f(x) \, dx$ between the limits A and B by using Gaussian quadrature, which is expressed as

$$\int_A^B f(x) \, dx = \frac{B - A}{2} \sum_{i=1}^N w_i f \left(\frac{(B - A)t_i + (B + A)}{2} \right),$$

where $w_1, w_2, \ldots, w_N$ are weighting coefficients and $t_1, t_2, \ldots$ t_N are the roots of the Legendre polynomial $P_N(t) = 0$. The value of N ranges from 3 to 6 in this program. The computation starts with $N = 3$. The program will first compare the result based on $N = 3$ with that based on $N = 4$. The results must satisfy the criterion

$$\epsilon \geq \frac{A_{n+1} - A_n}{A_n}, \tag{12.61}$$

where A_{n+1} is the answer based on $N + 1$ points, A_n is the answer based on N points, and ϵ is arbitrarily taken (in this program) to be 10^{-4}. If the result fails to pass the above test, the value of N will be increased by one. The maximum value of N is set to 6 in this program.

Nine data cards are provided here to be read in as the weighting coefficients and the roots of Legendre polynomials. They should not be used during the compilation stage, but only during execution. The subroutine GAUSS and its flow chart are shown in Figs. 12.15 and 12.16, respectively.

The output from this subroutine is in one of two forms:

1) If the result passes the test shown in Eq. (12.61), an answer having the following format will be printed:

BY CONVERGENCE, 5 POINT GAUSS. QUADRATURE GIVES ANSWER= -.49523118E+01

2) If it does not pass the test when $N = 6$, the following answer will be printed:

BY LIMITS OF PROGRAM, 6 POINT GAUSS. QUADRATURE GIVES ANS.=   -.49521140E+01

Example

Using Gaussian quadrature, write a main program to approximate the integral

$$\int_0^3 x^2 \cos x \, dx.$$

This main program will be used to call the subroutine GAUSS, shown in Fig. 12.16; it will also serve to express the upper and lower limits and the integrand $f(x) = x^2 \cos x$.

```
          SUBROUTINE GAUSS (A,B,X,F,KOUNT)
    C
    C  INTEGRATION OF F(X).DX BY GAUSSIAN QUADRATURE BETWEEN THE
    C  LIMITS A AND B.
    C
    C  NOMENCLATURE FOR THE ARGUMENTS
    C  A=THE LOWER LIMIT OF INTEGRATION.
    C  B=THE UPPER LIMIT OF INTEGRATION.
    C  X=INDEPENDENT VARIABLE OF FUNCTION F(X).
    C  F=F(X) UNDER INTEGRAL SIGN.
    C  KOUNT=AN INTEGER USED TO CONTROL THE WAY OF EXECUTION.
    C
          DATA IREAD,IWRITE/5,6/
          DIMENSION W(4,6), T(4,6)
          IF(KOUNT.EQ.0) GO TO 10
        8 GO TO 30
       10 ANS=1.0
          IPOINT=2
    C
    C  EPSILON FOR CONVERGENCE
    C
          EPS=10.E-05
    C
    C  STORE WEIGHTING COEFFICIENTS AND LEGENDRE ROOTS.
    C
          DO 2 I=1,4
          IP2=I+2
        2 READ(IREAD,12) (W(I,K),K=1,IP2),(T(I,K),K=1,IP2)
       12 FORMAT (5F15.10)
    C
    C  CHANGE INTEGRATION LIMITS TO (-1 TO +1).
    C
    C  THE NEXT STATEMENT IS TO EVALUATE THE COEFFICIENT OF THE
    C  NEW FUNTION.
    C
          C=(B-A)/2.
       18 IPOINT=IPOINT+1                                      (cont.)
          TEMP=ANS
```

Fig. 12.15. Subroutine GAUSS. The nine data cards must be used during the computing stage, but not during the compiling stage.

```
      ANS=0.
      IPM2=IPOINT-2
      KOUNT=1
C
C EVALUATE NEW VARIABLES WHICH ARE EXPRESSED IN TERM OF
C THE LEGENDRE ROOT.
C
   20 X=C*T(IPM2,KOUNT)+(B+A)/2.
      RETURN
C
C CARRY OUT INTEGRATION BY CALCULATING
C ANS=C*(WI*F(XI)+W2*F(X2)+---).
C
   30 ANS=ANS+C*W(IPM2,KOUNT)*F
      KOUNT=KOUNT+1
      IF(KOUNT.LE.IPOINT) GO TO 20
      IF(IPOINT.LE.3) GO TO 18
C NEXT THREE STATEMENTS FOR DETERMINING WHETHER THE DEVIATION
C OF ANSWER IS WITHIN THE LIMIT.
C IF IT IS NOT, TAKE ONE MORE POINT GAUSSIAN INTEGRATION.
C
   50 DELT=ABS(ANS-TEMP)
      RATIO=DELT/ABS(TEMP)
    7 IF(RATIO.GT.EPS) GO TO 80
C
C WRITE OUT THE ANSWER IF THE DEVIATION COMES WITHIN THE LIMIT.
C
   70 WRITE(IWRITE,72) IPOINT,ANS
   72 FORMAT('1',' BY CONVERGENCE, ',I2,' POINT GAUSS. QUADRATURE GIVES
     1 ANSWER =',E16.8//)
      KOUNT=7
      RETURN
   80 IF(IPOINT.LT.6) GO TO 18
  100 WRITE(IWRITE,102) IPOINT,ANS
C
C THE ANSWER IS WRITE OUT AFTER SIX POINT GAUSSIAN INTEGRATION
C HAS BEEN EXECUTED AND THE DEVIATION IS STILL NOT WITHIN THE LIMIT.
C
  102 FORMAT(///5X,'BY LIMITS OF PROGRAM ',I2,' POINT GAUSS. QUADRATURE
     1GIVES ANSWER =',E16.8//)
      KOUNT=7
      RETURN
C
C DATA FOR WEIGHTING COFFICIENTS AND LEGENDRE ROOTS
C
      END
```

.55555555	.88888888	.555555555	.774596669	.0
-.774596669				
.347854845	.652145154	.652145154	.347854845	.861136311
.339981043	-.339981043	-.861136311		
.236926885	.478628670	.568888888	.478268670	.236926885
.906179845	.538469310	.0	-.538469310	-.906179845
.171324492	.360761573	.467913934	.467913934	.360761573
.171324492	.932469514	.661209386	.238619186	-.238619186
-.661209386	-.932469514			

Fig 12.15   (*concl.*)

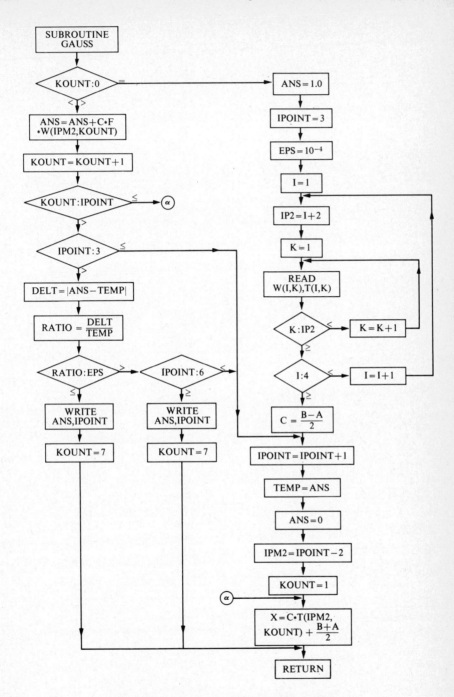

Fig. 12.16. Flow chart for subroutine GAUSS.

```
C MAIN PROGRAM FOR GAUSSIAN INTEGRATION OF Z(X).DX BETWEEN THE
C LIMITS A AND B.
C YOU MUST CHANGE THE STATEMENTS 1.2 AND 5. FOR A DIFFERENT PROBLEM.
C STATEMENT 1 IS FOR THE LOWER LIMIT OF INTEGRATION.
C STATEMENT 2 IS FOR THE UPPER LIMIT OF INTEGRATION.
C STATEMENT 5 IS FOR THE FUNCTION UNDER THE INTEGRAL SIGN.
C NEXT TWO STATEMENT FOR LOWER AND UPPER LIMITS.
C YOU CHANGE THE RIGHT HAND SIDE.
C
    1 A=0.
    2 B=3.
C
C NEXT STATEMENT FOR INITIALIZATION
C
   10 KOUNT=0
      X=9.999
C
C NEXT STATEMENT IS THE FUNCTION UNDER INTEGRAL SIGN.
C
    5 F=X*X*COS(X)
    6 CALL GAUSS (A,B,X,F,KOUNT)
   14 GO TO (5,5,5,5,5,5,17), KOUNT
   17 STOP
      END
```

Fig. 12.17. Main program to call SUBROUTINE GAUSS.

Figure 12.17 illustrates a sample main program. Statements 1 and 2 serve to indicate the lower and upper limits of integration, respectively. Statement 5 is used to express the integrand $f(x)$. Statement 14 provides an easy method of taking care of multiple exits by using a computed GO TO statement. This statement,

$$\text{GO TO } (5,5,5,5,5,5,17), \text{KOUNT}$$

will transfer operations to statement 5 if KOUNT $= 1$, to statement 5 if KOUNT $= 2, \ldots$, to statement 17 if KOUNT $= 7$. The output is

BY CONVERGENCE, 5 POINT GAUSS. QUADRATURE GIVES ANSWER=−.49523118E+01

It should be noted that this main program can be used for any integrand and for any given limits with only a slight modification. A change of only three statements is involved, namely, statements 1, 2, and 5. If, for example, we wish to approximate the integral

$$\int_{0.1}^{10} (x/\sqrt{1 + x}) \, dx,$$

then these three statements become

```
C     . . .
C     . . .
    1 A = 0.1
    2 B = 10.
      :
    5 F = X/SQRT(1.+X)
      :
      END
```

BIBLIOGRAPHY

General

Abramowitz, M., "On the practical evaluation of integrals," *J. Soc. Ind. Appl. Math.*, **2,** 20–35 (1954).

Burlirsch, R., "Numerical calculation of elliptic integrals and elliptic functions," *Num. Math.*, **7,** 78–90 (1965).

Clenshaw, C. W., and A. R. Curtis, "A method for numerical integration on an automatic computer," *Num. Math.*, **2,** 197–205 (1960).

Eberlein, W. F., "Theory of numerical integration, I, preliminary report," *Bull. Am. Math. Soc.*, **60,** 366–367 (1954).

Fettis, H. E., "Numerical calculation of certain definite integrals by Poisson's summation formula," *Math. Tab. Wash.*, **9,** 85–92 (1955).

Gautschi, W., "Recursive computation of certain integrals," *J. Assoc. Comput. Mach.*, **8,** 21–40 (1961).

Hammerlin, G., "Zur numerischen Integration periodischer Funktionen," *Z. Angew. Math. Mech.*, **39,** 80–82 (1959).

Hartree, D. R., "The evaluation of a diffraction integral," *Proc. Cambridge Phil. Soc.*, **50,** 567–576 (1954).

Havie, T., "On a modification of the Clenshaw-Curtis quadrature formula," *BIT*, **9,** no. 4, 338–350 (1969).

Krylov, V. J., *Approximate Calculation of Integrals.* Macmillan, New York, 1962, pp. 100–111, 337–340.

Luke, Y. L., "Evaluation of an integral arising in numerical integration near a logarithmic singularity," *Math. Tab., Wash.*, **10,** 14–21 (1956).

Stroud, A. H., "Approximate multiple integration," *Mathematical Models for Digital Computers*, Vol. 2, pp. 145–155, A. Ralston and H. S. Wilf, eds. Wiley, New York, 1967.

Todd, J., "Evaluation of the exponential integral for large complex arguments," *J. Res. Natn. Bur. Stand.*, **52,** 313–317 (1956).

Romberg's Quadrature

Bauer, F. L., H. Rutishauser, and E. Stiefel, "New aspects in numerical quadrature," *Proc. Symp. Appl. Math., XV. Am. Math. Soc.*, 199–218 (1963).

Burlirsch, R., "Bemerkungen zur Romberg Integration," *Num. Math.*, **6**, 61–66 (1964).

Fairweather, G., "Modified Romberg quadrature," *Commun. Assoc. Comput. Mach.*, **12**, 324 (1969).

Filippi, S., "Das Verfahren von Romberg-Stiefel-Bauer als Spezialfall des allgemeinen Prinzips von Richardson," *Mathematik Tech. Wirt.*, **11**, no. 2, 49–54 (1964).

Havie, T., "On a modification of Romberg's algorithm," *BIT*, **6**, 24–30 (1966).

Krasun, A. M., and W. Prager, "Remark on Romberg quadrature," *Commun. Assoc. Comput. Mach.*, **8**, 236–237 (1965).

Kubik, R. N., "Havie integrator," *Commun. Assoc. Comput. Mach.*, **8**, 381 (1965).

Miller, J. C. P., "Neville's and Romberg's processes—a fresh appraisal with extensions," *Phil. Trans. R. Soc.*, Ser. A, **263**, 525–562 (1968).

Gaussian Quadrature

Davis, P., and P. Rabinowitz, "Abscissas and weights for Gaussian quadratures of high order," *J. Res. Natn. Bur. Stand.*, **56**, 35–37 (1956).

Davis, P., and P. Rabinowitz, "Additional abscissas and weights for Gaussian quadratures of high order . . . values for *N*-64, 80, and 96," *J. Res. Natn. Bur. Stand.*, **60**, 613–614 (1958).

Fourier Series and Transform

Cooley, V. W., and J. W. Tukey, "An algorithm for the machine calculation of complex Fourier series," *Math. Comp.*, **19**, 297–301 (1965).

Gentleman, W. M., and G. Sande, "Fast Fourier transform for fun and profit," *Fall Joint Computer Conference, AFIPS, Proceedings*, **29**, 563–578 (1966).

PROBLEMS

Write a complete FORTRAN program to evaluate the following integrals by using the Gaussian quadrature formula for three points and the Romberg method:

1. $\int_0^3 x^2 \sin x^2 \, dx.$ 2. $\int_0^2 \frac{t \sin t \, dt}{27 + t^3}.$ 3. $\int_0^4 \frac{e^{-x^2/2}}{\sqrt{2\pi}}.$

4. $\int_0^1 \frac{dx}{1 + x^2}.$ 5. $\int_0^\pi \frac{\sin x}{x} \, dx.$

6. It is desired to generate a certain waveform in an electric circuit. The waveform $f(t)$ is defined by

$$f(t) = 50 \cos (2\pi/200)t, \quad -50 \le t \le 50;$$

$$f(t) = 0, \quad -100 \le t \le -50, \quad 50 \le t \le 100.$$

Thus the period T is seen to be 200. The waveform will be represented by the first 10 terms in the Fourier series expansion of f(t). We seek

$$f(t) = A_0 + \sum_{n=1}^{9} [2A_n \cos (n\omega t)],$$

where $\omega = 2\pi/T$. Given that

$$A_n = (2/T) \int_0^{T/2} f(t) \cos (n\omega t) \, dt; \qquad n = 0, 1, 2, \ldots,$$

find $A_0, A_1, A_2, \ldots, A_9$.

7. Evaluate the following integral

$$\int_a^b \frac{dx}{1 + 10x^2}$$

for the following cases:

Case	I	II	III	VI
a	1	0	0	0
b	10	1	10	1000

8. Evaluate the integral

$$\int_0^2 \frac{x}{27 + x^3} \, dx$$

by (a) using the closed-form solution; (b) using the Gaussian quadrature formula for three points, (c) using the Gaussian quadrature formula for four points, and (d) using Romberg method. It is known that the following closed-form expression exists:

$$\int_0^x \frac{t}{a^3 + t^3} \, dt = \frac{1}{\sqrt{3a}} \tan^{-1} \frac{\sqrt{3}x}{2a - x} + \frac{1}{6a} \log_e \frac{a^2 - ax + x^2}{(a + x)^2}, \qquad -a < x < 2a.$$

The output should be in the following form:

```
ANSWER FROM CLOSED-FORM EXPRESSION = XXX.XXXXXX
ANSWER FROM THREE-POINT GAUSSIAN QUAD. EQ. = XXX.XXXXXX
ANSWER FROM FOUR-POINT GAUSSIAN QUAD. EQ. = XXX.XXXXXX
ANSWER FROM ROMBERG QUADRATURE = XXX.XXXXXX
```

9. Given the integral $\int_0^3 3x^2 \sin x \, dx$, and the closed solution

$$\int 3x^2 \sin x \, dx = 3[2x \sin x - (x^2 - 2) \cos x],$$

write a program in FORTRAN to evaluate the given integral by using the Gaussian quadrature formula both for four points and for five points. The output should be in the following form:

```
   NAME                  DATE                  EDITION 3
4-POINT GAUSSIAN QUADRATURE
              T(0) = X.XXXXXXXX   W(0) = X.XXXXXXXX
              T(1) = X.XXXXXXXX   W(1) = X.XXXXXXXX
                ⋮                   ⋮
              T(3) = X.XXXXXXXX   W(3) = X.XXXXXXXX
5-POINT GAUSSIAN QUADRATURE
              T(0) = X.XXXXXXXX   W(0) = X.XXXXXXXX
                ⋮                   ⋮
              T(4) = X.XXXXXXXX   W(4) = X.XXXXXXXX
ANSWER FROM CLOSED-FORM EXPRESSION = X.XXXXX
ANSWER FROM 4-POINT GAUSSIAN QUADRATURE = X.XXXXX
ANSWER FROM 5-POINT GAUSSIAN QUADRATURE = X.XXXXX
```

CHAPTER 13

Errors

13.1 PRELIMINARY REMARKS

Aside from possible mistakes in the initial data, there may also be round-off and truncation errors in machine computations. Round-off errors stem from a finite number of digits in a computer word or from initial data, while truncation errors are due mainly to finite approximations of limiting processes.

This chapter summarizes the elements of these two types of errors. Error analysis for complicated computations is subject to continuing research. The interested reader is referred to the bibliography at the end of this chapter for further information.

13.2 ROUND-OFF ERRORS

If we are given a decimal number which contains a fractional part and we attempt to convert it to its binary equivalent, a conversion error due to the finite word length of the computer may be introduced, particularly if there is not *exact* binary equivalent.

For example, the decimal number 0.625 can be represented as a binary number without conversion error since it has an exact binary equivalent:†

$$0.625 = \tfrac{5}{8} = (\tfrac{1}{2})^1 + (\tfrac{1}{2})^3$$

$$= \boxed{.\ |\ 1\ |\ 0\ |\ 1}$$

---

† See Appendix A for a discussion of decimal-binary conversion.

However, if we attempt to convert the decimal number 0.626 to its binary equivalent, we find that an infinite series is needed:

$$0.626 = (\tfrac{1}{2})^1 + (\tfrac{1}{2})^3 + (\tfrac{1}{2})^{10} + (\tfrac{1}{2})^{16} + (\tfrac{1}{2})^{17} + (\tfrac{1}{2})^{21} + \cdots$$

$$= \; . \; | 1 | 0 | 1 | 0 | 0 | 0 | 0 | 0 | 0 | 1 | 0 | 0 | 0 | 0 | 0 | 1 | 1 | 0 | 0 | 0 | 1 | . | . | .$$

Assuming that the binary machine has twenty bits available for representing the binary mantissa, we can read 0.626 as either

$$. \; | 1 | 0 | 1 | 0 | 0 | 0 | 0 | 0 | 0 | 1 | 0 | 0 | 0 | 0 | 0 | 1 | 1 | 0 | 0 | 0 \qquad \text{(without rounding),}$$

or

$$. \; | 1 | 0 | 1 | 0 | 0 | 0 | 0 | 0 | 0 | 1 | 0 | 0 | 0 | 0 | 0 | 1 | 1 | 0 | 0 | 1 \qquad \text{(with rounding).}$$

When either of these values is reconverted to an exact decimal equivalent with eight-digit accuracy, the error introduced by the initial decimal-binary conversion is clearly shown:

$$. \; | 1 | 0 | 1 | 0 | 0 | 0 | 0 | 0 | 0 | 1 | 0 | 0 | 0 | 0 | 0 | 1 | 1 | 0 | 0 | 0 \quad = 0.62599945,$$

$$. \; | 1 | 0 | 1 | 0 | 0 | 0 | 0 | 0 | 0 | 1 | 0 | 0 | 0 | 0 | 0 | 1 | 1 | 0 | 0 | 1 \quad = 0.62600040.$$

Even though no decimal-binary conversion may be necessary, another source of round-off error may be introduced if the calculation requires more digits than available through a machine or compiler. For example, if a decimal computer has a capacity of eight significant digits and we attempt to add the number 0.33333333 to itself four times, we find that the last significant digit will be truncated:

$$
\begin{array}{r}
0.33333333 \\
+0.33333333 \\
\hline
0.66666666 \\
+0.33333333 \\
\hline
0.99999999 \\
+0.33333333 \\
\hline
1.33333332 \\
\end{array}
$$

Truncated to

True value $\longrightarrow$ 1.33333332 $\longrightarrow$ 1.3333333

Error = 0.00000002

After 3000 such additions, the error becomes 0.00908:

Expected value: 999.99999
Rounded-off value: 999.99091
Error: 0.00908

The study of round-off errors is important in high-speed digital computations and may be needed to:

1) Estimate the final round-off error made in solving a given problem by a specific numerical method.

2) Compare the round-off errors made in solving a given problem by different numerical methods.

3) Examine the effect which the round-off errors of a particular numerical method may have upon the given problem in order to determine whether or not computtation by that numerical method is worth while.

13.3 PROPAGATION OF ERROR FROM THE INITIAL DATA

In the previous section, we discussed the round-off errors due to a finite number of digits in a computer word. We turn now to a brief discussion of the second type of rounding error. This type of error arises solely from the initial data; no rounding errors of the first type are involved. As an illustrative example, let us consider the integrals

$$J_n = \int_0^1 x^n e^{x-1} \, dx \qquad n = 1, 2, 3, \ldots \tag{13.1}$$

If we integrate the right-hand side of Eq. (13.1) by parts, we have

$$\int_0^1 x^n e^{x-1} \, dx = [x^n e^{x-1}]_0^1 - \int_0^1 n x^{n-1} e^{x-1} \, dx$$

$$= 1 - n \int_0^1 x^{n-1} e^{x-1} \, dx$$

or,

$$J_n = 1 - n J_{n-1}. \tag{13.2}$$

The special case of Eq. (13.1) for $n = 1$ is

$$J_1 = \int_0^1 x e^{x-1} \, dx = e^{-1}. \tag{13.3}$$

If we take $e^{-1} = 0.367879$, the exact number e^{-1} is rounded to its six-digit approximation. On the basis of this initial value, we can calculate $J_2, J_3, \ldots$ from Eq. (13.2) The results are tabulated in column 2 of Table 13.1. The correct values of J_1 through J_{10} are shown in column 3 (only six significant digits are tabulated). The propagation of error is quite obvious. The errors arise solely from the inaccuracy in J_1, which has been taken to six significant digits. No rounding errors of the first type are involved.

Table 13.1 Errors in recursive use of the equation $J_n = 1 - nJ_{n-1}$

(1) n	(2) J based on $e^{-1} = 0.367879$	(3) Correct answer for J_n
1	0.367879	0.367879
2	0.264242	0.264241
3	0.207274	0.207276
4	0.170904	0.170893
5	0.145480	0.145532
6	0.127120	0.126803
7	0.110160	0.112375
8	0.118720	0.100932
9	−0.068480	0.0916123
10	1.684800	0.0838770

13.4 TRUNCATION ERRORS

Round-off errors were discussed in the previous section. We now consider the second type of errors—truncation errors—which stem from finite approximations of limiting processes. We will also discuss propagation of errors.

The truncation error and its propagation will be treated by means of an initial-value problem (see Chapter 7). In this type of problem it may be given that at $x = x_0$, $y = y_0$; and knowing that $dy/dx = f(y, x)$, we may be asked to find the value of y corresponding to $x = x_m$. We might attempt to solve this problem by using a finite-difference approach in which a succession of ordinates $(y_1^*, y_2^*, \ldots, y_k^*, y_{k+1}^*, \ldots, y_m^*)$ corresponding to a set of abscissas $(x_1, x_2, \ldots, x_k, x_{k+1}, \ldots, x_m)$, spaced at equal intervals of length h, are approximated according to the relation

$$y_{k+1}^* - y_k^* = h \cdot f(y_k^*, x_k). \tag{13.4}$$

Corresponding to each approximated ordinate y_k^*, there is a true value y_k on the given curve, as shown in Fig. 13.1. The difference of the values can be expressed by

$$\Delta y_k = y_k - y_k^*. \tag{13.5}$$

If we attempt to apply Eq. (13.4) to the true ordinates y_k and y_{k+1}, we find that an additional error, which may be called δ_{k+1}, is introduced because we are approximating the slope of the given curve by a straight line:

$$y_{k+1} - y_k = h \cdot f(y_k, x_k) + \delta_{k+1}. \tag{13.6}$$

To determine the total error Δy_{k+1} at step $k + 1$, we can subtract Eq. (13.4) from Eq. (13.6) and obtain:

$$(y_{k+1} - y_k) - (y_{k+1}^* - y_k^*) = h \cdot f(y_k, x_k) + \delta_{k+1} - h \cdot f(y_k^*, x_k).$$

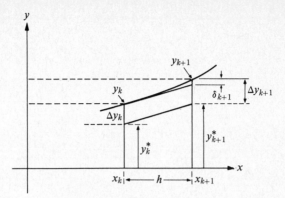

Fig. 13.1. Truncation errors.

After rearranging terms, we obtain

$$(y_{k+1} - y_{k+1}^*) - (y_k - y_k^*) = h[f(y_k, x_k) - f(y_k^*, x_k)] + \delta_{k+1};$$

and applying Eq. (13.5), we get

$$\Delta y_{k+1} - \Delta y_k = h[f(y_k^* + \Delta y_k, x_k) - f(y_k^*, x_k)] + \delta_{k+1}$$

or

$$\Delta y_{k+1} = \Delta y_k + h[f(y_k^* + \Delta y_k, x_k) - f(y_k^*, x_k)] + \delta_{k+1}. \qquad (13.7)$$

We have thus obtained an expression for Δy_{k+1}, the total error accumulated after $k + 1$ steps, in terms of the total error Δy_k, after k steps, the error caused by Δy_k in making the $(k + 1)$-step (the second term on the right-hand side), and the error introduced strictly because of the truncation in the finite series approximation at step $k + 1$, or δ_{k+1}.

We recall that the Taylor series expansion can be written

$$f(y + k, x + h) = \sum_{n=0}^{N-1} \frac{1}{n!} \left(k \frac{\partial}{\partial y} + h \frac{\partial}{\partial x} \right)^n f(y, x) + R_N,$$

where the remainder after N terms is

$$R_N = \frac{1}{N!} \left(k \frac{\partial}{\partial y} + h \frac{\partial}{\partial x} \right)^N f(y + \theta k, x + \theta h), \qquad 0 \le \theta \le 1.$$

In applying this expansion we may rewrite the following function as two terms and a remainder:

$$f(y_k^* + \Delta y_k, x_k) = f(y_k^*, x_k) + f_y(y_k^*, x_k)(\Delta y_k) + f_{yy}(y_k^* + \theta \Delta y_k, x_k)(\Delta y_k)^2/2,$$
$$0 \le \theta \le 1.$$

Substituting this expansion into Eq. (13.7), we obtain

$$\Delta y_{k+1} = \Delta y_k + h[f(y_k^*, x_k) + f_y(y_k^*, x_k)(\Delta y_k) \\ + f_{yy}(y_k^* + \theta \Delta y_k, x_k)(\Delta y_k)^2/2 - f(y_k^*, x_k)] + \delta_{k+1} \\ = \Delta y_k + h \Delta y_k[f_y(y_k^*, x_k)] + h(\Delta y_k)^2 f_{yy}(y_k^* + \theta \Delta y, x_k)/2 + \delta_{k+1}.$$

Since the second derivative term involves the small value $(\Delta y)^2$, it can often be ignored; thus the total error at the $(k + 1)$-step is expressed as

$$\Delta y_{k+1} = \Delta y_k + h \Delta y_k f_y(y_k^*, x_k) + \delta_{k+1}. \tag{13.8}$$

It should be noted that in our discussion we have considered the error introduced by, and solely due to, the $(k + 1)$-step as simply the truncation error δ_{k+1}. At each step, however, there are other errors introduced; e.g., round-off errors. Although we have ignored these other factors in our calculations, the total error introduced by the $(k + 1)$-step is actually different from δ_{k+1}.

In calculating the value of Δy_m in terms of the errors $\delta_1, \delta_2, \ldots, \delta_i, \ldots, \delta_m$, we begin at the given initial point and use Eq. (13.8) with $k = 0$ to evaluate successively:

$$\Delta y_0 = 0, \quad \text{at initial point } y = y_0 = y_0^*,$$

$$\Delta y_1 = \delta_1,$$

$$\Delta y_2 = \Delta y_1[1 + hf_y(y_1^*, x_1)] + \delta_2 \\ = \delta_1[1 + hf_y(y_1^*, x_1)] + \delta_2,$$

$$\Delta y_3 = \Delta y_2[1 + hf_y(y_2^*, x_2)] + \delta_3 \\ = \delta_1[1 + hf_y(y_1^*, x_1)][1 + hf_y(y_2^*, x_2)] + \delta_2[1 + hf_y(y_2^*, x_2)] + \delta_3,$$

$$\vdots$$

$$\Delta y_m = \delta_1[1 + hf_y(y_1^*, x_1)][1 + hf_y(y_2^*, x_2)] \cdots [1 + hf_y(y_{m-1}^*, x_{m-1})] \\ + \delta_2[1 + hf_y(y_2^*, x_2)] \cdots [1 + hf_y(y_{m-1}^*, x_{m-1})] + \cdots \\ + \delta_i[1 + hf_y(y_i^*, x_i)] \cdots [1 + hf_y(y_{m-1}^*, x_{m-1})] + \cdots + \delta_m. \tag{13.9}$$

We can see that the error contributed by δ_i to Δy_m can be expressed as

$$\delta_{im} = \delta_i \prod_{n=i}^{m-1} [1 + hf_y(y_n^*, x_n)], \quad i = 1, \ldots, m - 1. \tag{13.10}$$

Thus Eq. (13.9) may be rewritten as

$$\Delta y_m = \sum_{i=1}^{m-1} \delta_{im} + \delta_m. \tag{13.11}$$

Example 13.1

We wish to find the value of y at $x = 1$ for the initial-value problem where

$$dy/dx = f(y, x) = -y \tag{13.12}$$

and $y_0 = 1$ at $x_0 = 0$. Euler's methods will be used to reach an approximate answer with $h = 0.001$. We can compare this answer with the analytic solution and obtain the *actual error*, which in turn can then be contrasted to the error Δy_m as derived above.

We approximate δ_i by the following term in the Taylor series expansion:

$$\delta_i = \frac{h^2}{2}\left(\frac{d^2 y}{dx^2}\right)_{x=x_{i-1}}$$

Using the given condition in Eq. (13.12), we have

$$\delta_i = \frac{h^2}{2}\left[\frac{d}{dx}\left(\frac{dy}{dx}\right)_{x=x_{i-1}}\right] = \frac{h^2}{2}\left[\frac{d}{dx}(-y)\right]_{x=x_{i-1}}$$

$$= h^2 \frac{y_{i-1}^*}{2}. \tag{13.13}$$

In order to obtain an expression for y_{i-1}^* in terms of the given initial conditions, we can apply the finite-difference formula, Eq. (13.4):

$$y_1^* = y_0 + hf(y_0, x_0) = y_0 + h(-y_0) = y_0(1 - h),$$
$$y_2^* = y_1^* + hf(y_1^*, x_1) = y_1^* + h(-y_1^*)$$
$$= y_1^*(1 - h) = y_0(1 - h)^2,$$
$$y_3^* = y_0(1 - h)^3,$$
$$\vdots$$
$$y_{i-1}^* = y_0(1 - h)^{i-1}.$$

Substituting this expression into Eq. (13.13) we obtain

$$\delta_i = y_0(1 - h)^{i-1}(h^2/2). \tag{13.14}$$

Equation (13.9) now becomes

$$\delta_{im} = y_0(1 - h)^{i-1}(h^2/2)\prod_{n=i}^{m-1}[1 + hf_y(y_n^*, x_n)]. \tag{13.15}$$

From Eq. (13.12) we have

$$f_y(y_n^*, x_n) = (d/dy)[f(y, x)]_{x=x_n} = (d/dy)(-y)_{x=x_n} = -1,$$

and so Eq. (13.15) can be further reduced to

$$\delta_{im} = y_0(1 - h)^{i-1}(h^2/2)\prod_{n=i}^{m-1}(1 - h) \qquad i = 1, \ldots, m - 1,$$

$$= y_0(1 - h)^{i-1}(h^2/2)(1 - h)^{m-i}$$

$$= y_0(h^2/2)(1 - h)^{m-1}. \tag{13.16}$$

Substituting Eq. (13.14) with $i = m$ and Eq. (13.16) in Eq. (13.11), we have

$$\Delta y_m = \sum_{i=1}^{m-1} y_0 \left(\frac{h^2}{2}\right)(1 - h)^{m-1} + y_0 \left(\frac{h^2}{2}\right)(1 - h)^{m-1}$$

$$= \frac{my_0(h^2/2)(1 - h)^m}{1 - h}. \tag{13.17}$$

If we define the interval length A as

$$A = mh = x_m - x_0 = 1 - 0 = 1,$$

then Eq. (13.17) becomes

$$\Delta y_m = \frac{y_0(h/2)(1 - h)^{1/h}}{1 - h}. \tag{13.18}$$

Since

$$\lim_{h \to 0} \frac{1}{1 - h} = 1 \quad \text{and} \quad (1 - h)^{1/h} = e^{-1},$$

Eq. (13.18) becomes

$$\Delta y_m = \frac{y_0 h e^{-1}}{2}.$$

Since we chose $h = 0.001$ and were given that $y_0 = 1$, we have

$$\Delta y_m = \frac{1 \times 0.001 e^{-1}}{2} = 0.00018394.$$

Table 13.2 Truncation errors

Row	Precision Interval h	Single-precision		Double-precision	
		0.001	0.0001	0.001	0.0001
1	e^{-1}	0.36787946	0.36787946	0.36787946	0.36787946
2	Eulerian approximation on IBM 360	0.36768100	0.36769542	0.36769542	0.36786105
3	Actual error row 1–row 2	0.00019846	0.00019575	0.00018404	0.00001841
4	Theoretical error Δy_m	0.00018394	0.00001839	0.00018394	0.00001839
5	Difference between rows 3 and 4	0.00001452	0.00017736	0.00000010	0.00000002

The Euler approximation was based on the double-precision calculation with $h = 0.001$. The results for this example and the same example with $h = 0.0001$, using both single- and double-precision, are summarized in Table 13.2. It is noted that in the single-precision computation, the difference between the actual error (truncation, round-off, and other errors) and the theoretical error (truncation only) increases with a smaller interval ($h = 0.0001$). On the other hand, this difference decreases in the case where double precision was used.

BIBLIOGRAPHY

Error in Interpolation

Ostrowski, A. M., "On the rounding-off of difference tables for linear interpolation," *Math. Tables Aids Comput.*, **6**, 212–214 (1952).

Error in Numerical Integration

Hull, T. E., and A. C. R. Newberry, "Error bounds for a family of three-point integration procedures," *J. Soc. Ind. Appl. Math.*, **7**, 402–412 (1959).

Huskey, H. D., "On the precision of a certain process of numerical integration" (with an appendix by D. R. Hartree), *J. Res. Natn. Bur. Stand.*, **42**, 57–62 (1949).

Lotkin, M., "The propagation of error in numerical integration," *Proc. Am. Math. Soc.*, **5**, 869–887 (1954).

Rademacher, H., "On the accumulation of errors in processes of integration on high-speed calculating machines," *Ann. Comput. Lab. Harvard Univ.*, **16**, 176–187 (1948).

Thompson, R. J., "Improving round-off in Runge-Kutta computations with Gill's method," *Commun. Assoc. Comput. Mach.*, **13**, 739–740 (1970).

Error in the Approximation of Analytical Functions

Davis, P., "Errors of numerical approximation for analytic functions," *J. Rat. Mech. Analysis*, **2**, 303–313 (1953).

Davis, P., and P. Rabinowitz, "On the estimation of quadrature errors for analytic functions," *Math. Tables Aids Comput.*, **8**, 193–202 (1954).

Error in Matrix Calculation

Turing, A. M., "Rounding-off errors in matrix processes," *Quart. J. Mech. Appl. Math.*, **1**, 287–308 (1948).

Wilkinson, J. H., "Error analysis of direct methods of matrix inversion," *J. Assoc. Comput. Mach.*, **8**, 281–330 (1961).

Error in the Numerical Integration of Differential Equations

Carr, J. W., III, "Error bounds for the Runge-Kutta single-step integration process," *J. Assoc. Comput. Mach.*, **5**, 39–44 (1958).

Henrici, P., *Error Propagation for Difference Methods*. Wiley, New York, 1963.

Lotkin, M., "On the accuracy of Runge-Kutta's method," *Math. Tables Aids Comput.*, **5**, 128–133 (1951).

Moore, R. E., "The automatic analysis and control of error in digital computation based on the use of interval numbers," *Error in Digital Computation*, Vol. 1, pp. 61–130. Wiley, New York, 1965.

Scraton, R. E., "Estimation of the truncation error in Runge-Kutta and allied processes," *Comp. J.*, **7**, 246–248 (1964).

Sterne, T. E., "The accuracy of numerical solutions of ordinary differential equations," *Math. Tab., Wash.*, **7**, 159–164 (1953).

Wasow, W., "On the truncation error in the solution of Laplace's equation by finite differences," *J. Res. Natn. Bur. Stand.*, **48**, 345–348 (1952).

Other References on Errors

Davis, P. J., and P. Rabinowitz, "On the estimation of quadrature errors for analytic functions," *Math. Tables Aids Comput.*, **8**, 193–203 (1954).

Macon, N., and M. Baskerville, "On the generation of errors in digital evaluation of continued fractions," *J. Assoc. Comp. Mach.*, **3**, 199–202 (1956).

Wasow, W., "The accuracy of difference approximations to plane Dirichlet problems with piecewise analytic boundary values," *Quart. Appl. Math.*, **15**, 53–63 (1957).

Weeg, G. P., "Truncation error in the Graeffe root squaring method," *J. Assoc. Comp. Mach.*, **7**, 69–71 (1960).

Wilkinson, J. H., "Error analysis of floating-point computation," *Num. Math.*, **2**, 319–340 (1960).

Wilkinson, J. H., *Rounding Errors in Algebraic Processes*. Prentice-Hall, Englewood Cliffs, N.J., 1963.

Zondek, B., and J. W. Sheldon, "On the error propagation in Adams' extrapolation method," *Math. Tables Aids Comput.*, **13**, 52–55 (1959).

PROBLEMS

1. A unit sphere is placed at the base of a wall, not necessarily in the corner, in a rectangular room.[†] It is known that radius r of the largest sphere S that can be passed between the unit sphere, the wall, and the floor is in Fig. 13.2,

$$r = \frac{\sqrt{2} - 1}{\sqrt{2} + 1},$$

and the volume of this sphere S is

$$V = \frac{4\pi r^3}{3}.$$

[†] This problem was originally suggested by Professor R. V. Andree of the University of Oklahoma.

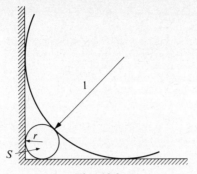

Fig. 13.2

From the two equations above, the following five expressions for V may be obtained:

$$V_1 = \frac{4}{3}\pi(\sqrt{2} - 1)^6,$$

$$V_2 = \frac{4}{3}\pi/(\sqrt{2} + 1)^6,$$

$$V_3 = \frac{4}{3}\pi(99 - 70\sqrt{2}),$$

$$V_4 = \frac{4}{3}\pi/(99 + 70\sqrt{2}),$$

$$V_5 = \frac{4}{3}\pi\left(\frac{\sqrt{2} - 1}{\sqrt{2} + 1}\right)^3.$$

Compute V_1 through V_5, using $\sqrt{2} = 1.0$, $\sqrt{2} = 1.4$, $\sqrt{2} = 1.41$, $\sqrt{2} = 1.414$, $\sqrt{2} = 1.4142$, and $\sqrt{2} = 1.41421$ respectively, and fill in the accompanying table.

$\sqrt{2}$	V_1	V_2	V_3	V_4	V_5
1.0	0.0	0.065450	121.4749	0.024786	0.0
1.4					
1.41					
1.414					
1.4142					
1.41421	0.021155	0.021156	0.022201	0.021156	0.021156

```
            SUBROUTINE RUTIS(N,A,B,ANS)
            ZN=N
            COFF=(B-A)/ZN
      C        INITIALIZATION
            IF (N-16) 10,10,15
     10  N1=N
            GO TO 20
     15  N1=16
            IF(N-256) 20,20,25
     20  N2=N
            GO TO 30
     25  N2=256
            IF(N-4096) 30,30,35
     30  N3=N
            GO TO 40
     35  N3=4096
      C        MAKING SUMS
     40  DO 200 K3=1,N,4096
              SUM3=0.
              J=K3+N3-1
              DO 150 K2=K3,J,256
                SUM2=0.
                JJ=K2+N2-1
                DO 100 K1=K2,JJ,16
                  SUM1=0.
                  JJJ=K1+N1-1
                  DO 90 K=K1,JJJ,2
                    ZK=K
     90             SUM1=SUM1+F(A+ZK*COFF)
    100          SUM2=SUM1+SUM2
    150        SUM3=SUM2+SUM3
    200  ANS=ANS+SUM3
            RETURN
            END
```

Fig. 13.3. Subroutine RUTIS.

2. In the evaluation of the following expression,

$$R = \sum_{j=1}^{n} f\left[a + (j - \tfrac{1}{2})\,\frac{b-a}{n}\right],$$

the repeated addition of small terms to a large partial sum can lead to serious roundoff error. Rutishauser[†] suggested a procedure for the evaluation of the value R which significantly reduces this error. This procedure is shown in the subroutine RUTIS (Fig. 13.3). Draw a flowchart for this subroutine. The function f, and the variables a, b, n are given.

[†] Rutishauser, H., "Description of Algol 60," *Handbook for Automatic Computation*, Vol. 1, Part a. Springer-Verlag, Berlin, 1968.

MODERN METHODS

Many numerical methods devised before the advent of high-speed computers and merely adapted for computer usage were discussed in Part II. However, application of some new methods has grown so extensively in recent years that an understanding of the underlying principles of the techniques is indispensable in handling certain scientific and engineering problems. Two of these methods, the Monte-Carlo method and linear programming, have been selected for discussion in PART III.

CHAPTER 14

Monte Carlo Method
and Random Numbers

14.1 MONTE CARLO METHOD

One of the most powerful features of the digital computer is its high speed, which makes possible the performance of many repeated arithmetic operations within a short period of time. The Monte Carlo method is a procedure which takes advantage of this speed in solving complex problems in science and engineering where analytical formulation is not available and experimental procedure is not possible. The Monte Carlo method is mainly used to predict the final consequence of a series of occurrences, each having its own probability. We shall illustrate this procedure by considering the simple example of obtaining the shaded area under the curve $f(x) = L \sin x$, where $L = 0.876$ (see Fig. 14.1).

To "solve" this problem by the Monte Carlo method, we begin by asking, If a series of darts are thrown at random onto a rectangular board having dimensions $d \times \pi$, what is the probability P_a that the darts will hit the shaded area?

$$P_a = \frac{\text{Number of darts hitting the shaded area}}{\text{Total number of darts thrown}}. \qquad (14.1)$$

This probability is clearly related to the shaded area itself. For example, a low probability of hitting the shaded area implies that this area is relatively small:

$$\text{The shaded area} = P_a \cdot (\text{total area OABC}). \qquad (14.2)$$

A hit or a miss can readily be expressed by the following expressions: if

$$Y_i \geq y_i, \quad \text{it is a miss}, \qquad (14.3)$$

and if

$$Y_i < y_i, \quad \text{it is a hit}, \qquad (14.4)$$

327

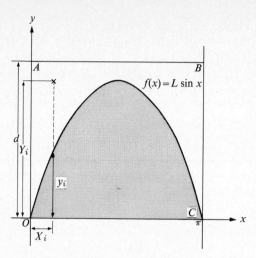

Fig. 14.1. Area under a given curve.

where

y_i = ordinate of the curve at a given X_i,

Y_i = ordinate of the point of impact (see Fig. 14.1), $0 \le Y_i \le d$.

In order to calculate the shaded area by the Monte Carlo method, we follow these steps:

1) We adopt two random input values for X_i and Y_i representing the abscissa and ordinate, respectively, for an impact point ($0 \le X_i \le \pi$ and $0 \le Y_i \le d$).

2) We compare Y_i with the value of y_i corresponding to X_i. If Eq. (14.4) is satisfied, we register a hit.

3) We calculate P_a, using Eq. (14.1).

4) We repeat steps 1, 2, and 3 N times, where N is some large integer.

It should be mentioned that in order to repeat Step 1 many times, we must be able to generate a large supply of random numbers, and since this in itself is a rather involved process, we will defer its discussion to the next section. We may state here, however, that many subprograms have been written to generate floating-point random numbers between 0 and 1, and that by a simple transformation these subprograms can be made to generate numbers randomly in some other interval, such as $(0, \pi)$ or $(0, d)$. For example, if we wanted to generate numbers randomly between 0 and π and if a subprogram called RANDU had already been written to generate random numbers between 0 and 1,

$$Z = \text{RANDU(IX,IY,YFL)} \qquad 0 \le Z \le 1,$$

then we could simply multiply by the factor π to obtain

$$X = \text{RANDU(IX,1Y,YFL)} * \pi, \qquad 0 \le X \le \pi.$$

Table 14.1 Area under a sine curve by the Monte Carlo method

IX=   123   XL=   0.876000   D=   1.122999

N	PROBABILITY	AREA
100	0.510000	1.799282
200	0.515000	1.816922
300	0.486667	1.716962
400	0.502500	1.772821
500	0.492000	1.735778
600	0.493333	1.740481
700	0.492857	1.738802
800	0.497500	1.755181
900	0.492222	1.736562
1000	0.489000	1.725194
1100	0.490000	1.728722
1200	0.485000	1.711082
1300	0.484615	1.709724
1400	0.482143	1.701001
1500	0.487333	1.719314
1600	0.485000	1.711082
1700	0.485294	1.712119
1800	0.482778	1.703241
1900	0.485789	1.713867
2000	0.485500	1.712846
.		
.		
.		
.		
8100	0.489012	1.725237
8200	0.488902	1.724850
8300	0.489639	1.727447
8400	0.490238	1.729562
8500	0.489412	1.726646
8600	0.490698	1.731183
8700	0.490805	1.731561
8800	0.490341	1.729924
8900	0.490112	1.729118
9000	0.490111	1.729114
9100	0.491209	1.732986
9200	0.492391	1.737158
9300	0.492473	1.737447
9400	0.493191	1.739982
9500	0.493789	1.742091
9600	0.493437	1.740849
9700	0.494124	1.743270
9800	0.493775	1.742042
9900	0.494545	1.744758
10000	0.495200	1.747067

An actual run on a digital computer has been made to compute the shaded area of Fig. 14.1. The associated FORTRAN IV program for the Monte Carlo method is listed in Fig. 14.2. The results, as shown in Table 14.1, were obtained by following the four steps mentioned previously. We begin by enclosing the sine curve in a rectangle of length π and of arbitrarily chosen height $d = 1.123$. After every 100 darts thrown, the results for the probability P_a and the computed area are printed out. Thus, after the first 100 throws there are 51 hits; $P_a = 51/100 = 0.51$, and the first estimate for

```
C     PROGRAM TO FIND THE AREA UNDER SINE CURVE
C     WITH RANDU SUBROUTINE
C     IX = INITIAL RANDOM NUMBER,   XL = COEFFICIENT OF SIN(X)
C
      DATA IREAD,IWRITE/5,6/
      READ(IREAD,1) IX,XL,D
      WRITE(IWRITE,2) IX,XL,D
      WRITE(IWRITE,10)
      HIT=0.
      N=0
      PI=3.14159265
      TA=PI*D
      DO 100 K=1,100
      DO 25 I=1,100
C
C     FOLLOWING 6 STATEMENTS TO FIGURE OUT THE ABSCISSA AND
C     ORDINATE OF THE POINT THAT DART HITS
C
      CALL RANDU(IX,IY,YFL)
      IX=IY
      Y=YFL*D
      CALL RANDU(IX,IY,YFL)
      IX=IY
      Z=XL*SIN(YFL*PI)
      IF(Y.GE.Z) GO TO 25
      HIT=HIT+1.
   25 CONTINUE
C
C     FOLLOWING 6 STATEMENTS FOR FINDING THE PROBABILITY AND
C     THE AREA AFTER EVERY 100 SHOTS
C
      N=N+100
      ZN=N
      P=HIT/ZN
      SA=TA*P
      WRITE(IWRITE,95) N,P,SA
  100 CONTINUE
    1 FORMAT(I10,2F10.5)
    2 FORMAT(' IX=',I10,5X,'XL=',F10.6,5X,'D=',F10.6/)
   10 FORMAT(6X,'N',22X,'PROBABILITY',22X,'AREA'/)
   95 FORMAT(3X,I5,21X,F8.6,21X,F9.6)
 1000 STOP
      END

      123    .876    1.123
```

Fig. 14.2. Area under a sine curve.

the shaded area is $1.123\pi \times 0.51 = 1.7992$. This value for the area is far from the true area 1.752; however, from Table 14.1, we note that the results may fluctuate in the beginning but tend to converge to 1.752 as N gets large. This fluctuation can be observed in Fig. 14.3.

14.2 BUFFON NEEDLE PROBLEM

As a second illustration of the Monte Carlo method, let us consider the classic *buffon needle* problem. The physical situation is quite simple. We are given a level

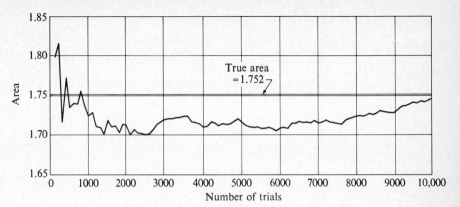

Fig. 14.3. Area under a sine curve by Monte Carlo procedure.

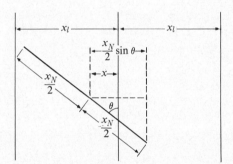

Fig. 14.4. Buffon needle problem.

plane surface with many parallel lines drawn a distance x_l apart and a thin needle of
length x_N. The question is, If we drop the needle on the plane at random a great
many times (Fig. 14.4), what is the ratio of the number of times it hits one of the
lines to the number of times it is dropped? According to statistical theory[†] the answer
is known as $2x_N/\pi x_l$; we would like to check this result by the Monte Carlo procedure.
    From Fig. 14.4 it is obvious that if

$$x \le \frac{x_N}{2} \sin \theta,$$

then the needle is touching the line; and if $x > (x_N/2) \sin \theta$, it is not touching. We
know that if the needle were actually dropped on the plane, then the values of x would
occur at random. This act of dropping the needle may be simulated by picking random input values for θ and x. Clearly, in this case θ would lie between 0 and π,

---

[†] For example, see Gredenko, B. V., *The Theory of Probability*. Chelsea, New York,
pp. 41–43, 1962.

while x would lie between 0 and $x_l/2$. After dropping the needle many times we are able to calculate the following ratio,

$$\frac{\text{Number of hits}}{\text{Number of times dropped}},$$

which should be approximately equal to $2x_N/\pi x_l$.

Figures 14.5 and 14.6 are, respectively, a flow chart and a FORTRAN IV program for solving the buffon needle problem, i.e., determining the ratio of the number of trials to hits. We take

$$x_l = 2.5 \quad \text{and} \quad x_N = 1.48.$$

A recommended output format is shown below where N is the number of random sets tested.

BUFFON NEEDLE PROBLEM

(Skip 1 line)

```
   IX = XXXXXXXXX              XL = XX.XXXXX              XN = XX.XXXXX
```

(Skip 1 line)

```
   N                        PROBABILITY                      TRUE PROB.
```

(Skip 1 line)

```
   100                       X.XXXXXX                        X.XXXXXX
   :
   5000                      X.XXXXXX                        X.XXXXXX
```

The relation of the number of throws to the corresponding probability of hitting a line based on the computer output is depicted in Fig. 14.7. It is seen that the computer answers converge well to the true answer, which is known to be $2x_N/\pi x_l = 0.376878$. The time for computing 10,000 throws on an IBM System 360 Model 50 is about 10 seconds.

We have illustrated how the Monte Carlo method is used to evaluate *approximately* a definite integral and to solve the buffon needle problem. The method can be used to find approximate answers for problems in such simulation studies as a job shop or automobile traffic flow, and for problems in nuclear physics (see references listed at the end of this chapter). It cannot be overemphasized that answers obtained by the Monte Carlo method are not very accurate when the sample size is small; increased accuracy may come only with a large number of trials.

A need for random numbers is evidenced by the two examples in the preceding sections. We shall proceed to discuss the available methods for generation of such random numbers, present a FORTRAN IV program for implementing it, and touch briefly on other possible procedures for generation of random numbers.

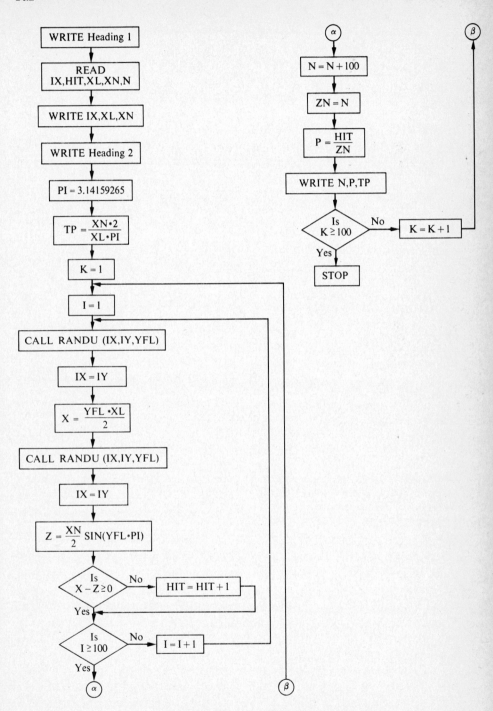

Fig. 14.5. Flow chart for buffon needle problem.

```
C      PROGRAM TO SOLVE BUFFON NEEDLE PROBLEM
C      USING RANDU SUBPROGRAM
C
C      XL=DISTANCE BETWEEN PARALLEL LINES OF THE GRID
C      XN=LENGTH OF THE NEEDLE
C
       DATA IREAD,IWRITE/5,6/
       WRITE(IWRITE,20)
C
C      READ IN THE INITIAL RANDOM NUMBER AND OTHER DATA
C
       READ(IREAD,2) IX,HIT,XL,XN,N
       WRITE(IWRITE,25) IX,XL,XN
       WRITE(IWRITE,30)
       PI=3.14159265
C
C      CALCULATING THE TRUE ANSWER BY STATISTICAL THEORY
C
       TP=XN*2./(XL*PI)
C
C
       DO 18 K=1,60
       DO 15 I=1,250
       CALL RANDU(IX,IY,YFL)
       IX=IY
       X=YFL*XL/2.
       CALL RANDU(IX,IY,YFL)
       IX=IY
       Z=(XN/2.)*SIN(YFL*PI)
C
C      COMPARE X WITH Z.IF X IS LESS THAN Z,IT MEANS THAT
C      THE NEEDLE IS TOUCHING THE LINE
C
       IF(X.GE.Z) GO TO 15
   10  HIT=HIT+1.
   15  CONTINUE
       N=N,250
       ZN=N
C
C      AFTER DROPPING EVERY 100 TIMES, CALCULATE THE PROBABILITY
C      WHICH SHOULD BE APPROXIMATELY EQUAL TO TP(TRUE ANSWER)
C
       P=HIT/ZN
C
C      PRINTING OUT THE NUMBER OF DROPPINGS, THE PROBABILITY
C      OF THE NEEDLE DROPPED ON THE LINE BY MONTE CARLO METHOD,
C      AND THE TRUE ANSWER BY STATISTICAL THEORY
C
       WRITE(IWRITE,50) N,P,TP
   18  CONTINUE
    1  FORMAT(I1)
    2  FORMAT(I10,5X,F15.3,5X,F10.5,5X,F10.5,5X,I5)
   20  FORMAT('1',30X,'BUFFON NEEDLE PROBLEM'//)
   25  FORMAT(5X,'IX=',I9,5X,'XL=',F8.5,5X,'XN=',F8.5/)
   30  FORMAT(/6X,'N',22X,'PROBABILITY',19X,'TRUE PROB.'/)
   50  FORMAT(3X,I5,21X,F8.6,21X,F8.6)
   90  STOP
       END

        135        0.                      2.5            1.48            0
```

Fig. 14.6. FORTRAN program and data for buffon needle problem.

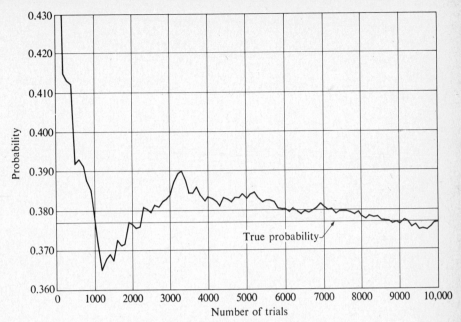

Fig. 14.7. Buffon needle problem.

14.3 GENERATION OF RANDOM NUMBERS—POWER RESIDUE METHOD

Of the many methods attempted in the past for generating random numbers with the aid of computers, the most widely used one is believed to be the power-residue method. Before this method is presented, a brief introduction to some notations and concepts in number theory is necessary.

When the difference of two numbers s and t is *evenly* divisible by an integer M, then s is said to be "congruent to t modulo M," or

$$s = t \bmod M. \tag{14.5}$$

For example, in the expression

$$19 = 9 \bmod 5,$$

$s = 19$, $t = 9$, and their difference, 10, is evenly divisible by 5. In other words, the quotient $(s - t)/M$ is an integer:

$$(19 - 9)/5 = 2.$$

If s, t and M satisfy Eq. (14.5), then the following two relations exist:

a) s and t must have the same remainder after division by M. For example,

$$19 = 9 \bmod 5,$$

$$\tfrac{19}{5} = 3\tfrac{4}{5} \quad \text{and} \quad \tfrac{9}{5} = 1\tfrac{4}{5}.$$

b) $s - t = i \cdot M$, where i = integer. For example,

$$19 = 9 \bmod 5, \qquad 19 - 9 = 2 \cdot 5.$$

In relation b) it is seen that given t and M, s may have many possible solutions which are known as a *collection* (see Table 14.2).

Table 14.2 A collection of s for the relation $s = 26 \bmod 6$

i	1	0	-1	-2	-3	-4	-5
s	32	26	20	14	8	2	-4

In the following, we choose s to be the smallest positive number in the collection, known as the *least positive residue*. Thus in Table 14.2, the least positive residue is equal to 2.

If the value of t and M are given, then the least positive residue s can readily be evaluated by the relation

$$s = t - (t/M)M, \qquad (14.6)$$

where (t/M) is a truncated-integer division; that is, any fractional part of the quotient (t/M) is ignored, *leaving only the integral part*. As a practical detail, the value of s is equal to the remainder of the division t/m.

Suppose, for example, that we wish to evaluate the value of s when $t = 10$ and $M = 4$. Then

$$s = 10 \bmod 4,$$

or

$$s = 10 - (\tfrac{10}{4})4 = 10 - (2)4 = 10 - 8 = 2.$$

Note that only the integral part, that is, 2, of the quotient 2.5 is used; the decimal part (0.5) of the quotient is discarded.

It is possible to look at the value of s another way; s is simply equal to the remainder of the division 10/4, which is 2.

We also define the set of power residues s_n as

$$s_n = t \pmod{M} \qquad \text{for} \quad n = 1, 2, 3, \ldots$$

For example, if $t = 5$ and $M = 31$, then

$$
\begin{aligned}
s_1 &= 5^1 \pmod{31} = 5 - (\tfrac{5}{31})31 = 5 - 0 = 5, \\
s_2 &= 5^2 \pmod{31} = 25 - (\tfrac{25}{31})31 = 25 - 0 = 25, \\
s_3 &= 5^3 \pmod{31} = 125 - (\tfrac{125}{31})31 = 125 - 124 = 1, \\
s_4 &= 5^4 \pmod{31} = 625 - (\tfrac{625}{31})31 = 625 - 620 = 5, \\
&\;\;\vdots
\end{aligned}
$$

(this sequence repeats)

Here we see that 5, 25, and 1 are the power residues.

In using the power-residue concept to obtain random numbers with a binary computer, we frequently generate the random number u_{n+1} by the following relation, known as the *multiplicative congruence method*:

$$u_{n+1} = xu_n (\text{mod } 2^d), \tag{14.7}$$

where u_n is the previous random number, d represents the number of significant bits in a word, and x is a constant multiplier chosen so as to obtain the longest possible sequence of random number without repetition. The choice of x is determined by a more involved power-residue theory than presented above. If we take d as 31, and x as $2^{16} + 3$, Eq. (14.7) becomes

$$u_{n+1} = u_n(2^{16} + 3)(\text{mod } 2^{31}), \tag{14.8}$$

or,

$$u_{n+1} = 65{,}539u_n(\text{mod } 2^{31}). \tag{14.9}$$

For a beginner, it is convenient to consider that Eq. (14.9) is merely a shorthand version of the following equation:

$$u_{n+1} = 65{,}539u_n - \left[\frac{65{,}539u_n}{2^{31}}\right]2^{31}, \tag{14.10}$$

where the division inside the square brackets [ ] indicates a truncated integer division.

We shall now illustrate the usage of Eq. (14.10) by two specific examples. In either example, the value of u_n is arbitrarily given, and we must compute u_{n+1}.

Example 14.1

Assuming that $u_n = 262{,}145$, compute u_{n+1}.

The truncated-integer division in Eq. (14.10) may be performed as follows:

$$\left[\frac{65{,}539u_n}{2^{31}}\right] = \left[\frac{65{,}539 \times 262{,}145}{2{,}147{,}483{,}648}\right] = 8.$$

Substituting this value into Eq. (14.10), we have

$$u_{n+1} = 65{,}539 \times 262{,}145 - 8 \times 2{,}147{,}483{,}648,$$

or

$$u_{n+1} = 851{,}971. \tag{14.11}$$

It is important to note that the value of u_{n+1} as obtained in Eq. (14.11) is simply the remainder of the following division:

$$\frac{65{,}539u_n}{2^{31}} = \frac{65{,}539 \times 262{,}145}{2{,}147{,}483{,}648}.$$

Example 14.2

Compute u_{n+1} with u_n taken as 1,044,702,915. The truncated-integer division may be performed as follows:

$$\left[\frac{65{,}539u_n}{2^{31}}\right] = \left[\frac{65{,}539 \times 1{,}044{,}702{,}915}{2{,}147{,}483{,}648}\right] = 31{,}883.$$

Substituting this value into Eq. (14.10), we obtain

$$u_{n+1} = 65{,}539 \times 1{,}044{,}702{,}915 - 31{,}883 \times 2{,}147{,}483{,}648$$

$$= 68{,}468{,}784{,}346{,}185 - 68{,}468{,}221{,}149{,}184,$$

or

$$u_{n+1} = 563{,}197{,}001, \tag{14.12}$$

which is also the remainder for the following division:

$$\frac{65{,}539u_n}{2^{31}} = \frac{65{,}539 \times 1{,}044{,}702{,}915}{2{,}147{,}483{,}648}.$$

14.4 FORTRAN SUBROUTINE FOR GENERATION OF RANDOM NUMBERS

A subroutine RANDU written in FORTRAN IV language is designed to generate random numbers according to Eq. (14.8). As shown in Fig. 14.8, the calculations for each random number are made in fixed-point arithmetic; and the answer IY is then converted to a floating-point number YFL between 0 and 1. The argument IX must be initially defined in the main program as an odd integer with nine digits or less.

```
      SUBROUTINE RANDU(IX,IY,YFL)
      IY=IX*65539
      IF(IY.GE.0) GO TO 6
    5 IY=IY+2147483647+1
    6 YFL=IY
      YFL=YFL*.4656613E-9
      RETURN
      END
```

Fig. 14.8. SUBROUTINE RANDU.

Although there are only a few statements in this subroutine, a clear understanding of its action demands a basic knowledge of the following three important topics:

1. Representation of a positive fixed-point binary integer in a fullword.

2. Representation of a negative fixed-point binary integer in a fullword.

3. Conversion of a binary integer to and from its decimal equivalence.

We shall discuss the first two topics in this and next sections, leaving the third one in Appendix A.

A fixed-point binary integer can be stored either in a halfword or in a fullword. As mentioned in Section 2.6, a halfword has 16 bits and a fullword, 32 bits. The 0 bit, or the sign bit, is used to indicate the sign of the number in consideration. Consider, for example, the representation of a positive integer $+9$ in a fullword as shown in Fig. 14.9. When the number is positive, the zero-bit contains a 0 and bits 1–31 inclusive contain a string of 0's and 1's, the binary representation of the number. It should be noted that the maximum possible positive number in a fullword is $(2^{31} - 1)$. This can be seen with the aid of Fig. 14.10. Since the sign bit is occupied by a zero, the largest positive 32-bit number consists of 31 one bits and a zero bit, which is equal to $2^{31} - 1$, or decimal 2,147,483,647.

Bit position

0	4	8	12	16	20	24	28
0000	0000	0000	0000	0000	0000	0000	1001

Fig. 14.9. Representation of $+9$ in a fullword.

We now turn to some statements used in the subroutine RANDU (See Fig. 14.8). The first statement IY = IX × 65,539 serves to compute the value of $65,539u_n$. For the sake of comparison with the calculations carried out in the Example 14.1 in the previous section, we shall also take IX = 262,145 as the initial random number. We have IY = IX × 65,539 = 17,180,721,155.

Bit position

0	4	8	12	16	20	24	28
0111	1111	1111	1111	1111	1111	1111	1111

Fig. 14.10. Representation of the maximum possible positive number $(2^{31} - 1)$ in a fullword.

This number can be readily converted into its binary equivalence (See Appendix A, decimal integer to binary operation); and its result is shown in Fig. 14.11. This 64-bit product IY is now divided by 2^{31} and the remainder is taken as the next random number. Instead of performing such a division, we need only truncate the leftmost 32 bits of the product IY. In other words, the 32 rightmost bits retained in the fullword, as shown in Fig. 14.12, represent the remainder of the division, and are therefore the required answer. Its decimal equivalence is 851,971, which is identical with the answer u_{n+1} calculated in Example 14.1 in the previous section (p. 337).

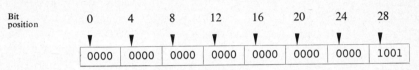

0	0100	0000	0000	0000	0011	0100	0000	0000	0011

Fig. 14.11. Binary representation of the decimal number 17,180,721,155.

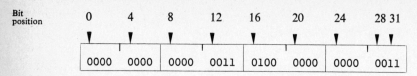

Fig. 14.12. Contents in the rightmost 32 bits, representing decimal number 851,971 (note that sign bit is 0).

In this example, the answer shown in Fig. 14.12 turns out to be a positive number, and the statement IF(IY.GE.0) GO TO 5 requires that we go to Statement 5, which is used to convert the integer IY to a floating-point number YFL. The value of this YFL is then normalized in Statement 6 so that the final answer is between 0 and 1. The normalization is performed by dividing the value of YFL obtained in Statement 5 by $2^{31} - 1$, the maximum positive number in a fullword. Note that $1/(2^{31} - 1)$ is approximately equal to 0.4656613×10^{-9}.

14.5 A STATEMENT IN SUBROUTINE RANDU

In the previous section, we studied subroutine RANDU by taking the value of the initial random number u_n as 262,145. Its result checks well with the value of u_{n+1} as computed in Example 14.1 of Section 14.3. However, Statement 4 in the subroutine (see Fig. 14.8) has not been discussed, because the contents in the rightmost 32 bits, as shown in Fig. 14.12, represent a positive quantity. In this section, we shall study this important statement by taking a different value of the initial random number: IX = 1,044,702,915.

Let us start with the Statement 2 in Fig. 14.8. We have

$$\text{IY} = \text{IX} * 65539 = 68,468,784,346,185.$$

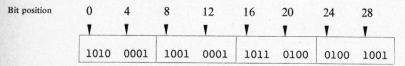

Fig. 14.13. Binary equivalence of the decimal number 68,468,784,346,185.

The binary representation of this decimal number is shown in Fig. 14.13. The product IY, 64 bits long, should now be divided by 2^{31} and the remainder used as the next random number. Instead of actually performing such division, it is easier to truncate the leftmost 32 bits of the product IY. The 32 rightmost bits, as shown in Fig. 14.14, represent the remainder of the division and are therefore the required answer. Their decimal equivalent is $-1,584,286,647$.

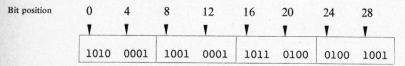

Fig. 14.14. Contents in the fullword after truncation, representing decimal number $-1,584,286,647$.

Since the value IY is now a negative number, we proceed to the Statement 4: IY = IY + 2147483647 + 1. The final IY value is

$$-1,584,286,647 + 2,147,483,648 = 563,197,001.$$

It should be noted that the addition of 2^{31}, or $2,147,483,648$, to the original negative number makes the IY a positive number, representing the *least positive residue s*. This final IY value checks identically with the answer u_{n+1} calculated in Example 14.2 of Section 14.3 (p. 338).

We shall conclude this section by showing how the decimal equivalence of the binary integer shown in Fig. 14.14 was obtained.

As mentioned before, in the IBM System/370 or IBM System/360 computers, a fixed-point binary integer can be stored either in a halfword or a fullword. The zero-bit is the sign bit. When it contains a 1, the binary integer is negative. For example, a negative number -9, represented in a fullword, is shown in Fig. 14.15. The bits 1 through 31 contain the so-called two's complement of the number 9.

Bit position	0	4	8	12	16	20	24	28
	1111	1111	1111	1111	1111	1111	1111	0111

Fig. 14.15. Representation of -9 in a fullword.

The complement of a given number is a number which, when added to the given number, yields zero as the answer and carries from the leftmost position. For example, it is known that the complement of a decimal number 653 is 347. Here:

$$
\begin{array}{rl}
653 & \longleftarrow \text{Given number} \\
+\,347 & \longleftarrow \text{Its complement} \\
\hline
\boxed{1}000 & \\
\uparrow\!\!\!\text{\underline{\hspace{3cm}}} & \text{Carry}
\end{array}
$$

A practical way to find the complement of a decimal number is as follows: subtract the number from the highest number of that degree, then add 1 to the result. For example, if we wish to find the complement of 653, we can perform the following operation:

$$
\begin{array}{rl}
999 & \longleftarrow \text{Highest of all three-digit numbers} \\
-)\,653 & \longleftarrow \text{Given number} \\
\hline
346 & \longleftarrow \text{Nine's complement} \\
+)\ \ 1 & \longleftarrow \text{Add 1} \\
\hline
347 & \longleftarrow \text{The required complement (ten's complement)}
\end{array}
$$

where the answer 347 is known as the ten's complement of 653.

The complements in the binary system work in a similar manner. A binary number may have its one's or two's complement. The one's complement of a binary number is found by subtracting each digit from 1; and two's complement, by adding 1 to the one's complement. For example, the complement of the binary number 1101 can be calculated as follows:

$$
\begin{array}{ll}
1111 & \longleftarrow \text{Highest of all four-digit binary numbers} \\
-)1101 & \longleftarrow \text{Given binary number} \\
\hline
0010 & \longleftarrow \text{One's complement} \\
+)\ \ \ 1 & \longleftarrow \text{Add 1} \\
\hline
0011 & \longleftarrow \text{The required complement (two's complement)}
\end{array}
$$

As a practical detail, the result of the subtraction, or 0010, in the above example can be obtained simply by reversing the sense of each binary digit of the given number 1101.

It is important to note that all negative binary integers (fixed point numbers) in the IBM System/370 or System/360 are carried in the two's complement form. In order to find the *true* answer, we must first recomplement and then change sign. Consider, for example, the negative binary number as shown in Fig. 14.14. To find its *true* form, the complement is first obtained (see Fig. 14.16); its decimal equivalence is 1,584,286,647. Then we change its sign, yielding $-1,584,286,647$ as the correct answer.

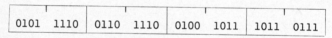

Fig. 14.16. The two's complement notation of the binary number shown in Fig. 14.14.

BIBLIOGRAPHY

Random Numbers

Coveyou, R. R., and R. D. MacPherson, "Fourier analysis of uniform random number generators," *J. Assoc. Comput. Mach.*, **14**, no. 1, 100–119 (1967).

Downham, D. Y., and F. D. K. Roberts, "Multiplicative congruential pseudo-random number generators," *Comput. J.*, **10**, no. 1, 74–77 (1967).

Hutchinson, D. W., "A new uniform pseudo-random number generator," *Commun. Assoc. Comput. Mach.*, **9**, no. 6, 432–433 (1966).

Jansson, B., *Random Number Generators*, p. 205. Almquist and Wiksell, Stockholm, 1966.

Jucosa, M. L., "Random number generation on the BRL high-speed computing machines," *BRL Report No. 855* (1953).

Knuth, D. E., *The Art of Computer Programming*, Vol. II, Ch. 3. Addison-Wesley, Reading, Mass., 1969.

Lewis, P. A. W., A. S. Goodman, and J. M. Miller, "A pseudo-random number generator for the System/360," *IBM System Journal*, **8**, no. 2, 136–146 (1969).

Marsaglia, G., "Random numbers fall mainly in the planes," *Proc. Natn. Acad. Sci. U.S.A.*, **61**, no. 2, 25–28 (1968).

Whittlesey, J. R. B., "A comparison of the correlational behavior of random number generators for the IBM System/360," *Commun. Assoc. Comput. Mach.*, **11**, no. 9, 641–644 (1968).

Monte Carlo Method

Bauer, W. F., "The Monte Carlo method," *J. Soc. Ind. Appl. Math.*, **6**, 438–451 (1958).

Beckenback, E. F. (ed.), *Modern Mathematics for the Engineer*, Ch. 12.   University of California, Engineering Extension Series. McGraw-Hill, New York, 1956.

Hammersley, J. M., and K. W. Morton, "Poor man's Monte Carlo," *J. Roy. Statist. Soc., Ser. B*, **16**, 23–38 (1954).

Householder, A. S., G. E. Forsythe, and H. H. Germond (eds.), "Monte Carlo method," *Natn. Bur. Stand. Appl. Math. Ser.*, **12** (1951).

McCracken, D. D., "The Monte Carlo method," *Scient. Am.*, **192**, 90 (May 1955).

Meyer, H. A. (ed.), *Symposium on Monte Carlo Methods*. Wiley, New York, 1956.

Shreider, U. A., *The Monte Carlo Method*. Pergamon Press, New York, 1966.

Zaremba, S. K., "The mathematical basis of Monte Carlo methods," *SIAM Rev.*, **10**, 303–314 (1968).

Application of Monte Carlo Method

Davis, P. J., and P. Rabinowitz, "Some Monte Carlo experiments in computing multiple integrals," *Math. Tables Aids Comput.*, **10**, 1–7 (1956).

Hammersley, J. M., "Monte Carlo methods for solving multivariable problems," *Ann. N.Y. Acad. Sci.*, **86**, 844–874 (1960).

King, G. T., "Monte Carlo method for solving diffusion problems," *Ind. Engng. Chem.*, **43**, no. 11, 2475–2478 (1951).

Matthes, W., "Model calculation of phase transactions with Monte Carlo," *J. Computat. Phys.*, **4**, no. 4, 431–450 (1969).

Richtmyer, R. D., "The evaluation of definite integrals and a quasi-Monte Carlo method based on the properties of algebraic numbers," *Los Alamos Scientific Laboratory Report* LA1342 (1951–1952).

Talley, W. K., and S. Whitaker, "Monte Carlo analysis of Knudsen flow," *J. Computat. Phys.*, **4**, no. 3, 389–410 (1969).

Todd, J., "Experiments in the solution of differential equations by Monte Carlo methods," *J. Wash. Acad. Sci.*, **44**, 377–381 (1954).

Other References

Dixon, W. J., and F. J. Massey, Jr., *Introduction to Statistical Analysis*, 2nd ed. McGraw-Hill, New York, 1957.

Hoel, P. G., *Introduction to Mathematical Statistics*. Wiley, New York, 1956.

Lehmer, D. H., "Mathematical methods in large-scale computing units," *Proceedings of a Second Symposium* (1949) *on Large-Scale Digital Calculating Machinery.* Harvard University Press, Cambridge, Mass., 1951.

Nagell, T., *Introduction to Number Theory.* Wiley, New York, 1951.

Stoneham, R. G., "On a new class of multiplicative pseudo-random generators," *BIT* **10**, no. 4, 481-500 (1970).

PROBLEMS

1. Using subroutine RANDU, generate 5000 numbers between 0 and 1, calculate their mean and compare the calculated mean with the expected mean (0.5) as a test of randomness.

2. Calculate the distribution in intervals of (0.1) for 5000 numbers generated by subroutine RANDU, i.e., find the percentage of numbers between 0 and 0.1, between 0.1 and 0.2, etc. Compare these results with the expected distribution of 10% in each interval.

3. Find the area under the bell-shaped curve $y = e^{-x^2/2}/\sqrt{2\pi}$ by using the Monte Carlo method with $0 \le x \le 4$.

4. Bernoulli's lemniscate is described by the equation $\rho^2 = a^2 \cos 2\theta$, where ρ and θ are polar coordinates, and the constant a denotes the size of the lemniscate (see Fig. 14.17). Use the Monte Carlo method to find the area under one loop of this curve and compare your result with the analytical solution.

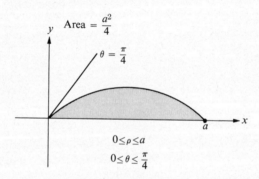

Fig. 14.17

5. Use the Monte Carlo method to calculate the volume of the sphere

$$x^2 + y^2 + z^2 = 1.$$

6. Two suburbanites plan to meet in front of the city hall at 6 p.m. Because of the traffic situation, they agree to arrive at the city hall sometime between 6 p.m. and 7 p.m. They also agree that each will stay only 10 minutes after arrival and then leave if the other has not shown up. Use the Monte Carlo method to find the probability of their actually meeting and compare your results with the correct answer, 11/36.

7. Three containers are available. Initially, container I is filled with numbered balls marked from 1 to N. (N is a large number). From the third container filled with N numbered

slips of paper, one picks a numeral at random and transfers the ball marked with that number from one container to the other. He then returns the slip of paper and repeats the process. In the beginning, the flow of balls will be strongly from container I to container II, as the former contains many more balls than the latter. This game is known as the Ehrenfest model and is used to illustrate a Markov chain. Play the Ehrenfest game with 16,384 balls and 150,000 drawings. Plot the number of balls in container I vs. container II every 1,000 drawings.

8. A drunk starts off at a lamppost and takes 16 steps, each of length 2 and each in a random direction θ, where $0° \leq \theta \leq 360°$. Use the Monte Carlo method to find his most probable distance from the lamppost after the 16 steps and compare your results with the correct answer, $2\sqrt{16} = 8$. What is the distance after 200 steps?

9. On an American roulette wheel there are 38 slots, 36 for the digits 1 to 36, one for 0, and one for 00. The slots corresponding to the odd numbers in the ranges 1 through 10 and 19 through 28 and to the even numbers in the ranges 11 through 18 and 29 through 36 are colored red. The 0 and 00 slots are colored green and the rest are black. Using random numbers, write a FORTRAN program to simulate the operation of the roulette wheel and to print out the number and color for a succession of spins.

10. In each play of a particular game, a pair of dice are thrown and the sum of the spots on the upper sides of the dice are counted. Using random numbers, find the probability that each of the different possible sums will occur.

11. By playing a simple card game, it is possible to use random numbers and multiply two proper functions n_1/d_1 and n_2/d_2. In two decks each containing d cards, n of them are marked. Each deck is then shuffled separately and one card from each is chosen at random. If both of the cards chosen are marked, a win is recorded; if not, a loss. The cards are then replaced, the decks reshuffled, and after the procedure has been repeated many times the number of wins would approximate the product

$$\frac{n_1}{d_1} \cdot \frac{n_2}{d_2}.$$

Approximate the product $\frac{2}{3} \times \frac{5}{9}$ by adapting this game to the Monte Carlo procedure.

CHAPTER 15

Linear Programming

15.1 CONSTRAINT, NONNEGATIVE CONDITION, AND OBJECTIVE FUNCTION

In Chapter 8 we introduced the Gauss-Jordan method for solving a system of simultaneous equations:

$$a_{11}x_1 + a_{12}x_2 + a_{13}x_3 = c_1,$$
$$a_{21}x_2 + a_{22}x_2 + a_{23}x_3 = c_2, \tag{15.1}$$
$$a_{31}x_3 + a_{32}x_2 + a_{33}x_3 = c_3,$$

or

$$[A][x] = [C], \tag{15.2}$$

and we observed that in each instance $[A]$ was a nonsingular square matrix. In other words, the number of unknowns was the same as the number of independent linear equations. In this chapter we shall extend our discussion to an important class of linear inequalities:

$$a_{11}x_1 + a_{12}x_2 + \cdots + a_{1n}x_n \geq c_1 \quad (\text{or } \leq c_1),$$
$$a_{21}x_1 + a_{22}x_2 + \cdots + a_{2n}x_n \geq c_2 \quad (\text{or } \leq c_2),$$
$$\vdots \tag{15.3}$$
$$a_{m1}x_1 + a_{m2}x_2 + \cdots + a_{mn}x_n \geq c_m \quad (\text{or } \leq c_m),$$

where we have $n \geq m$ or $n \leq m$, and an infinite number of solutions. In order to choose an optimal solution from all the possible solutions, we must introduce two additional conditions:

$$x_i \geq 0, \quad i = 1, 2, \ldots, n \tag{15.4}$$

and

$$z = b_1 x_1 + b_2 x_2 + \cdots + b_n x_n, \tag{15.5}$$

where all the coefficients, a, b, and c, are known. We also wish to minimize z in Eq. (15.5).

For the sake of discussion, we shall refer to Eq. (15.3) as the constraints of the system, Eq. (15.4) as the nonnegative condition, and Eq. (15.5) as the objective function. We note that Eqs. (15.3) and (15.5) are assumed to be linear. This fact should not be overlooked since the linear programming method is used to optimize a linear function, such as Eq. (15.5), is subject to a set of linear constraints, such as given in Eq. (15.3), and is restricted by a set of nonnegative conditions, as expressed in Eq. (15.4).

15.2 CONVERSION TO THE STANDARD MINIMIZATION FORM

We noted that constraints (15.3) are a set of linear equations involving inequalities. Each inequality sign can be eliminated simply by adding or subtracting a new variable which must be nonnegative. To illustrate this point, let us consider the two constraints

$$x_1 + 2x_2 \le 3, \tag{15.6}$$

$$2x_1 + x_2 \ge 4. \tag{15.7}$$

If we add a nonnegative variable x_3 to the left-hand side of the constraint (15.6), we have

$$x_1 + 2x_2 + x_3 = 3. \tag{15.6a}$$

Similarly, if the term x_4 is subtracted from the left-hand side of the constraint (15.7), we obtain

$$2x_1 + x_2 - x_4 = 4, \tag{15.7a}$$

where x_4 is a nonnegative variable. In the above example, we have introduced two new variables x_3 and x_4 which have to be determined. They are called *slack* variables. Thus the constraint (15.3) can be rewritten as

$$\begin{aligned}
a_{11}x_1 + a_{12}x_2 + \cdots + a_{1n}x_n + x_{n+1} &= c_1, \\
a_{21}x_1 + a_{22}x_2 + \cdots + a_{2n}x_n + x_{n+2} &= c_2, \\
\vdots \\
a_{m1}x_1 + a_{m2}x_2 + \cdots + a_{mn}x_n + x_{n+m} &= c_m.
\end{aligned} \tag{15.8}$$

Our problem of linear programming set forth in the previous section may now be stated in the following standard minimization form. Determine

$$x_i \ge 0, \qquad i = 1, 2, \ldots, n + m, \tag{15.9}$$

so that

$$z = [B] \cdot [x] \tag{15.10}$$

is a minimum and

$$[A][x] = [C], \tag{15.11}$$

where $[A]$ is a matrix of m rows and $n + m$ columns; $[x]$ and $[C]$ are column matrices of degree $n + m$ and m, respectively; and $[B]$ is a row matrix of degree $n + m$.

We now proceed to examine the cases in which the objective function is to be maximized, rather than minimized. In order to transform the given objective function to the standard minimization form, we can multiply it by -1; the resulting equation will have the precise form of Eq. (15.10). For example, consider the objective function z,

$$z = 3x_1 - 6x_2, \tag{15.12}$$

which is to be maximized. By changing the sign of each term, we can reduce this to the following equation, where we wish to determine x_1 and x_2 such that

$$-z = -3x_1 + 6x_2 \quad \text{is a minimum.}$$

15.3 TWO-VARIABLE PROBLEM—GRAPHICAL SOLUTION

Let us examine a specific problem using the graphical method, which involves two variables and three constraints. Determine

$$x_1 \geq 0, \quad x_2 \geq 0, \tag{15.13}$$

such that

$$z = 10x_1 + 11x_2 \tag{15.14}$$

is a maximum, and such that the two variables will satisfy the following three constraints:

$$3x_1 + 4x_2 \leq 9, \tag{15.15a}$$

$$5x_1 + 2x_2 \leq 8, \tag{15.15b}$$

$$x_1 - 2x_2 \leq 1. \tag{15.15c}$$

We shall now draw a graph representing the two nonnegative conditions and the three constraints. From Eq. (15.13) it is clear that any possible point representing solutions for x_1 and x_2 must be confined to the first quadrant (Fig. 15.1), including the points on both the x_1- and x_2-axes. Next we consider the three constraints (15.15). It is evident that the set of points satisfying Eq. (15.15a) must be below the line

$$3x_1 + 4x_2 = 9,$$

as shown in Fig. 15.2, where no solution is possible in the shaded portion. The other two constraints are readily plotted in the same way. An area bounded by the lines

$$x_1 = 0, \quad x_2 = 0,$$

$$3x_1 + 4x_2 = 9, \quad 5x_1 + 2x_2 = 8, \quad x_1 - 2x_2 = 1$$

is thus formed, as shown in Fig. 15.3.

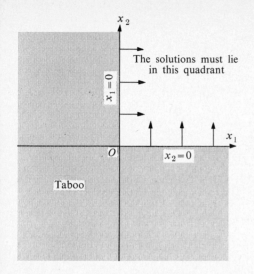

Fig. 15.1. Feasible solutions for $x_1 \geq 0$, $x_2 \geq 0$.

Fig. 15.2. Feasible solutions for $3x + 4x \leq 9$.

Our problem now is to select, from all points in the unshaded area (Fig. 15.3), one corresponding to the maximum z in Eq. (15.14). This task can be readily accomplished by plotting a family of straight lines representing

$$10x_1 + 11x_2 = \text{constant},$$

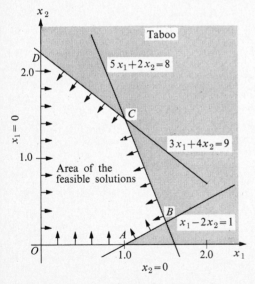

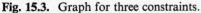

Fig. 15.3. Graph for three constraints.

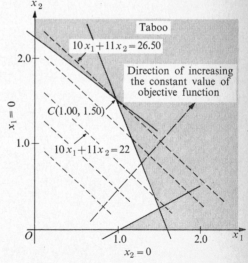

Fig. 15.4. The optimal solution.

as shown in Fig. 15.4. The solution of this example clearly occurs at point C, where

$$x_1 = 1.0, \qquad x_2 = 1.5.$$

Hence $z = 10x_1 + 11x_2 = 26.5$.

Before proceeding to the next section, it would be useful to introduce, with the aid of Fig. 15.3, the concepts of *feasible* and *optimum* solutions.

Feasible solutions are represented by all points confined inside the polygon $OABCD$ (including the points on the five sides). In other words, a feasible solution is one in which all variables satisfy the three constraints, Eqs. (15.15), and the non-negative condition, Eq. (15.13). The optimum solution is one of the feasible solutions which maximizes the objective function, Eq. (15.14).

15.4 BASIC FEASIBLE SOLUTIONS AND VERTICES

In this section, we shall introduce the concept of *basic feasible solutions* and discuss their relation with the vertices of a polygon (Fig. 15.3) in a two-variable case.

Let us reexamine the example discussed in the previous section. Determine

$$x_1 \geq 0, \qquad x_2 \geq 0, \tag{15.13}$$

so that

$$z = 10x_1 + 11x_2 \tag{15.14}$$

is a maximum, and so that

$$\begin{aligned} 3x_1 + 4x_2 &\leq 9, \\ 5x_1 + 2x_2 &\leq 8, \\ x_1 - 2x_2 &\leq 1. \end{aligned} \tag{15.15}$$

By introducing additional slack variables x_3, x_4, and x_5, we can convert the problem into the following standard form. Determine

$$x_i \geq 0, \qquad i = 1, 2, \ldots, 5, \tag{15.16}$$

so that the variables will maximize the objective function

$$z = 10x_1 + 11x_2, \tag{15.17}$$

and will satisfy the linear equations

$$\begin{aligned} 3x_1 + 4x_2 + x_3 \qquad\qquad &= 9, \\ 5x_1 + 2x_2 + \quad + x_4 \qquad &= 8, \\ x_1 - 2x_2 \qquad\qquad + x_5 &= 1. \end{aligned} \tag{15.18}$$

Equations (15.18) involve five unknowns and, therefore, have an infinite number of solutions. However, if we arbitrarily assign the value zero to two of the five variables and solve for the other three, then there are ten possible solutions. These ten solutions, known as *basic solutions*, are calculated and tabulated in Table 15.1, which will be found helpful in our discussion. An inspection of the table will indicate that the solutions numbered 3, 4, 5, 6, and 9 contain negative values of the variables and are therefore not feasible, because they do not satisfy the nonnegative requirement in Eq. (15.16).

The remaining five solutions are called *basic feasible solutions*, since they satisfy two conditions: the condition of nonnegativity, as expressed in Eq. (15.16), and the constraints, Eqs. (15.18). It is important to note that these five basic feasible solutions have a one-to-one correspondence to the five vertices shown in Fig. 15.3. This correspondence is properly marked in the last column of Table 15.1.

Table 15.1 Ten possible solutions for assigning two variables to be zero and solving for the other three

Number of possible solutions	Five variables					Vertex in Fig. 15.3
	x_1	x_2	x_3	x_4	x_5	
1	0	0	9.00	8.00	1.00	O
2	0	2.25	0	3.50	5.50	D
3	0	4.00	-7.00	0	9.00	—
4	0	-0.50	11.00	9.00	0	—
5	3.00	0	0	-7.00	-2.00	—
6	1.60	0	4.20	0	-0.60	—
7	1.00	0	6.00	3.00	0	A
8	1.00	1.50	0	0	3.00	C
9	2.20	0.60	0	-4.20	0	—
10	1.50	0.25	3.50	0	0	B

In retrospect, we have ten basic solutions. They are obtained by arbitrarily assigning the value zero to two variables and then solving for the other three. Out of these ten basic solutions, there exist only five basic feasible solutions which meet the nonnegativity requirement. Finally, these five basic feasible solutions correspond to the five vertices of the polygon $OABCD$ of Fig. 15.3; they are arranged in Table 15.2.

Table 15.2 Five basic feasible solutions

Vertex	x_1	x_2	x_3	x_4	x_5
O	0	0	9.00	8.00	1.00
A	1.00	0	6.00	3.00	0
B	1.50	0.25	3.50	0	0
C	1.00	1.50	0	0	3.00
D	0	2.25	0	3.50	5.50

15.5 SIMPLEX METHOD FOR AN OPTIMUM SOLUTION

One of the special features of all linear programming problems is the discrepancy between the number of constraints and the number of variables. Frequently, the number of variables is several times the number of constraints. When both m and n are large, the graphical method presented in Section 15.3 becomes useless and other methods must be used. Of the several computational methods available, Dantzig's Simplex method is probably the most widely used. We shall now introduce this method by means of the previous example, which involves two variables and three constraints; see Eqs. (15.16) through (15.18).

The first step of the Simplex method is to eliminate the inequality signs. As mentioned before, this can be accomplished readily by adding the so-called non-negative slack variables x_3, x_4, and x_5. The problem is then converted to the following form: Determine

$$x_i \geq 0, \qquad i = 1, 2, 3, 4, 5, \tag{15.19}$$

so that

$$-z = -10x_1 - 11x_2 + 0x_3 + 0x_4 + 0x_5 \tag{15.20}$$

is as small as possible and

$$3x_1 + 4x_2 + x_3 \qquad\qquad = 9, \tag{15.21a}$$

$$5x_1 + 2x_2 \qquad + x_4 \qquad = 8, \tag{15.21b}$$

$$x_1 - 2x_2 \qquad\qquad + x_5 = 1. \tag{15.21c}$$

The second step is to set x_1 and x_2 equal to zero and solve for the other three variables, x_3, x_4, and x_5. At this moment, x_1 and x_2 are called *nonbasic variables*; x_3, x_4, and x_5 are called *basic variables*. We know from Solution 1 of Table 15.1 that

$$x_3 = 9, \qquad x_4 = 8, \qquad x_5 = 1.$$

It should be noted that the value of $-z$ based on this basic feasible solution just computed is equal to zero. Clearly, it is not the best solution. We shall now attempt to improve this situation in the third step described below.

From Eq. (15.20), it is evident that the value of $-z$ can be decreased by increasing either x_1 or x_2. We then ask, Which one should be chosen? Since x_2 in Eq. (15.20) has a smaller (more negative) coefficient (that is, -11), it should be allowed to increase; x_1 should be kept at zero. Now that x_2 is no longer zero, we can replace it by x_3, x_4, or x_5. In other words, knowing that $x_1 = 0$ and $x_2 \neq 0$, we select x_3, x_4, or x_5 and set the chosen variable equal to zero in order to obtain a second set of basic feasible solutions.

We next select one of the three variables and set it equal to zero. We observe that this increase in the value of x_2 will have the following effects:

1) It will increase the value of z; that is, decrease the value of $-z$(see Eq. 15.20).
2) It will decrease the values of x_3, x_4, and x_5.

The second effect can be seen from Eqs. (15.21) which can now be written as

$$4x_2 + x_3 \qquad\qquad = 9, \tag{15.22a}$$

$$2x_2 \qquad + x_4 \qquad = 8, \tag{15.22b}$$

$$-2x_2 \qquad\qquad + x_5 = 1, \tag{15.22c}$$

where all the variables are nonnegative. In particular, x_3 in Eq. (15.22a) cannot be negative. This condition forces us to increase the value of x_2 by no more than $\frac{9}{4}$ or

$$x_2 \leq 2.25. \tag{15.23}$$

Similarly, from Eq. (15.22b) we have

$$x_2 \leq 4.00. \tag{15.24}$$

Since x_2 has a negative coefficient in Eq. (15.22c), x_5 will always be positive and no restriction is needed for x_2 in Eq. (15.22c).

Comparing Eq. (15.23) with Eq. (15.24), we find that the smaller value of x_2 is 2.25, and since this comes from Eq. (15.23), we now set $x_3 = 0$ (as the new nonbasic variable) in place of x_2. In other words, our new basic feasible solution is

$$x_1 = 0, \qquad x_2 = 2.25, \qquad x_3 = 0, \qquad x_4 = 3.50, \qquad x_5 = 5.50.$$

We shall now eliminate x_2 from Eq. (15.21b) and (15.21c). This task is readily accomplished by first converting Eq. (15.21a) to the form

$$\tfrac{3}{4}x_1 + x_2 + \tfrac{1}{4}x_3 = \tfrac{9}{4}, \tag{15.25a}$$

and then using it to eliminate x_2 from the other two equations. As a result, we have

$$\tfrac{3}{4}x_1 + x_2 + \tfrac{1}{4}x_3 \qquad\qquad = \tfrac{9}{4}, \tag{15.25a}$$

$$\tfrac{7}{2}x_1 \qquad - \tfrac{1}{2}x_3 + x_4 \qquad = \tfrac{7}{2}, \tag{15.25b}$$

$$\tfrac{5}{2}x_1 \qquad + \tfrac{1}{2}x_3 \qquad + x_5 = \tfrac{11}{2}. \tag{15.25c}$$

The remainder of this step is to find the value of $-z$. Substituting Eq. (15.25a) into Eq. (15.20), we obtain

$$-z = -24.75 - 1.75x_1 + 2.75x_3, \tag{15.26}$$

or

$$-z = -24.75.$$

We note that the value of $-z$ is now smaller than zero, which is a substantial improvement over the previous solution.

The procedure used in the third step described above will be used over and over again to further improve the z-value. Consider, for example, the fourth step. From Eq. (15.26) we see that x_3 must be set to zero and x_1 be increased in order to improve the z-value. In addition to the variable x_3, we must select one more variable among x_2, x_4, and x_5 which will be set to zero, in order to obtain the third set of a basic feasible solution. To determine which one is to be set to zero, let us examine Eq.

(15.25). The variable x_2 in Eq. (15.25a) cannot be negative; as a result, the value of x_1 cannot be increased by more than 3 or

$$x_1 \leq 3. \tag{15.27}$$

Similarly, from Eqs. (15.25b) and (15.25c) respectively, we have

$$x_1 \leq 1 \tag{15.28}$$

and

$$x_1 \leq 2.2. \tag{15.29}$$

The smallest value of x_1 is 1, which comes from Eq. (15.28). As a result, we set $x_4 = 0$ as the new nonbasic variable in place of x_1. At this point, the new basic feasible solution is

$$x_1 = 1.00, \qquad x_2 = 1.50, \qquad x_3 = 0, \qquad x_4 = 0, \qquad x_5 = 3.00.$$

We shall now solve x_1, x_2, and x_5, each in terms of two variables x_3 and x_4. The resulting equations are

$$x_1 \quad - 0.143x_3 + 0.286x_4 \qquad = 1 \tag{15.30a}$$

$$x_2 + 0.357x_3 - 0.214x_4 \qquad = 1.5 \tag{15.30b}$$

$$0.857x_3 - 0.714x_4 + x_5 = 3 \tag{15.30c}$$

The remainder of this fourth step is to find the value of $-z$. Substituting Eq. (15.30a) into Eq. (15.20), we obtain

$$-z = -26.5 - 2.5x_3 - 0.5x_4, \tag{15.31}$$

or

$$-z = -26.5.$$

Since any increase of the values of x_3 or x_4 in Eq. (15.31) will decrease the value of z, no further improvement is possible for the value of $-z$. Thus our answers are

$$x_1 = 1.0, \qquad x_2 = 1.5, \qquad \text{max } z = 26.5.$$

In order to simplify the discussion in the previous sections, we chose a well-behaved problem to illustrate the Simplex method. There are, however, some problems which introduce further complications in the method. Among these exceptional cases is *degeneracy*, in which one or more of the basic variables become zero at the same time that the objective function is zero.

In addition to the problem of degeneracy, other complications may arise, such as (1) no obvious feasible solution or contradictory constraint equations, or (2) multiple solutions or an infinite number of solutions. For a more detailed explanation of these exceptional cases, we refer the interested reader to the references listed at the end of this chapter.

15.6 GAUSS-JORDAN PROCEDURE AND SIMPLEX TABLEAU

In Chapter 8, we introduced the Gauss-Jordan elimination method to solve a set of simultaneous equations. We shall now extend this method to the solution of a set of linear algebraic equations where the number of unknowns is not equal to the number of the equations; this extension is important to the computer solution of linear programming problems.

To illustrate this extension, let us consider the following three equations in five unknowns:

$$3x_1 + 4x_2 + x_3 + x_4 + x_5 = 9,$$
$$5x_1 + 2x_2 + x_3 + x_4 + x_5 = 8, \tag{15.32}$$
$$x_1 - 2x_2 + x_3 + x_4 + x_5 = 1.$$

It should be noted that this set of equations has no relation whatsoever with Eqs. (15.18) or Eqs. (15.21). We wish to solve for x_1, x_2, and x_3 in terms of x_4 and x_5. In other words, we wish to transform Eqs. (15.32) into the form

$$x_1 \qquad + a'_{14}x_4 + a'_{15}x_5 = c'_1,$$
$$\qquad x_2 \quad + a'_{24}x_4 + a'_{25}x_5 = c'_2, \tag{15.33}$$
$$\qquad x_3 + a'_{34}x_4 + a'_{35}x_5 = c'_3,$$

and to determine the nine coefficients (a's and c's) in Eqs. (15.33). This transformation can be readily accomplished by the procedures discussed in Chapter 8:

1) interchange of rows if the coefficient of a pivot element is zero,
2) normalization (see Section 8.4),
3) elimination (see Section 8.3).

And Eqs. (15.32) are then transformed into the following form:

$$x_1 \qquad + 0x_4 + 0x_5 = \tfrac{5}{8},$$
$$\qquad x_2 \quad + 0x_4 + 0x_5 = \tfrac{9}{8}, \tag{15.34}$$
$$\qquad x_3 + \quad x_4 + \quad x_5 = \tfrac{21}{8}.$$

Hence, the solutions for x_1, x_2, and x_3 in terms of x_4 and x_5 are

$$x_1 = \tfrac{5}{8}, \qquad x_2 = \tfrac{9}{8}, \qquad x_3 = \tfrac{21}{8} - x_4 - x_5.$$

We shall now incorporate the above extension of the Gauss-Jordan elimination procedure in the Simplex method. Special attention will be paid to the selection of a pivotal element. We consider once more the sample problem from the previous sections in which we attempted to determine

$$x_i \geq 0, \qquad i = 1, 2, 3, 4, 5, \tag{15.16}$$

so that

$$-10x_1 - 11x_2 + 0x_3 + 0x_4 + 0x_5 + z = 0, \qquad (15.17)$$

where z is the value to be maximized, and

$$3x_1 + 4x_2 + x_3 \qquad\qquad = 9, \qquad (15.18a)$$

$$5x_1 + 2x_2 \qquad + x_4 \qquad = 8, \qquad (15.18b)$$

$$x_1 - 2x_2 \qquad\qquad + x_5 = 1. \qquad (15.18c)$$

The above four equations can be written in the following matrix form, known as a *Simplex tableau*:

$$A = \begin{bmatrix} -10 & -11 & 0 & 0 & 0 & 1 & 0 \\ 3 & 4 & 1 & 0 & 0 & 0 & 9 \\ 5 & 2 & 0 & 1 & 0 & 0 & 8 \\ 1 & -2 & 0 & 0 & 1 & 0 & 1 \end{bmatrix}. \qquad (15.35)$$

To initiate the computation, we set x_1 and x_2 equal to zero (nonbasic variables). Then, by inspecting the tableau, we arrive at the first basic feasible solution:

$$x_1 = 0, \qquad x_2 = 0, \qquad x_3 = 9 - 3x_1 - 4x_2 = 9,$$

$$x_4 = 8 - 5x_1 - 2x_2 = 8, \qquad x_5 = 1 - x_1 + 2x_2 = 1,$$

$$z = 0 + 10x_1 + 11x_2 = 0.$$

Since z is equal to zero, this is clearly not the optimal solution.

As discussed in the second step of the Simplex method (Section 15.5), the value of x_2 is the first to be increased since it is the nonbasic variable with the most negative coefficient in Eq. (15.17) and therefore has the most pronounced effect in the maximization of z. At this point the Gauss-Jordan elimination procedure can be used advantageously to obtain a set of basic feasible solutions. Each basic feasible solution is obtained by arbitrarily setting two variables equal to zero and solving for the other three. Since we chose x_2 as the variable to be increased, we must determine the pivot element from column 2 of the Simplex tableau (Eq. 15.35). This pivot element can be found by (1) dividing each positive element in column 2 into the rightmost element of the row in which it appears and (2) selecting as the pivot element the one which after division yields the smallest quotient (excluding row 1).

In our example we note that only the second and third elements in this column are positive. We therefore compute only the two quotients

$$c_2/a_{22} = \tfrac{9}{4} = 2.25, \qquad c_3/a_{32} = \tfrac{8}{2} = 4,$$

and observe that the smaller quotient, that is, 2.25, is associated with the element a_{22}. Here the coefficients, a's and c's, are defined in Eq. (15.8).

We thus choose the element a_{22}, whose value is 4, as the pivot element in the next step of the procedure, normalization. Dividing each element in the second row by the value of the pivot element, we find that Eq. (15.35) becomes

$$A = \begin{bmatrix} -10 & -11 & 0 & 0 & 0 & 1 & 0 \\ 0.75 & ① & 0.25 & 0 & 0 & 0 & 2.25 \\ 5 & 2 & 0 & 1 & 0 & 0 & 8 \\ 1 & -2 & 0 & 0 & 1 & 0 & 1 \end{bmatrix} \quad (15.36)$$

The pivot element is indicated above by a circle and is used to eliminate the other elements in column 2. At the end of this first elimination step the tableau becomes

$$A = \begin{bmatrix} -1.75 & 0 & 2.75 & 0 & 0 & 1 & 24.75 \\ 0.75 & 1 & 0.25 & 0 & 0 & 0 & 2.25 \\ 3.50 & 0 & -0.50 & 1 & 0 & 0 & 3.50 \\ 2.50 & 0 & 0.50 & 0 & 1 & 0 & 5.50 \end{bmatrix}. \quad (15.37)$$

The new basic feasible solution obtained from the above matrix is

$$x_1 = 0, \quad x_3 = 0,$$
$$x_2 = 2.25 - 0.75x_1 - 0.25x_3 = 2.25,$$
$$x_4 = 3.50 - 3.50x_1 + 0.50x_3 = 3.50,$$
$$x_5 = 5.50 - 2.50x_1 - 0.50x_3 = 5.50,$$
$$z = 24.75 + 1.75x_1 - 2.75x_3 = 24.75.$$

Although the value of z has been increased from 0 to 24.75, it can be made still greater by increasing the value of x_1, since the coefficient of x_1 is positive, while at the same time x_3 is kept equal to zero. To perform this operation, we simply select the appropriate pivot element from column 1 and repeat the normalization and elimination procedures as before to obtain

$$A = \begin{bmatrix} 0 & 0 & 2.50 & 0.50 & 0 & 1 & 26.5 \\ 0 & 1 & 0.35 & -0.21 & 0 & 0 & 1.5 \\ ① & 0 & -0.14 & 0.28 & 0 & 0 & 1.0 \\ 0 & 0 & 0.85 & -0.71 & 1 & 0 & 3.0 \end{bmatrix}. \quad (15.38)$$

We have now arrived at the optimal solution:

$$x_3 = 0, \quad x_4 = 0,$$
$$x_2 = 1.50 - 0.35x_3 + 0.21x_4 = 1.50,$$
$$x_1 = 1.00 + 0.14x_3 - 0.28x_4 = 1.00,$$
$$x_5 = 3.00 - 0.85x_3 + 0.71x_4 = 3.00,$$
$$z = 26.50 - 2.50x_3 - 0.50x_4 = 26.50.$$

Table 15.3 Simplex tableau

Pass	Selection of the pivot element	Matrix

Pass	Selection of the pivot element	x_1	x_2	x_3	x_4	x_5	z	
Given augmented matrix	$\frac{9}{4} = 2.25$ $\frac{8}{2} = 4.00$ $\rightarrow$	-10 3 5 1	-11 $④$ 2 -2 $\uparrow$	0 1 0 0	0 0 1 0	0 0 0 1	1 0 0 0	0 9 8 1

First pass

New row 2 = row 2/4

$$\rightarrow \begin{pmatrix} -10 & -11 & 0 & 0 & 0 & 1 & 0 \\ 0.75 & ① & 0.25 & 0 & 0 & 0 & 2.25 \\ 5 & 2 & 0 & 1 & 0 & 0 & 8 \\ 1 & -2 & 0 & 0 & 1 & 0 & 1 \end{pmatrix}$$
$\uparrow$

New row 1 = row 1 − row 2 × (−11)

$$\rightarrow \begin{pmatrix} -1.75 & 0 & 2.75 & 0 & 0 & 1 & 24.75 \\ 0.75 & ① & 0.25 & 0 & 0 & 0 & 2.25 \\ 5 & 2 & 0 & 1 & 0 & 0 & 8 \\ 1 & -2 & 0 & 0 & 1 & 0 & 1 \end{pmatrix}$$
$\uparrow$

New row 3 = row 3 − row 2 × 2

$$\rightarrow \begin{pmatrix} -1.75 & 0 & 2.75 & 0 & 0 & 1 & 24.75 \\ 0.75 & ① & 0.25 & 0 & 0 & 0 & 2.25 \\ 3.50 & 0 & -0.50 & 1 & 0 & 0 & 3.50 \\ 1 & -2 & 0 & 0 & 1 & 0 & 1 \end{pmatrix}$$
$\uparrow$

New row 4 = row 4 − row 2 × (−2)

$\frac{2.25}{0.75} = 3.00$	-1.75	0	2.75	0	0	1	24.75
$\frac{3.50}{3.50} = 1.00$ $\rightarrow$	0.75	1	0.25	0	0	0	2.25
	(3.50)	0	-0.50	1	0	0	3.50
$\frac{5.50}{2.50} = 2.20$	2.50 $\uparrow$	0	0.50	0	1	0	5.50

		New row 3 $=$ row 3/3.50

$$\rightarrow \begin{pmatrix} -1.75 & 0 & 2.75 & 0 & 0 & 1 & 24.75 \\ 0.75 & 1 & 0.25 & 0 & 0 & 0 & 2.25 \\ ① & 0 & -0.143 & 0.286 & 0 & 0 & 1.00 \\ 2.50 & 0 & 0.50 & 0 & 1 & 0 & 5.50 \end{pmatrix}$$

New row 1 $=$ row 1 $-$ row 3 $\times$ (-1.75)

$$\rightarrow \begin{pmatrix} 0 & 0 & 2.50 & 0.50 & 0 & 1 & 26.50 \\ 0.75 & 1 & 0.25 & 0 & 0 & 0 & 2.25 \\ ① & 0 & -0.143 & 0.286 & 0 & 0 & 1.00 \\ 2.50 & 0 & 0.50 & 0 & 1 & 0 & 5.50 \end{pmatrix}$$

Second pass

New row 2 $=$ row 2 $-$ row 3 $\times$ 0.75

$$\rightarrow \begin{pmatrix} 0 & 0 & 2.50 & 0.50 & 0 & 1 & 26.50 \\ 0 & 1 & 0.357 & -0.214 & 0 & 0 & 1.50 \\ ① & 0 & -0.143 & 0.286 & 0 & 0 & 1.00 \\ 2.50 & 0 & 0.50 & 0 & 1 & 0 & 5.50 \end{pmatrix}$$

New row 4 $=$ row 4 $-$ row 3 $\times$ 2.50

$$\rightarrow \begin{pmatrix} 0 & 0 & 2.50 & 0.50 & 0 & 1 & 26.50 \\ 0 & 1 & 0.357 & -0.214 & 0 & 0 & 1.50 \\ ① & 0 & -0.143 & 0.286 & 0 & 0 & 1.00 \\ 0 & 0 & 0.857 & -0.714 & 1 & 0 & 3.00 \end{pmatrix}$$

We can see that we have reached the maximum value for z, since the coefficients of both x_3 and x_4 are negative and an increase in either variable would only decrease z. On the other hand, if we decreased x_3 or x_4, then they would be less than zero—in violation of the nonnegative requirement. Table 15.3 summarizes the two passes in the elimination procedure. The computations made to select the two pivot elements are listed, and the pivot elements at each step of the procedure are circled. In addition, the pivotal rows and columns at each step are indicated by arrows.

15.7 COMPUTER APPLICATION IN LINEAR PROGRAMMING

In recent years, the use of linear programming methods has grown rather extensively. One of the key factors contributing to this remarkable development has been the availability of computers and computer programs. Many case studies of practical applications of linear programming are listed in the references at the end of this chapter. The main purpose of this section is to introduce to the reader one of the possible FORTRAN IV programs for the Simplex method.

The FORTRAN IV program shown in Fig. 15.5 can be used to solve a linear programming problem containing up to 34 constraints and 85 variables.

```
C        SIMPLEX METHOD FOR SOLVING LINEAR PROGRAMMING PROBLEMS
C        FORTRAN IV MUST BE USED
C
C        THE LIMITS OF THE PROGRAM ARE
C            MAX. NUMBER OF VARIABLES INCLUDING SLACK VARIABLES = 84
C            MAX. NUMBER OF EQUATIONS INCLUDING THE OBJECTIVE FUNCTION = 34
C        THE INITIAL SIMPLEX TABLEAU CORRESPONDING TO THE GIVEN EQUATIONS
C        MUST BE PUT INTO THE FORM
C
C            I--                                          --I
C            I   A(1,1)   A(1,2) . . . . . . .   A(1,JJ)   I
C            I                                             I
C            I   A(2,1)   A(2,2) . . . . . . .   A(2,JJ)   I
C            I        .        .                   .       I
C            I        .        .                   .       I
C            I        .        .                   .       I
C            I        .        .                   .       I
C            I        .        .                   .       I
C            I   A(II,1)   A(II,2) . . . . . . .  A(II,JJ)  I
C            I                                             I
C            I   A(III,1) A(III,2) . . . . . .  A(III,JJ)  I
C            I--                                        --I
C
C        WHERE THE ELEMENTS WERE DEFINED IN EQ.(15.39).
C
         DIMENSION A(35,85), W(35), L(35)
C
C        READ IN
C            II = TOTAL NUMBER OF THE GIVEN EQUATIONS INCLUDING THE
C                 OBJECTIVE FUNCTION.
C               = TOTAL NUMBER OF ROWS OF THE SIMPLEX TABLEAU - 1.
C            JJ = TOTAL NUMBER OF COLUMNS OF THE TABLEAU.
C
         DATA IREAD,IWRITE/5,6/
         WRITE(IWRITE,2)
       2 FORMAT('1')
     108 READ(IREAD,1) II,JJ
       1 FORMAT (18I4)
         III = II + 1
         DO 10 I=1, III
         W(I) = 0.
      10 L(I) = 0
C
C        READ IN THE ELEMENTS OF THE MATRIX ROW BY ROW.
C
         READ(IREAD,4) ((A(I,J),J=1,JJ),I=1,III)
       4 FORMAT (7F10.4)
C
C        READ IN THE SUBSCRIPT FOR THE SLACK VARIABLE ON ROW(I), WHERE
C        I IS NOT EQUAL TO 1 OR III.
C
         READ(IREAD,1) (L(I),I=2,II)
C
C        NEXT STATEMENT FOR INITIALIZATION
C
```

Fig. 15.5. FORTRAN program for linear programming problems.

```
      KKK = 0
C
C     LOOK FOR THE ROW IN WHICH THERE IS NO SLACK VARIABLE
C     (EXCLUDING FIRST ROW)
C
   22 I = 1
   23 I = I + 1
      IF(I.GE.III) GO TO 40
      IF(L(I).NE.0) GO TO 23
C
C     CALCULATE
C          NEW LAST ROW = LAST ROW - THE ROW WITHOUT SLACK VARIABLE
C
   25 DO 27 J=1, JJ
      IF(A(I.J).EQ.0.) GO TO 27
   26 A(III.J) = A(III.J) - A(I.J)
   27 CONTINUE
      GO TO 23
C
C     NEXT STATEMENTS FOR SEARCHING FOR THE COLUMN IN WHICH THE MOST
C     NEGATIVE ENTRY APPEARS EITHER IN THE FIRST(OBJECTIVE FUNCTION)
C     OR LAST(FORM P) ROW.
C
   40 K = III
   44 J = 0
      W(K) = 0.
      L(K) = 0
   42 J = J + 1
      IF(J.GE.JJ) GO TO 45
      IF((A(K.J).GE.0.).OR.(W(K).LE.A(K.J))) GO TO 42
   47 W(K) = A(K.J)
      L(K) = J
      GO TO 42
C
C     TEST FOR L(K). IF L(K) IS EQUAL TO ZERO, THAT IS, ALL THE
C     ENTRIES EXCEPT THE EXTREME RIGHT ONE EITHER IN THE FIRST OR LAST
C     ROW ARE POSITIVE, GO TO ST. 62 FOR FURTHER EXAMINATION.
C
   45 IF(L(K).EQ.0) GO TO 62
C
C     SEARCH FOR PIVOT COLUMN
C
   46 KJ = L(K)
C
C     TEST EACH ENTRY IN THE PIVOT COLUMN TO SEE IF IT IS POSITIVE
C     OR NOT. IF TI IS, GO TO ST. 121 TO COMPUTE THE RATIO DEFINED IN
C     THE LAST SECTION.
C
      DO 120 I=2, II
      IF(A(I.KJ).GT.0.) GO TO 121
  120 CONTINUE
C
C     IF ALL THE ENTRIES IN THE PIVOT COLUMN ARE ZERO OR NEGATIVE
C     NUMBERS, UNBOUNDED WILL BE TYPED.
C
      WRITE(IWRITE,130)
  130 FORMAT(5X,'UNBOUNDED')
```

(cont.)

Fig. 15.5 *(concl.)*

```
      GO TO 70
C
C     THE FOLLOWING STATEMENTS ARE FOR COMPUTING THE RATIO DEFINED
C     IN SECTION 15.4, AND FOR DETERMINING THE LOCATION OF THE PIVOT.
C
  121 I = 1
      JK = 0
   50 I = I + 1
      IF(I.GT.II) GO TO 56
      IF(A(I,KJ).LE.0.) GO TO 50
   51 X = A(I,JJ)/A(I,KJ)
      IF(JK.EQ.0) GO TO 53
      IF(X.GE.XMIN) GO TO 50
   53 XMIN = X
      JK = I
      GO TO 50
C
C     THE NEXT STATEMENT INDICATES THE PIVOT ELEMENT BEFORE NORMALIZATION.
C
   56 X = A(JK,KJ)
      L(JK) = KJ
C
C     NEXT STATEMENTS FOR CALCULATING THE NEW ROWS ABOVE THE PIVOT ROW
C
      DO 57 I=1, III
   57 W(I) = A(I,KJ)
      IJ = JK - 1
      DO 59 I=1, IJ
      DO 59 J=1, JJ
      IF((A(JK,J).EQ.0.).OR.(W(I).EQ.0.)) GO TO 59
  580 A(I,J) = A(I,J) - W(I)*(A(JK,J)/X)
   59 CONTINUE
C
C     NEXT STATEMENTS FOR CALCULATING THE NEW ROWS BELOW THE PIVOT ROW
C
      IJ = JK + 1
      DO 61 I=IJ, III
      DO 61 J=1, JJ
      IF((A(JK,J).EQ.0.).OR.(W(I).EQ.0.)) GO TO 61
  600 A(I,J) = A(I,J) - W(I)*(A(JK,J)/X)
   61 CONTINUE
C
C     NEXT STATEMENTS FOR NORMALIZATION
C
      DO 205 J=1, JJ
  205 A(JK,J) = A(JK,J)/X
      KKK = KKK + 1
      WRITE(IWRITE,105) KKK,A(K,JJ),L(JK)
  105 FORMAT (1X, I4, 6X, F15.2, 10X, I4)
      GO TO 44
C
C     NEXT STATEMENT FOR TESTING TO SEE IF IT IS THE FIRST ROW ON
C     WHICH ALL THE ENTRIES ARE POSITIVE EXCEPT THE EXTREME RIGHT ONE.
C     IF IT IS, THAT MEANS, NO FURTHER IMPROVEMENT ON THE SOLUTION
C     CAN BE MADE. GO TO ST. 70 AND THE ANSWER WILL BE TYPED OUT.
C
   62 IF(K.LE.1) GO TO 70
   63 IJ = JJ - 1
C
```

```
C       TEST TO SEE WHETHER ALL THE ELEMENTS ON THE LAST ROW(NOT
C       INCLUDING THE EXTREME RIGHT ONE) ARE CLOSE TO ZERO.
C       IT IS DEFINED IN THE NEXT STATEMENTS THAT THE PROBLEM IS
C       INFEASIBLE IF ONE (OR MORE) OF THEM IS LARGER THAN 0.0001.
C
        DO 65 J=1, IJ
        IF(A(K,J).GT.0.0001) GO TO 66
     65 CONTINUE
        WRITE(IWRITE,103)
    103 FORMAT(5X,'FEASIBLE')
        WRITE(IWRITE,101)
    101 FORMAT(1X,'ITERATION      OBJ. FUNCTION      NEW BASIC VAR.')
C
C       IF, AFTER ITERATIONS, ALL THE ELEMENTS IN THE LAST ROW HAVE
C       BECOME POSITIVE BUT NEAR ZERO, DEFINE ALL OF THEM TO BE ZERO.C
C
        DO 140 J=1, JJ
    140 A(III,J) = 0.
C
C       IN CASE OF NONARTIFICIAL PROBLEMS, DEFINE K = 1, AND GO TO ST. 44
C       TO SEARCH FOR THE PIVOT COLUMN.
C
        K = 1
        KKK = 0
        GO TO 44
C
C       TYPE OUT THE SOLUTION
C
     66 WRITE(IWRITE,6)
      6 FORMAT(5X,'INFEASIBLE')
     70 WRITE(IWRITE,8) A(1,JJ)
      8 FORMAT(///5X,'OBJ. FUNCTION',F20.8/)
        WRITE(IWRITE,7)
      7 FORMAT(1X,'VARIAVLE          VALUE')
        DO 71 I=2, II
     71 WRITE(IWRITE,5) L(I),A(I,JJ)
      5 FORMAT (1X, I4, F20.8)
C
C       NEXT STATEMENTS FOR PRINTING THE FINAL MATRIX
C
        WRITE(IWRITE,100)
    100 FORMAT(////1X,'THE FINAL MATRIX')
        DO 78 I=1, III
        WRITE(IWRITE,150) I
    150 FORMAT(//35X,'ROW ',I2/)
     78 WRITE(IWRITE,4) (A(I,J),J=1,JJ)
        STOP
        END
```

The Simplex tableau mentioned in the previous section is followed very closely in this program. A detailed flow chart is shown in Fig. 15.6.

The first step is to form the following matrix:

$$[\mathscr{A}] = \begin{bmatrix} a_{11} & a_{12} & \cdots & a_{1,n+1} \\ a_{21} & a_{22} & \cdots & a_{2,n+1} \\ \vdots & & & \\ a_{m+1,1} & a_{m+1,2} & \cdots & a_{m+1,n+1} \\ 0 & 0 & \cdots & 0 \end{bmatrix}. \tag{15.39}$$

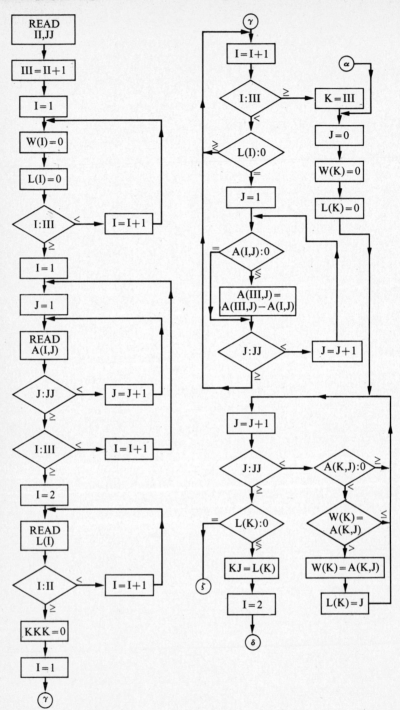

Fig. 15.6. Flow chart for simplex method.

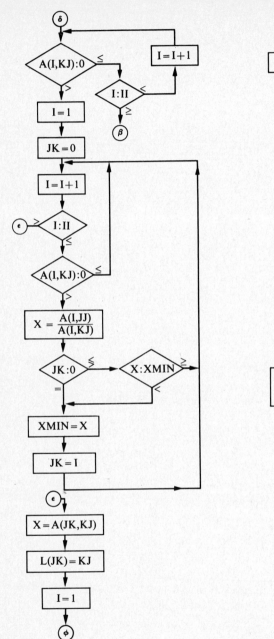

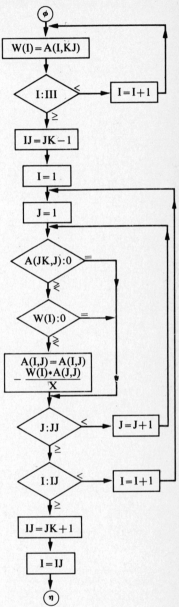

(cont.)

Fig. 15.6 *(concl.)*

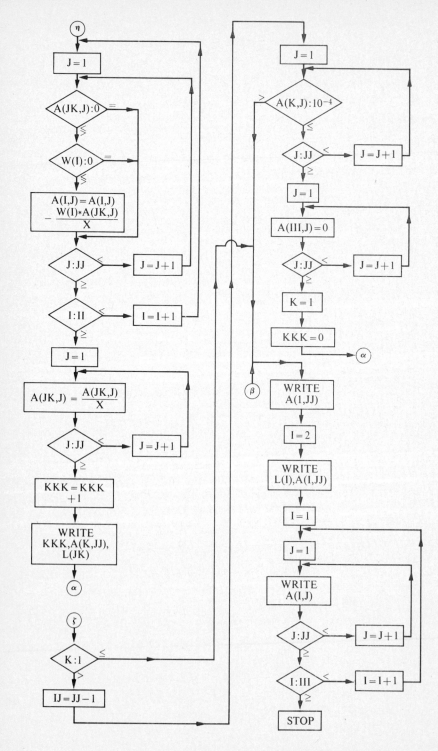

The matrix $[\mathscr{A}]$ can be obtained from $[A]$ by simply deleting the column for z in Eq. (15.35). The column for z is the second column from the right. We can make the following observations.

1) The elements in the first represent the coefficients of the variables and the constant in the objective function.

2) The elements in the second row through the $m + 1$ row represent the coefficients of the variables and the constant in the constraints.

3) The elements in the last row (all zeros) are for artificial variable problems which will be discussed in the next section.

Example 15.1

As a sample problem we may apply this program to the example developed in Section 15.6. We would thus read in the data in the form shown in Fig. 15.7.

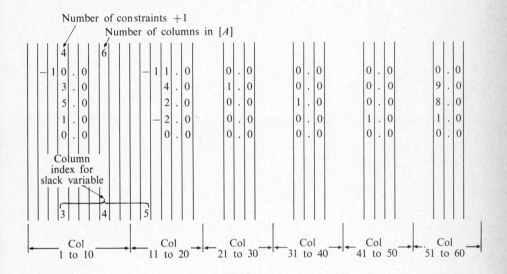

Fig. 15.7

The required input are as follows.

Card 1
    Columns 1 to 4 inclusive: Number of the constraints + 1 (FORMAT 14).
    Columns 5 to 8 inclusive: Number of columns in Eq. (15.39) (FORMAT 14).

Starting with Card 2
    Matrix $[\mathscr{A}]$ as shown in Eq. (15.39), row by row in FORMAT(7F10.4).

Table 15.4 Output for Example 15.1

```
FEASIBLE
ITERATION        OBJ. FUNCTION      NEW BASIC VAR.
    1                24.75                 2
    2                26.50                 1


OBJ. FUNCTION            26.50000000

VARIABLE            VALUE
    2            1.50000000
    1            1.00000000
    5            3.00000000



THE FINAL MATRIX
```

ROW   1

| 0.0000 | 0.0000 | 2.5000 | .4999 | 0.0000 | 26.5000 |

ROW   2

| 0.0000 | 1.0000 | .3571 | -.2142 | 0.0000 | 1.5000 |

ROW   3

| 1.0000 | 0.0000 | -.1428 | .2857 | 0.0000 | 1.0000 |

ROW   4

| 0.0000 | 0.0000 | .8571 | -.7142 | 1.0000 | 3.0000 |

ROW   5

| 0.0000 | 0.0000 | 0.0000 | 0.0000 | 0.0000 | 0.0000 |

Last Card

Column index for the slack variable in each row (ignore the first and last rows) in FORMAT(18I4). We assume here that there are less than 18 slack variables.

Here the number of columns in $[\mathscr{A}]$ is 6, not 7, because the elements in the z-column, such as shown in Table 15.3, are not included in $[\mathscr{A}]$. Depending on the problem, one of the following three messages will be printed out as output for the program:

FEASIBLE,   INFEASIBLE,   or   UNBOUNDED.

In our particular example, the solution is feasible and the FEASIBLE message is printed out. Table 15.4 lists the complete output for Example 15.1.

15.8 ARTIFICIAL VARIABLES

In the previous section we used the Simplex procedure to solve a maximization problem involving constraints containing only inequalities of the "less than or equal to" type. The initial basic feasible solution that started off the iterations could be easily found by assigning x_1 and x_2 as nonbasic variables, and then setting them equal to zero. However, not all linear programming problems are quite so simple. We shall now consider a more difficult example in which the objective function is subject to a set of constraints containing $\geq$-inequalities. This type of problem is known as the *artificial-variable* problem. In order to explain the artificial-variable technique, let us begin by stating the following example, which is actually the same as the sample problem in Section 15.7 except that the inequality sign in the second constraint has been reversed.

Determine

$$x_1 \geq 0, \qquad x_2 \geq 0, \tag{15.40}$$

so that

$$z = 10x_1 + 11x_2 \tag{15.41}$$

is a maximum and

$$3x_1 + 4x_2 \leq 9, \tag{15.42}$$

$$5x_1 + 2x_2 \geq 8, \tag{15.43}$$

$$x_1 - 2x_2 \leq 1. \tag{15.44}$$

By introducing the slack variables x_3, x_4, and x_5 as before, the problem is reformulated as follows. Determine

$$x_i \geq 0, \qquad i = 1, 2, 3, 4, 5, \tag{15.45}$$

so that

$$-z = -10x_1 - 11x_2 \tag{15.46}$$

is a minimum and

$$3x_1 + 4x_2 + x_3 \qquad\qquad = 9, \tag{15.47}$$

$$5x_1 + 2x_2 \qquad - x_4 \qquad = 8, \tag{15.48}$$

$$x_1 - 2x_2 \qquad\qquad + x_5 = 1. \tag{15.49}$$

The problem now is to evaluate the initial basic feasible solution. It is obvious that the origin, $x_1 = 0$, $x_2 = 0$, is not one of the basic feasible solutions to the above constraints, for it makes x_4 equal to -8, and this clearly violates the nonnegativity condition of Eq. (15.45). These equations must therefore be rearranged so that we can easily find a first feasible solution and the iterations will finally lead to an optimal solution satisfying the original as well as the rearranged set of equations. For this purpose, we introduce in Eq. (15.48) a new variable, x_6, which is called an *artificial variable*, since it has no direct connection with the original constraints. Having made this addition, we see that the constraints become

$$x_i \geq 0, \qquad i = 1, 2, 3, 4, 5, 6, \tag{15.50}$$

$$3x_1 + 4x_2 + x_3 \qquad\qquad\qquad = 9, \tag{15.51}$$

$$5x_1 + 2x_2 \qquad - x_4 \qquad + x_6 = 8, \tag{15.52}$$

$$x_1 - 2x_2 \qquad\qquad + x_5 \qquad = 1. \tag{15.53}$$

We immediately have a basic feasible solution by setting x_1, x_2, and x_4 equal zero:

$$x_1 = 0, \qquad x_2 = 0, \qquad x_4 = 0, \qquad x_3 = 9, \qquad x_5 = 1, \qquad x_6 = 8. \tag{15.54}$$

It should be mentioned, however, that this first basic feasible solution of Eqs. (15.51) through (15.53) is not necessarily a basic feasible solution to the original set of constraints, Eqs. (15.47) through (15.49). Any basic feasible solution of the constraints, Eqs. (15.51) through (15.53), which is also a basic feasible solution to the constraints, Eqs. (15.47) through (15.49), must have $x_6 = 0$. In other words, the artificial valiable must be assigned as a nonbasic variable.

Our next step is to use the Simplex method to search for a basic feasible solution common to both sets of constraints. In searching for this solution for which $x_6 = 0$, we may ask, Can we improve the value of the objective function while the search is being made? The answer to this question is yes; we can by introducing in the objective function a quantity px_6, where p is chosen as an unspecified, yet extremely large, positive number. Let us define

$$-z' = -z + px_6 \tag{15.55a}$$

$$= -10x_1 - 11x_2 + 0x_3 + 0x_4 + 0x_5 + px_6, \tag{15.55b}$$

and then proceed to seek a minimum value for $-z'$ instead of $-z$. As a result, we now have the following problem. Determine

$$x_i \geq 0, \qquad i = 1, 2, 3, 4, 5, 6, \tag{15.50}$$

so that

$$-z' = -10x_1 - 11x_2 + 0x_3 + 0x_4 + 0x_5 + px_6 \tag{15.55b}$$

is a minimum and

$$3x_1 + 4x_2 + x_3 \qquad\qquad\qquad = 9, \tag{15.51}$$

$$5x_1 + 2x_2 \qquad - x_4 \qquad + x_6 = 8, \tag{15.52}$$

$$x_1 - 2x_2 \qquad\qquad + x_5 \qquad = 1. \tag{15.53}$$

If there is a solution at all to the original constraints, Eqs. (15.47) through (15.49), we shall eventually arrive at such a basic feasible solution with $x_6 = 0$, by applying the Simplex procedure to the above modified system. The large coefficient p will prevent the appearance of the artificial variable x_6 in the optimal solution to Eqs. (15.50) through (15.53) and (15.55b); and the value of $-z'$ will not reach its minimum while the search is being made for a basic feasible solution with $x_6 = 0$. It should be noted here that any optimal solution with $x_6 = 0$ for Eqs. (15.50) through (15.53) and (15.55b) is not necessarily the optimal solution to the original set of constraints, Eqs. (15.45) through (15.49); it is just one of the basic feasible solutions of the original constraints subject to the objective function $-z$. Having attained a basic feasible solution with $x_6 = 0$, we can therefore drop the artificial variable x_6 from the objective function and continue with the original variables, applying the Simplex procedure until either the minimum value of $-z'$ is found or $-z'$ is determined to have an infinite minimum.

The first Simplex tableau (see Section 15.6) can be written as

$$A = \begin{array}{c} \begin{array}{ccccccc} x_1 & x_2 & x_3 & x_4 & x_5 & x_6 & z' \end{array} \\ \left[\begin{array}{ccccccc|c} -10 & -11 & 0 & 0 & 0 & p & 1 & 0 \\ 3 & 4 & 1 & 0 & 0 & 0 & 0 & 9 \\ 5 & 2 & 0 & -1 & 0 & 1 & 0 & 8 \\ 1 & -2 & 0 & 0 & 1 & 0 & 0 & 1 \end{array} \right]. \end{array} \qquad (15.56)$$

It is convenient to separate the first row into two parts. The element p is transferred to a new row (the last one) in the following matrix:

$$\begin{array}{c} \begin{array}{ccccccc} x_1 & x_2 & x_3 & x_4 & x_5 & x_6 & z' \end{array} \\ \left[\begin{array}{ccccccc|c} -10 & -11 & 0 & 0 & 0 & 0 & 1 & 0 \\ 3 & 4 & 1 & 0 & 0 & 0 & 0 & 9 \\ 5 & 2 & 0 & -1 & 0 & 1 & 0 & 8 \\ ① & -2 & 0 & 0 & 1 & 0 & 0 & 1 \\ 0 & 0 & 0 & 0 & 0 & p & 0 & 0 \end{array} \right]. \end{array} \qquad (15.57)$$

Note that the first and last rows in the above tableau still represent a single objective function, Eq. (15.55b). We know that in this problem there are only three variables which can be taken as the nonbasic variables. Equation (15.57) does not yield a basic feasible solution because it contains four nonbasic variables, that is, x_1, x_2, x_4, and x_6. This difficulty can be easily solved by multiplying each entry of the third row in Eq. (15.57) by $-p$, and then adding the results to the fifth row in order to make x_6 a basic variable. The resulting matrix is

$$\begin{array}{c} \begin{array}{ccccccc} \quad\quad\quad x_1 & x_2 & x_3 & x_4 & x_5 & x_6 & z' \end{array} \\ \begin{array}{c} \\ \frac{9}{3} = 3 \\ \frac{8}{5} = 1.6 \\ \frac{1}{1} = 1 \end{array} \left[\begin{array}{ccccccc|c} -10 & -11 & 0 & 0 & 0 & 0 & 1 & 0 \\ 3 & 4 & 1 & 0 & 0 & 0 & 0 & 9 \\ 5 & 2 & 0 & -1 & 0 & 1 & 0 & 8 \\ \text{→}① & -2 & 0 & 0 & 1 & 0 & 0 & 1 \\ -5p & -2p & 0 & p & 0 & 0 & 0 & -8p \\ & \uparrow & & & & & & \end{array} \right]. \end{array} \qquad (15.58)$$

We have now reached the first basic feasible solution for Eqs. (15.51) to (15.53) and (15.55b) which can be taken from Eq. (15.58) as

$$x_1 = 0, \qquad x_2 = 0, \qquad x_4 = 0,$$
$$x_3 = 0, \qquad x_5 = 1, \qquad x_6 = 8, \qquad z' = -8p. \tag{15.59}$$

Since the first and last row of the artificial tableau denote a single objective function, we have from Eq. (15.58),

$$-z' = 8p - (10 + 5p)x_1 - (11 + 2p)x_2 + px_4. \tag{15.60}$$

This equation suggests that an increase in either x_1 or x_2 will decrease the value of $-z'$. From Section (15.5) it is known that the variable which has the smaller (more negative) coefficient should be increased first. Therefore, it is necessary to determine which variable has the smaller coefficient. Since the coefficient p is defined as an extremely large number, the coefficients -10 and -11 become negligible in comparison with $-5p$ and $-2p$, respectively. This leads us to the conclusion that the coefficient of x_1 is more negative than that of x_2, and that we should search in column 1 to locate the pivot element. By comparing the ratios of the rightmost elements to their corresponding entries in the pivot column, we choose element 1, encircled in Eq. (15.58), as the pivot element. Once this choice has been made, we use the Gauss-Jordan elimination method to perform the first iteration, the results of which are indicated in the following matrix:

$$
\begin{array}{c}
\begin{array}{cccccccc}
\quad x_1 & x_2 & x_3 & x_4 & x_5 & x_6 & z'
\end{array} \\
\begin{array}{l}
\frac{6}{10} = 0.6 \\
\frac{3}{12} = 0.25 \\
 \\
 \\

\end{array}
\left[
\begin{array}{ccccccc}
0 & -31 & 0 & 0 & 10 & 0 & 1 & 10 \\
0 & 10 & 1 & 0 & -3 & 0 & 0 & 6 \\
0 & \boxed{12} & 0 & -1 & -5 & 1 & 0 & 3 \\
1 & -2 & 0 & 0 & 1 & 0 & 0 & 1 \\
0 & -12p & 0 & p & 5p & 0 & 0 & -3p
\end{array}
\right],
\end{array}
\tag{15.61}
$$

with a basic feasible solution

$$x_2 = 0, \qquad x_4 = 0, \qquad x_5 = 0,$$
$$x_1 = 1, \qquad x_3 = 6, \qquad x_6 = 3, \qquad z' = 10 - 3p. \tag{15.62}$$

The large negative entry, $-12p$, in the last row of Eq. (15.61) implies that $-z'$ may be further improved. The second iteration is therefore performed by using the element 12 as a pivot, which is marked by a circle in Eq. (15.61). The result is

$$
\begin{array}{c}
\begin{array}{ccccccc}
x_1 & x_2 & x_3 & x_4 & x_5 & x_6 & z'
\end{array} \\
\left[
\begin{array}{ccccccc}
0 & 0 & 0 & -\frac{31}{12} & -\frac{35}{12} & \frac{31}{12} & 1 & \frac{71}{4} \\
0 & 0 & 1 & \frac{5}{6} & \frac{7}{6} & -\frac{7}{6} & 0 & \frac{7}{2} \\
0 & 1 & 0 & -\frac{1}{12} & -\frac{5}{12} & \frac{1}{12} & 0 & \frac{1}{4} \\
1 & 0 & 0 & -\frac{1}{6} & \frac{1}{6} & \frac{1}{6} & 0 & \frac{3}{2} \\
0 & 0 & 0 & 0 & 0 & p & 0 & 0
\end{array}
\right],
\end{array}
\tag{15.63}
$$

with a basic feasible solution

$$x_4 = 0, \qquad x_5 = 0, \qquad x_6 = 0, \qquad x_1 = \tfrac{3}{2} = 1.50,$$

$$x_2 = \tfrac{1}{4} = 0.25, \qquad x_3 = \tfrac{7}{2} = 3.50 \qquad z' = \tfrac{71}{4} = 17.75. \tag{15.64}$$

The artificial variable x_6 has now become nonbasic, namely, it is zero, and from Eq. (15.55) it follows that z' becomes equal to z. The objective function may be now written as

$$-z = -\tfrac{71}{4} - \tfrac{31}{12}x_4 - \tfrac{35}{12}x_5 + \tfrac{31}{12}x_6, \tag{15.65}$$

which signifies that $-z$ can be improved further by increasing the value of x_5.

We now come to the final phase of this problem in which we attempt to find the optimal solution for the original constraint. The last row in Eq. (15.63) may be ignored in this phase, because the large number p no longer influences the value of z. By omitting the last row from Eq. (15.63) we have

$$
\begin{array}{c}
\frac{7}{2} = 3.0 \\[4pt]
\frac{7}{6} \\[10pt]
\frac{3}{2} = 9.0 \\[4pt]
\frac{1}{6} \\[10pt]
\\
\\
\end{array}
\begin{array}{c}
x_1 \quad x_2 \quad x_3 \quad x_4 \quad x_5 \quad x_6 \quad z'
\end{array}
$$

$$
\begin{bmatrix}
0 & 0 & 0 & -\frac{31}{12} & -\frac{35}{12} & \frac{31}{12} & 1 & \frac{71}{4} \\
0 & 1 & 0 & \frac{5}{6} & \boxed{\frac{7}{6}} & -\frac{7}{6} & 0 & \frac{7}{2} \\
0 & 1 & 0 & -\frac{1}{12} & -\frac{5}{12} & \frac{1}{12} & 0 & \frac{1}{4} \\
1 & 0 & 0 & -\frac{1}{6} & \underset{\uparrow}{\frac{1}{6}} & \frac{1}{6} & 0 & \frac{3}{2}
\end{bmatrix}. \tag{15.63a}
$$

The next two iterations are

$$
\begin{array}{c}
\frac{3}{\frac{5}{7}} = 4.2 \\[12pt]
\frac{\frac{2}{3}}{\frac{3}{14}} = 7.0 \\[12pt]
\\
\\
\end{array}
\begin{array}{c}
x_1 \quad x_2 \quad x_3 \quad x_4 \quad x_5 \quad x_6 \quad z'
\end{array}
$$

$$
\begin{bmatrix}
0 & 0 & \frac{5}{2} & -\frac{1}{2} & 0 & \frac{1}{2} & 1 & \frac{53}{2} \\
\rightarrow 0 & 0 & \frac{6}{7} & \boxed{\frac{5}{7}} & 1 & -\frac{7}{5} & 0 & 3 \\
0 & 0 & \frac{5}{14} & \frac{3}{14} & 0 & -\frac{3}{14} & 0 & \frac{3}{2} \\
1 & 0 & -\frac{1}{7} & \underset{\uparrow}{-\frac{2}{7}} & 0 & -\frac{2}{7} & 0 & 1.0
\end{bmatrix}, \tag{15.66}
$$

$$
\begin{bmatrix}
0 & 0 & \frac{31}{10} & 0 & \frac{7}{10} & 0 & 1 & \frac{143}{5} \\
0 & 0 & \frac{6}{5} & 1 & \frac{7}{5} & -1 & 0 & \frac{21}{5} \\
0 & 1 & \frac{1}{10} & 0 & -\frac{3}{10} & 0 & 0 & \frac{3}{5} \\
1 & 0 & \frac{1}{5} & 0 & \frac{2}{5} & 0 & 0 & \frac{11}{5}
\end{bmatrix}, \tag{15.67}
$$

from which the final solution to this artificial-variable problem is found to be

$$x_3 = 0, \qquad x_5 = 0,$$

$$x_6 = 0, \qquad x_1 = 2.2, \qquad (15.68)$$

$$x_2 = 0.6, \qquad x_4 = 4.2,$$

$$\max z = 28.6$$

Figure 15.8 indicates the graphical solution to the same problem.

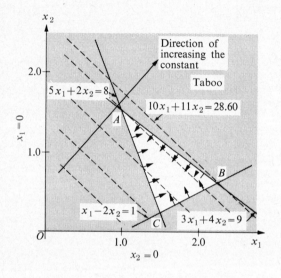

Fig. 15.8. Artificial variable problem.

Col 1 to 10				Col 10 to 20				Col 20 to 30			Col 30 to 40			Col 40 to 50			Col 50 to 60			Col 60 to 70	
	4		7																		
-1	0 . 0			-1	1 . 0			0 . 0			0 . 0			0 . 0			0 . 0			0 . 0	
	3 . 0				4 . 0			1 . 0			0 . 0			0 . 0			0 . 0			9 . 0	
	5 . 0				2 . 0			0 . 0			-1 . 0			0 . 0			1 . 0			8 . 0	
	1 . 0			-2	2 . 0			0 . 0			0 . 0			1 . 0			0 . 0			1 . 0	
	0 . 0				0 . 0			0 . 0			0 . 0			0 . 0			0 . 0			0 . 0	
	3		6		5																

Fig. 15.9.

```
FEASIBLE
ITERATION        OBJ. FUNCTION    NEW BASIC VAR.
    1                24.75               2
    2                26.50               1
    3                28.59               4


OBJ. FUNCTION         28.59999900

VARIABLE             VALUE
    2              .60000010
    1             2.19999990
    4             4.20000000
```

THE FINAL MATRIX

ROW   1

```
 0.0000    0.0000    3.1000    0.0000    .6999    0.0000    28.5999
```

ROW   2

```
 0.0000    1.0000    .1000     0.0000   -.2999    0.0000    .6000
```

ROW   3

```
 1.0000    0.0000    .1999     0.0000    .3999    0.0000    2.1999
```

ROW   4

```
 0.0000    0.0000    1.2000    1.0000   1.4000   -1.0000    4.2000
```

ROW   5

```
 0.0000    0.0000    0.0000    0.0000   0.0000    0.0000    0.0000
```

Fig. 15.10. Output for Example 15.2.

Example 15.2

If we were to apply the FORTRAN IV program listed in Fig. 15.5 to the example discussed above, the input would be read in as shown in Fig. 15.9. The output for this example is listed in Fig. 15.10.

BIBLIOGRAPHY

General

Bellman, R., "The theory of dynamic programming," *Modern Mathematics for the Engineer*, Ch. 11, E. F. Beckenbach, ed. McGraw-Hill, New York, 1956.

Berge, C., and A. Ghouila-Houri, *Programming, Games and Transportation Networks*. Methuen, London, 1965.

Charnes, A., and W. W. Cooper, *Management Models and Industrial Application of Linear Programming*, I and II. Wiley, New York, 1960.

Fletcher and Powell, "A rapidly convergent descent method for minimization," *Comput. J.*, **6**, no. 2, 163–168 (1963).

Gass, S., *Linear Programming Methods and Applications*, 3rd ed. McGraw-Hill, New York, 1969.

Harrison, J. O., Jr., "Linear programming and operations research," *Operations Research for Management*, J. F. McCloskey and F. N. Trefethen, eds., **1**, 217–237. Johns Hopkins Press, Baltimore, 1954.

Koopmans, T. C. (ed.), *Activity Analysis of Production and Allocation*, Cowles Commission Monograph 13. Wiley, New York, 1951.

Kunzi, H. P., H. G. Tzschach, and C. A. Zehnder, *Numerical Methods of Mathematical Optimization with ALGOL and FORTRAN Programs*. Translated by W. C. Rheinboldt and C. J. Rheinboldt. Academic Press, New York, 1968.

Application of Linear Programming

Bowman, E. H., "Production scheduling by the transportation method of linear programming," *Ops. Res.*, **4** (1956).

Charnes, A., W. W. Cooper, and D. Farr, "Linear programming and profit preference scheduling for a manufacturing firm," *J. Ops. Res. Soc. Am.* (now *Ops. Res.*), **1** (1953).

Charnes, A., W. W. Cooper, and B. Mellon, "Blending aviation gasolines—a study in programming interdependent activities in an integrated oil company," *Econometrica*, **20** (1952).

Eisemann, K., "The trim problem," *Mgmt. Sci.*, **3**, no. 3 (1957).

Eisemann, K., and J. R. Lourie, *The Machine Loading Problem*. IBM Applications Library, New York, 1959.

Gilmore, P. C., and R. E. Gomory, "A linear programming approach to the cutting stock problem—Part II," *Ops. Res.*, **11**, 863–887 (1963).

Joseph, J. A., "The application of linear programming to weapon selection and target analysis," *Technical Memorandum 42*, Operations Analysis Division, Headquarters, U.S. Air Force, Washington, D.C. (1954).

Katzman, Irwin, "Solving feed problems through linear programming," *J. Fm. Econ.*, **38** (1956).

Munkres, J., "Algorithms for the assignment and transportation problems," *J. Soc. Ind. Appl. Math.*, **5**, no. 1 (1957).

Neumann, J. von, "A numerical method to determine optimum strategy," *Nav. Res. Logist. Q.*, **1** (1954).

Stanley, E. D., D. Honig, and L. Gainen, "Linear programming in bid evaluation," *Nav. Res. Logist. Q.*, **1** (1954).

Suzuki, G., "A transportation simplex algorithm for machine computation based on the generalized simplex method," *Report 959*, The David W. Taylor Model Basin, Washington, D.C. (1955).

Wolfe, P., "A technique for resolving degeneracy in linear programming," *Rand Report RM-2995-PR*. The Rand Corporation, Santa Monica, Calif. (May 1962).

Traveling Salesman Problem

Bellman, R., "Dynamic programming treatment of the Traveling Salesman problem," *J. Assoc. Comput. Mach.*, **9,** 61–63 (1962).

Dantzig, G. B., R. Fulkerson, and S. Johnson, "Solution of a large scale Traveling Salesman problem," *Ops. Res.*, **2,** 293–404 (1954).

Gilmore, P. C., and R. E. Gomory, "Sequencing a one-state variable machine: a solvable case of the Traveling Salesman problem," *Ops. Res.*, **12,** 655–679 (1964).

Krolak, P., W. Felts, and G. Marble, "A man-machine approach toward solving the Traveling Salesman problem," *Commun. Assoc. Comput. Mach.*, **14,** no. 5, 327–334 (May 1971).

Kuo, S. S., and W. K. Young, "Computer studies of the Traveling Salesman problem," *Q. Bull., Can. Inf. Process. Soc.*, **8,** no. 5, 31–36 (1968).

Kuo, S. S., "4-opt treatment of the Traveling Salesman problem," *Proceedings of Purdue University Centennial Symposium on Information Processing*, pp. 121–130 (1969).

Kuo, S. S., "Computer studies of the asymmetric Traveling Salesman problem," *Proceedings of International Symposium of Digital Computation*, pp. 245–258. Istanbul Technical University Press, Turkey, Oct. 1970.

Lin, S., "Computer solutions of the Traveling Salesman problem," *The Bell System Technical Journal*, **XLIV,** no. 10, 2245–2269 (1965).

Michie, D., J. G. Fleming, and J. V. Oldfield, "A comparison of heuristic, interactive, and unaided methods of solving a shortest-route problem," *Mach. Intell.*, **3,** 245, 255 (1968).

Obruca, A. K., "Spanning tree manipulation and the Traveling Salesman problem," *Br. Comput. J.*, 374–377 (Feb. 1968).

PROBLEMS

1. a) Use the graphical method to find

$$x_1 \geq 0, \qquad x_2 \geq 0,$$

so that $P = 6x_1 + 9x_2$ is a maximum and

$$2x_1 + 3x_2 \leq 12,$$

$$4x_1 + x_2 \leq 8.$$

b) Tabulate all basic solutions.

c) How many feasible basic solutions are there?

2. Use (a) the graphical method, (b) the simplex method and, (c) the simplex tableau to maximize the following objective function:

$$P = 10A + 10B$$

subject to the constraints:

$$4A + 2B \leq 500$$
$$A + 2B \leq 200$$
$$A \geq 50$$
$$B \geq 0$$

3. Due to a mistake in planning, a manufacturing firm finds that it has 120 hr of skilled labor and 140 hr of machine time available next week. Product A requires 5 hr of machine time and 2 hr of skilled labor per unit. Product B requires 3 hr of machine time and 3 hr of skilled labor per unit. Product A will give the company a profit of $5 per unit and product B yields a profit of $4 per unit. Use the Simplex method to find the amounts of A and B which should be produced for maximum profits.

4. The weekly production schedule for three plants of an automobile manufacturer must be prepared. Plant A has a maximum total weekly output of 9800 units of all models, plant B a capacity for all models of 9419 units, and plant C a total maximum plant capacity of 9996 units. A breakdown of plant capacity by model type is shown in Table 15.5. The associated manufacturing costs by model type for each plant are shown in Table 15.6. Assign assembly quotas by model type to each plant based on the demand outlined in Table 15.7, in such a fashion that (a) the demand by model type is satisfied;

Table 15.5 Maximum plant capacity by model type

Plant	Custom	J300	Wildcat	Special	Riviera	Total
A	7154	None	1568	None	1078	9800
B	3014	2543	1036	1884	0942	9419
C	3899	2099	1499	2499	None	9996
Total	14067	4642	4103	4383	2020	

Table 15.6 Unit cost by model

Plant	Custom	J300	Wildcat	Special	Riviera
A	$2281	None	$3220	None	$3367
B	2404	$3478	3421	$1954	3501
C	2197	3495	3217	1917	None

Table 15.7 Sales orders for the week of November 4th, 1964

Model type	Custom	Jumbo 300	Wildcat	Special	Riviera
Quantity	8990	3101	2421	3785	0980

(b) the plant capacity by model type is not exceeded; (c) the total manufacturing costs of all plants is a minimum, based on the cost data outlined in Table 15.6.

5. Two types of cakes may be made in a factory. The first item (cake A) requires 2 hr of furnace baking and 3 hr of finishing. The second item (cake B) needs 2 hr of baking and 4 hr of finishing. The profit for each cake A is $1 and that for each cake B is $3. The owner may put as many as 10 workers on the finishing job to work a regular eight-hour shift. He has one furnace which, to operate economically, should work 24 hr a day and which bakes one cake at a time. How many cakes A and how many cakes B should be made *per day* to produce a maximum profit? Use the Simplex method.

Appendixes

Number Systems

A.1 INTRODUCTION

The most commonly used number system is the decimal system based on powers of ten. This system requires ten symbols:

$$0, 1, 2, 3, 4, 5, 6, 7, 8, 9.$$

The symbols 0 to 9 can be used to represent more than ten numbers by writing them in a particular order. To obtain the representation of a written integer such as 353, the digit farthest to the right is multiplied by $(10)^\circ$ or 1, and the succeeding digits to the left are multiplied by successively increased powers of 10. The products are then summed to give the integer 354. Or,

$$354 = 3 \times 10^2 + 5 \times 10^1 + 4 \times 10^0.$$

When numbers involve a decimal point, digits to the right are multiplied by successively increasing negative powers of 10, beginning with -1 at the digit adjacent to the decimal point. Digits to the left of the decimal point are multiplied by positive powers of 10, beginning with zero. As in the case of integers, the sum of these products is a representation of the value of the set of digits in this particular order. For example,

$$0.354 = 3 \times 10^{-1} + 5 \times 10^{-2} + 4 \times 10^{-3}.$$

Number systems involving bases other than 10 are also used. For instance, the binary system uses base 2, the hexadecimal system, base 16. Since the binary system has base 2, only two symbols are used, namely 1 and 0. Binary numbers can be represented in a manner similar to that applying to the number system based on 10. In binary, the symbols are multiplied by correct powers of 2 instead of 10, depending on their location with respect to the binary point. The products are then added to

383

obtain the decimal representation in the base-10 system. This concept will be further discussed in Section A.2.

In the hexadecimal system, sixteen symbols are used: 0, 1, 2, 3, ..., A, B, ..., F. The principal purpose of this appendix is to describe the three number systems and their interrelations.

A.2 BINARY SYSTEM

The relation between the binary and decimal systems (base-10 system) can be best explained by Fig. A.1. The four dials imprinted with the binary symbols 0 and 1

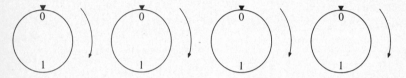

Fig. A.1. Binary number system.

represent the four digit locations to the left of the decimal point in a binary number. Each dial can rotate only half a revolution clockwise at one time. Thus the symbols can either occupy the positions shown or they can appear in the opposite position. Each time a dial moves through one complete revolution, the dial on its left rotates through half a revolution.

If we begin with all dials set at 0, which is equivalent to 0 in the decimal system, a comparison of the binary and decimal system can be summarized below:

Decimal system	Binary system	Decimal system	Binary system
0	0000	7	0111
1	0001	8	1000
2	0010	9	1001
3	0011	10	1010
4	0100	11	1011
5	0101	12	1100
6	0110		

It can be seen that the decimal system of numbers can be represented by certain logical arrangements of only two symbols. As the decimal numbers become larger, the corresponding binary numbers become extremely long and cumbersome.

Binary numbers can be converted into decimal form by following the rules set forth in Section A.1. For instance, to represent the binary number 11011 in decimal form, the steps indicated by the example below must be carried out.

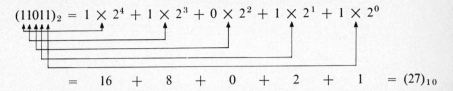

$$(11011)_2 = 1 \times 2^4 + 1 \times 2^3 + 0 \times 2^2 + 1 \times 2^1 + 1 \times 2^0$$

$$= \quad 16 \quad + \quad 8 \quad + \quad 0 \quad + \quad 2 \quad + \quad 1 \quad = (27)_{10}$$

Mathematical calculations in the binary system have their own rules corresponding to the familiar "carry" and "borrow" associated with the decimal system. These rules, along with simple examples, are given below. The decimal equivalents accompanying each example can be obtained by following the conversion process illustrated previously.

Rules for Binary Calculation

Addition

1. $\qquad\qquad (0)_2 + (0)_2 = (0)_2,$
2. $\qquad\qquad (0)_2 + (1)_2 = (1)_2,$
3. $\qquad\qquad (1)_2 + (1)_2 = (0)_2,$ carry $(1)_2$ to the left.

Example

Binary addition:

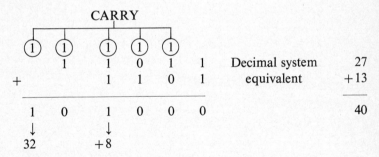

Subtraction

1. $\qquad\qquad (0)_2 - (0)_2 = (0)_2,$
2. $\qquad\qquad (1)_2 - (1)_2 = (0)_2,$
3. $\qquad\qquad (1)_2 - (0)_2 = (1)_2,$
4. $\qquad\qquad (0)_2 - (1)_2 = (1)_2,$ with $(1)_2$ borrowed from the left.

Example

Binary subtraction:

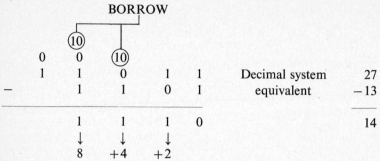

					Decimal system	27
					equivalent	-13
						14

Multiplication

1. $(0)_2 \times (0)_2 = (0)_2,$
2. $(0)_2 \times (1)_2 = (0)_2,$
3. $(1)_2 \times (1)_2 = (1)_2.$

Example

Binary multiplication:

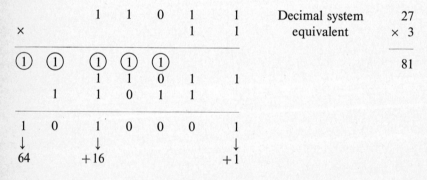

					Decimal system	27
					equivalent	$\times\ 3$
						81

Example

Binary division:

		1	0	0	1				9
11		1	1	0	1	1	Decimal system	3	27
		1	1				equivalent		27
			0	1	1				
				1	1				

A.3 HEXADECIMAL SYSTEM

As mentioned previously, the hexadecimal system uses ten numerical symbols and six alphabetic symbols. These are, in ascending order, 0, 1, 2, 3, 4, 5, 6, 7, 8, 9, A, B, C, D, E, F. All numbers in the hexadecimal system are represented by combinations of these symbols.

A comparison of the decimal, binary, and hexadecimal systems is shown below.

Decimal system	Binary system	Hexadecimal system
0	000	0
1	001	1
2	010	2
3	011	3
4	100	4
5	101	5
6	110	6
7	111	7
8	1000	8
9	1001	9
10	1010	A
11	1011	B
12	1100	C
13	1101	D
14	1110	E
15	1111	F
16	10000	10
$\vdots$	$\vdots$	$\vdots$

The meaning of a number in the hexadecimal system can be best understood by expanding the number in powers of 16. For example,

$$(BA6)_{16} = 11 \times 16^2 + 10 \times 16^1 + 6 \times 16^0 = (2982)_{10}.$$

Likewise, a hexadecimal number 5020 may be converted to a decimal number as follows:

$$5(16)^3 + 0(16)^2 + 2(16)^1 + 0(16)^0 = 20480 + 0 + 32 + 0 = 20512.$$

Table A.1 can be used to convert a hexadecimal number to its decimal equivalent. Each hexadecimal digit and the decimal equivalent for this digit are first located. These partial equivalents are added to obtain the desired decimal number. Consider again

Table A.1 Hexadecimal and decimal conversion

Column / Hex. no.	(1)	(2)	(3)	(4)	(5)	(6)
0	0	0	0	0	0	0
1	1	16	256	4,096	65,536	1,048,576
2	2	32	512	8,192	131,072	2,097,152
3	3	48	768	12,288	196,608	3,145,728
4	4	64	1,024	16,384	262,144	4,194,304
5	5	80	1,280	20,480	327,680	5,242,880
6	6	96	1,536	24,576	393,216	6,291,456
7	7	112	1,792	28,672	458,752	7,340,032
8	8	128	2,048	32,768	524,288	8,388,608
9	9	144	2,304	36,864	589,824	9,437,184
A	10	160	2,560	40,960	655,360	10,485,760
B	11	176	2,816	45,056	720,896	11,534,336
C	12	192	3,072	49,152	786,432	12,582,912
D	13	208	3,328	53,248	851,968	13,631,488
E	14	224	3,584	57,344	917,504	14,680,064
F	15	240	3,840	61,440	983,040	15,728,640

Example   F00 = 3,840; A0 = 160

$(5020)_{16}$. The decimal equivalent of the leftmost digit 5 is 20,480, which is located in column 4 of the table. The final sum 20,512 can be readily obtained as follows:

Hexadecimal digit	From Table A.1
0	0
2	32
0	0
5	20,480 +
	20,512

The hexadecimal number system is most often used to simplify working with the long and cumbersome binary data. A binary number is frequently shown in its hexadecimal representation. This conversion is fairly simple. Beginning at the right, separate the given binary number into groups of four. Each group is then represented by its equivalence of a symbol in the hexadecimal system.

Example

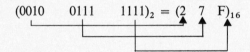

$$(0010 \quad 0111 \quad 1111)_2 = (2 \quad 7 \quad F)_{16}$$

A more complete discussion of conversion of integers and fractions from one system to another is given in Table A.4.

Like decimal numbers, hexadecimal numbers obey all the rules of arithmetic and can therefore be added, subtracted, multiplied, and divided. In hexadecimal addition, there is no carry until a multiple of $(16)_{10}$ is reached. Some illustrative examples follow:

$$F + 1 = 10.$$

$$A + E = 18.$$

$$\begin{array}{r} 75C \\ +4B6 \\ \hline C12 \end{array}$$

here

$$C = (12)_{10},$$

$$(12)_{10} + 6 = (18)_{10} = (12)_{16},$$

$$1_{\text{carry}} + 5 + (11)_{10} = (17)_{10} = (11)_{16},$$

$$1_{\text{carry}} + 7 + 4 = (12)_{10} = (C)_{16}.$$

5020	Decimal system	20512
+8FD	equivalent	2301+
591D		22813

As a practical detail, two hexadecimal numbers can be added by means of Table A.2. Locate one of the numbers to be added as the heading of a column and the other as the heading of a row. The sum is the intersection of the column and the row.

Example

3F4	Decimal system	1012
+23B	equivalent	+ 571
62F		1583

Table A.2 can also be used for subtraction. Go down the column with the heading of the subtrahend (the number to be subtracted) until the minuend (the number from which it is subtracted) is located. The heading of this row is the difference.

Example

3F4	Decimal system	1012
− 23B	equivalent	− 571
1B9		441

Table A.2 Hexadecimal addition and subtraction

	1	2	3	4	5	6	7	8	9	A	B	C	D	E	F	
1	02	03	04	05	06	07	08	09	0A	0B	0C	0D	0E	0F	10	1
2	03	04	05	06	07	08	09	0A	0B	0C	0D	0E	0F	10	11	2
3	04	05	06	07	08	09	0A	0B	0C	0D	0E	0F	10	11	12	3
4	05	06	07	08	09	0A	0B	0C	0D	0E	0F	10	11	12	13	4
5	06	07	08	09	0A	0B	0C	0D	0E	0F	10	11	12	13	14	5
6	07	08	09	0A	0B	0C	0D	0E	0F	10	11	12	13	14	15	6
7	08	09	0A	0B	0C	0D	0E	0F	10	11	12	13	14	15	16	7
8	09	0A	0B	0C	0D	0E	0F	10	11	12	13	14	15	16	17	8
9	0A	0B	0C	0D	0E	0F	10	11	12	13	14	15	16	17	18	9
A	0B	0C	0D	0E	0F	10	11	12	13	14	15	16	17	18	19	A
B	0C	0D	0E	0F	10	11	12	13	14	15	16	17	18	19	1A	B
C	0D	0E	0F	10	11	12	13	14	15	16	17	18	19	1A	1B	C
D	0E	0F	10	11	12	13	14	15	16	17	18	19	1A	1B	1C	D
E	0F	10	11	12	13	14	15	16	17	18	19	1A	1B	1C	1D	E
F	10	11	12	13	14	15	16	17	18	19	1A	1B	1C	1D	1E	F
	1	2	3	4	5	6	7	8	9	A	B	C	D	E	F	

Table A.3 can likewise be used to perform the multiplication of two hexadecimal numbers. In this table, locate one of the numbers to be multiplied as the heading of a column, and the other as the heading of a row. The product is the number at the intersection of the column and the row.

Table A.3 Hexadecimal multiplication

	2	3	4	5	6	7	8	9	A	B	C	D	E	F
2	04	06	08	0A	0C	0E	10	12	14	16	18	1A	1C	1E
3	06	09	0C	0F	12	15	18	1B	1E	21	24	27	2A	2D
4	08	0C	10	14	18	1C	20	24	28	2C	30	34	38	3C
5	0A	0F	14	19	1E	23	28	2D	32	37	3C	41	46	4B
6	0C	12	18	1E	24	2A	30	36	3C	42	48	4E	54	5A
7	0E	15	1C	23	2A	31	38	3F	46	4D	54	5B	62	69
8	10	18	20	28	30	38	40	48	50	58	60	68	70	78
9	12	1B	24	2D	36	3F	48	51	5A	63	6C	75	7E	87
A	14	1E	28	32	3C	46	50	5A	64	6E	78	82	8C	96
B	16	21	2C	37	42	4D	58	63	6E	79	84	8F	9A	A5
C	18	24	30	3C	48	54	60	6C	78	84	90	9C	A8	B4
D	1A	27	34	41	4E	5B	68	75	82	8F	9C	A9	B6	C3
E	1C	2A	38	46	54	62	70	7E	8C	9A	A8	B6	C4	D2
F	1E	2D	3C	4B	5A	69	78	87	96	A5	B4	C3	D2	E1

Table A.4

OPERATION: Hexadecimal to binary		OPERATION: Binary to hexadecimal	
INTEGERS	FRACTIONS	INTEGERS	FRACTIONS
Rule: Represent each symbol in the hexadecimal number by an equivalent four-digit symbol in the binary system as follows:	Rule: Represent each symbol in the hexadecimal number by an equivalent four digit symbol in the binary system, beginning at the decimal as shown.	Rule: Divide the binary number into groups of four digits, beginning at the right, and convert each group to its equivalent hexadecimal form.	Rule: Divide the binary number into groups of four digits, beginning at the decimal point, and convert each group into its equivalent hexadecimal form.
$(3F4)_{16}$ $(0011\ 1111\ 0100)_2$	$(0.3951)_{16}$ $(0.0011\ 1001\ 0101\ 0001)_2$	$(0011\ 1001\ 0100)_2$ $(394)_{16}$	$(0.0110\ 0111\ 0001)_2$ $(0.671)_{16}$

(cont.)

Table A.4

OPERATION: Decimal to binary		OPERATION: Binary to decimal	
INTEGERS	FRACTIONS	INTEGERS	FRACTIONS

OPERATION: Decimal to binary

INTEGERS

Rule: Divide the decimal number by 2 and proceed as shown.

$(421)_{10} = (?)_2$

```
2|421  remainder 1
2|210  remainder 1
2|105  remainder 1
2|52   remainder 0
2|26   remainder 0
2|13   remainder 1
2|6    remainder 0
2|3    remainder 1
2|1    remainder 1
  0
```

The binary number is read as directed by the arrow. Thus

$(421)_{10} = (110100101)_2$

FRACTIONS

Rule: Multiply the decimal number by 2 and develop as shown below.

$(0.321)_{10} = (?)_2$

```
    0.321
  ×     2
    0.642
  ×     2
    1.284
  ×     2
    0.568
  ×     2
    1.136
```

Only the portion of the number to the right of the decimal is multiplied by 2. The binary answer is made up of the single digits to the left of the decimal as directed by the arrow. Thus

$(0.321)_{10} = (0.0101)_2$

OPERATION: Binary to decimal

INTEGERS

Rule: Multiply the binary number by 2, and add as shown below.

$(010111)_2 = (?)_{10}$

```
0 1 0 1 1 1
      0
    × 2
      0
    + 1
      1
    × 2
      2
    + 0
      2
    × 2
      4
    + 1
      5
    × 2
     10
    + 1
     11
    × 2
     22
    + 1
     23
```

$(010111)_2 = (23)_{10}$

FRACTIONS

Rule: Express the binary number as powers of 2, then add and divide as shown.

$(0.11011)_2 = (?)_{10}$

$$= 1 \times 2^{-1} + 1 \times 2^{-2}$$
$$+ 0 \times 2^{-3} + 1 \times 2^{-4}$$
$$+ 1 \times 2^{-5}$$
$$= \tfrac{1}{2} + \tfrac{1}{4} + 0 + \tfrac{1}{16} + \tfrac{1}{32}$$
$$= 0.8437$$
$$(0.11011)_2 = (0.8437)_{10}$$

OPERATION: Decimal to hexadecimal

INTEGERS

Rule: Divide the decimal number by 16 and develop the hexadecimal number as shown below.

$(953)_{10} = (?)_{16}$

```
16│953  remainder  9  ↖
16│ 59  remainder 11  |
16│  3  remainder  3  |
      0
```

The hexadecimal number is read as directed by the arrow. Thus

$(953)_{10} = (3B9)_{16}$

Note: When converting from decimal to hexadecimal, it is easier to convert the decimal number to a binary number first and then to a hexadecimal number.

FRACTIONS

Rule: Multiply the decimal number by 16 and develop the hexadecimal number as shown.

$(0.953)_{10} = (?)_{16}$

```
  0.953
×    16
 15.248
×    16
  3.968
×    16
↓15.488
```

Only the portion of the number to the right of the decimal is multiplied by 16. The answer in hexadecimal is made up of the single digits to the left of the decimal point, as directed by the arrow. Thus

$(0.953)_{10} = (0.F3F)_{16}$

OPERATION: Hexadecimal to decimal

INTEGERS

Rule: Multiply the hexadecimal number by 16 and add as shown below.

$(953)_{16} = (?)_{10}$ $(BA6)_{16} = (?)_{10}$

```
  953            11
×  16          × 16
  144            66
+   5            11
  149           176
×  16          + 10
 2384           186
+   3          × 16
 2387          1116
                186
               2976
               +  6
            (2982)_{10}
```

$(953)_{16} = (2387)_{10}$

Note: Change all alphabetic symbols to their decimal equivalents before performing the operation and treat the two digit decimal number as a single symbol in the first step.

FRACTIONS

Rule: Express the hexadecimal number as powers of 16, add, and divide as shown.

$(0.953)_{16} = (?)_{10}$

$$= 9 \times 16^{-1} + 5 \times 16^{-2} + 3 \times 16^{-3}$$
$$= \frac{9}{16} + \frac{5}{256} + \frac{3}{4096}$$
$$= 0.56 + 0.031 + 0.00073$$
$$= 0.59173.$$
$$(0.953)_{16} = (0.59173)_{10}$$

Example

2F2	Decimal system	754
× 3A	equivalent	× 58
1D74		6032
8D6		3770
AAD4		43732

In practice, multiplication or division of two hexadecimal numbers can best be performed in the following manner: Convert the hexadecimal numbers to their binary equivalents, then perform the binary multiplication or division.

A.4 NUMBER CONVERSIONS

The conversion of integers and fractions from one system to either of the others may best be explained by a set of examples. Table A.4 (p. 391) illustrates these conversions between decimal, octal, and binary number systems.

Dumps

B.1 INTRODUCTION

A storage "dump" is a hard copy of the contents of all or part of main storage area. Dumps are useful both in debugging a program and in examining the intermediate results of a program. A sample layout of a storage area for a main program and three associated subroutines is shown in Fig. B.1. In the particular installation where these programs are executed, the first $(5020)_{16}$ bytes, known as the *relocatable quantity*, are reserved for the operation system. It is important to find out the value (in hexadecimal notation) of the relocatable quantity used in your installation.

This appendix deals primarily with the techniques of producing dumps and the steps to interpret them. We shall study a powerful subroutine PDUMP commonly used in a FORTRAN program.

B.2 PDUMP ROUTINE

To produce dump for a FORTRAN IV program, at least two routines can be used—DUMP and PDUMP routines. The FORTRAN call statement for the DUMP routine is:

$$\text{CALL DUMP } (a_1, b_1, f_1, \ldots, a_n, b_n, f_n)$$

where variables a and b indicate the limits of various storage areas to be dumped. Both a and b must be in the same program or in the common area; either a or b may be the lower or the upper limit of the storage area. The letter f indicates the dump format with the following permissible values:

$$0 - \text{hexadecimal}$$
$$4 - \text{integer}$$
$$5 - \text{real}$$
$$6 - \text{double precision}$$

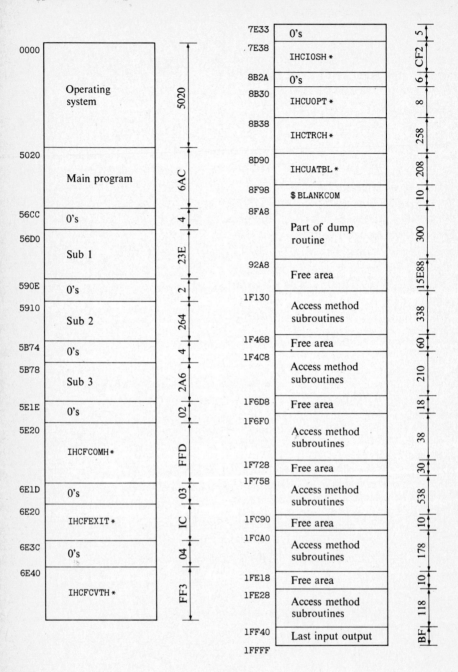

Fig. B.1. A sample layout of a core memory.

For example, the call statement:

CALL DUMP (T, ANS1, 0, B1, ANS2, 5)

would dump the contents of the storage area between T and ANS1 in hexadecimal format; also, from B1 to ANS2 in real format.

Likewise, the call statement for the PDUMP routine is as follows:

CALL PDUMP $(a_1, b_1, f_1, \ldots, a_n, b_n, f_n)$

where a's, b's, and f's have the same meanings as defined above.

There is only slight difference between a DUMP and a PDUMP routine. After a DUMP routine is executed the program is terminated, but after a PDUMP the program execution continues without interruption.

Consider, for example, the FORTRAN program shown in Fig. B.2 where a CALL PDUMP statement is used. This program dumps storage between J and ANS1 in hexadecimal, and between B1 and ANS2 in real format. The program also continues until its completion; hence its required output is also printed.

0001	READ(1,5)A,B
0002	5 FØRMAT(2F10.2)
0003	J = 6+A
0004	I = 40
0005	ANS1 = ((J*J)/I)+A−B
0006	B1 = B+A
0007	ANS2 = ((I*I)/J)+A−B
0008	CALL PDUMP(3,7)(J,ANS1,0,B1,ANS2,5)
0009	WRITE(3,7)(J,I,ANS1,B1,ANS2)
0010	7 FØRMAT(15,3X,15,3X,3F10.2)
0011	CALL EXIT
0012	END

Fig. B.2. A FORTRAN program using PDUMP routine.

0050B8	00000045 00000028 429830A4	⎫ Dump
0050C4	0.93049988E 02 0.561900C2E 02	⎬ answers
69	40 152.19 93.05 56.19	⎭

Fig. B.3. Dump and regular output.

FORTRAN IV G LEVEL 0, MOD 0   MAIN   DATE = 67291   11/01/35   PAGE 0002

SCALAR MAP

SYMBOL	LOCATION	SYMBOL	LOCATION	SYMBOL	LOCATION	SYMBOL	LOCATION	SYMBOL	LOCATION
A	90	B	94	J	98	I	9C	ANS1	A0
B1	A4	ANS2	A8						

SUBPROGRAMS CALLED

SYMBOL	LOCATION	SYMBOL	LOCATION	SYMBOL	LOCATION	SYMBOL	LOCATION
IBCOM=	AC	PDUMP	B0	EXIT	B4		

FORMAT STATEMENT MAP

SYMBOL	LOCATION	SYMBOL	LOCATION	SYMBOL	LOCATION
5	D0	7	D7		

TOTAL MEMORY REQUIREMENTS 00026C BYTES

Fig. B.4. Scalar map.

A sample dump output is shown in the first two lines in Fig. B.3, with the values of A and B taken respectively as 63.12 and 29.93. The first line shows the contents (in hexadecimal notation) of the storage area beginning at the byte location $(0050B8)_{16}$. Also, we learn from the scalar map (Fig. B.4) that the values of J, I and ANS1 are contained in this storage area. According to this map, the value of J is at location 98 (in hexadecimal). However, a relocatable quantity of 5020 bytes must be added to obtain its absolute address. Thus the value of J actually begins from the location 5020 + 98 = 50B8, which is printed on the first line of Fig. B.3. Also, the value of J takes a fullword or 32 bits in storage. The contents in these 4 bytes are 00000045, which equals $(69)_{10}$. This checks well with the value of J in the answer, which is shown on the third line of Fig. B.3.

Again by examining the scalar map, we learn that I is the variable located after J. The value of I takes a fullword and it is $(00000028)_{16}$, as shown in Fig. B.3. It should be noted that $(00000028)_{16}$ equals $(40)_{10}$, which checks exactly with the answer for I (see the last line in Fig. B.3).

In like manner we find that the next value at location 50C0 reads 429830A4. This is the hexadecimal representation of the floating point variable ANS1. The first byte, $42 = (66)_{10}$, is the characteristic byte. It is formed by adding 64 to the actual exponent. Hence, the exponent is equal to 2. Now the hexadecimal point is placed and the hexadecimal number 98.30A4 can be readily converted into its decimal equivalent:

$$(98)_{16} = 9 \times 16^1 + 8 \times 16^0 = (152)_{10}$$

$$30A4 = 3 \times 16^{-1} + 0 \times 16^{-2} + A \times 16^{-3} + 4 \times 16^{-4} = 0.19$$

Thus, the contents of location 50C0 in decimal is 152.19. This coincides with the result of ANS1 shown in the third line of Fig. B.3.

The second line of the dump output lists the contents of the storage area between B1 and ANS2 in real format. According to the storage map (Fig. B.4), B1 begins at location A4. Its absolute address is then 0050C4, and its contents are 93.049 in decimal. The last number in the second row of Fig. B.3 represents the floating number 56.19 in decimal. Its byte address is 0050C8. This again checks well with the result of ANS2, which is listed as the last number in Fig. B.3.

Derivation of Bairstow's iterative equations

Consider a nonlinear algebraic (polynomial) equation of the symbolic form

$$y^n + A_1 y^{n-1} + A_2 y^{n-2} + \cdots + A_{n-1} y + A_n = 0, \tag{C1}$$

where $A_1, \ldots, A_n$ are real. As was mentioned in Section 6.5, the problem is to find p and q in a quadratic factor of the form $(y^2 + py + q)$, by an iterative technique. If a polynomial of the form of Eq. (C1) is divided by the quadratic factor $(y^2 + py + q)$, it yields a polynomial of the form $y^{n-2} + B_1 y^{n-3} + B_2 y^{n-4} + \cdots + B_{n-3} y + B_{n-2}$ and a remainder of the form $Ry + S$.

Our objective is to reduce the remainders to zero, that is, R = S = 0. To do this we first establish the relationships between p, q, R, S, the A's, and the B's. This is a simple matter of rewriting Eq. (C1) in the factored form, expanding, and then equating coefficients of like powers.

If we let k denote the general term and n denote the last term, we have

$$y^n + A_1 y^{n-1} + \cdots + A_k y^{n-k} + \cdots + A_{n-1} y + A_n$$
$$= (y^2 + py + q)(y^{n-2} + B_1 y^{n-3} + B_2 y^{n-4} + \cdots + B_{k-2} y^{n-(k)}$$
$$+ B_{k-1} y^{n-(k+1)} + B_k y^{n-(k+2)} + \cdots + B_{n-3} y + B_{n-2})$$
$$+ Ry + S \tag{C2}$$

$$= y^n + (B_1 + p) y^{n-1} + (B_2 + B_1 p + q) y^{n-2} + \cdots$$
$$+ (B_k + pB_{k-1} + qB_{k-2}) y^{n-(k)} + \cdots$$
$$+ (pB_{n-2} + qB_{n-3} + R) y + (qB_{n-2} + S). \tag{C3}$$

Equating coefficients of like powers, we obtain

$$A_k = B_k + pB_{k-1} + qB_{k-2}, \qquad k = 1, 2, \ldots, n-2; \tag{C4}$$

$$A_{n-1} = R + pB_{n-2} + qB_{n-3}, \qquad k = n-1; \tag{C5}$$

$$A_n = S + qB_{n-2}, \qquad k = n. \tag{C6}$$

There is no y^{n-1}-term on the right-hand side of Eq. (C2) since this would lead to a y^{n+1}-term when expanded, and hence $B_{-1} = 0$ and $B_0 = 1$. All other B's, however, vary with p and q. Therefore, B_{n-1} and B_n, although they do not exist in Eq. (C2), can be evaluated and are used to express R and S as functions of p, q, and the B's. Extending Eq. (C4) for $k = n - 1$, we have

$$A_{n-1} = B_{n-1} + pB_{n-2} + qB_{n-3},$$

and for $k = n$ we have

$$A_n = B_n + pB_{n-1} + qB_{n-2}.$$

From Eqs. (C5) and (C6) we now have

$$R = A_{n-1} - pB_{n-2} - qB_{n-3} = B_{n-1}, \tag{C7}$$

$$S = A_n - qB_{n-2} = B_n + pB_{n-1}. \tag{C8}$$

Thus our requirement becomes $R = f_1(p, q) = 0$ and $S = f_2(p, q) = 0$, where f_1 and f_2 are B_{n-1} and $B_n + pB_{n-1}$, respectively.

Now that we have R and S as functions of p and q, we need only find the particular values of p and q which make $R = S = 0$. This is done by expressing the functions of R and S in terms of their Taylor series, setting the Taylor series equal to zero, and then finding the particular values of p and q that satisfy this equality.

We shall next discuss the general form of the Newton-Raphson iteration for a function of two variables. Taylor's theorem for a function with one variable can be written compactly as[†]

$$f(p) = \sum_{k=0}^{\infty} \frac{f^{(k)}(p_0)(p - p_0)^k}{k!}, \tag{C9}$$

where $f^{(k)}(p_0)$ denotes the kth derivative of $f(p)$ evaluated at p_0, the point about which the expansion takes place. This theorem may be extended, and for a function with two variables we have[‡]

$$f(p, q) = f(p_0, q_0) + [(p - p_0)f_p(p_0, q_0) + (q - q_0)f_q(p_0, q_0)]$$

$$+ \frac{1}{2!} [(p - p_0)^2 f_{pp}(p_0, q_0) + 2(p - p_0)(q - q_0)f_{pq}(p_0, q_0)$$

$$+ (q - q_0)^2 f_{qq}(p_0, q_0)] + \cdots, \tag{C10}$$

where $f_p \equiv \partial f/\partial p$, $f_{pp} \equiv \partial^2 f/\partial p^2$ and $f_{pq} \equiv \partial^2 f/\partial p\, \partial q$, and all are evaluated at the point (p_0, q_0).[§]

[†] For example, see Thomas, G. B., *Calculus and Analytic Geometry*, 3rd ed. Addison-Wesley, Reading, Mass., p. 787, 1960.
[‡] For example, see Hildebrand, F. B., *Advanced Calculus for Applications*. Prentice-Hall, Englewood Cliffs, N.J., p. 350, 1962.
[§] In the case of one variable, p_0 is the p-coordinate of the point $(p_0, f(p_0))$ about which the expansion takes place, and we say, "the function is expanded about the point p_0." Similarly, for two variables we are expanding about $(p_0, q_0, f(p_0, q_0))$ we say, "the function is expanded about the point (p_0, q_0)."

We consider the Taylor series expansion about a point (p_k, q_k). If P and Q are the exact values of p and q for which $f(P, Q) = 0$, we have

$$f(P, Q) = 0 = f(p_k, q_k) + (P - p_k)f_p(p_k, q_k) + (Q - q_k)f_q(p_k, q_k) + \cdots . \tag{C11}$$

We see that $P = p_k$ and $Q = q_k$ would satisfy Eq. (C11) nicely, but since we have no idea of the values of P and Q, we have no idea of the location of (p_k, q_k) about which we should expand the series. Letting $\Delta p = P - p_k$ and $\Delta q = Q - q_k$, we see that if Δp and Δq are made small, then nonlinear terms may be dropped because the series rapidly converges. We may, therefore, get an approximation by dropping nonlinear terms as follows:

$$f(P, Q) = 0 \cong f(p_k, q_k) + (P - p_k)f_p(p_k, q_k) + (Q - q_k)f_q(p_k, q_k). \tag{C12}$$

Letting p_k, q_k be initial "guesses" or estimates of the expansion point coordinates, we have for some other point (p_{k+1}, q_{k+1}),

$$f_1(P, Q) = 0 \approx f_1(p_k, q_k) + \Delta p f_{1p}(p_k, q_k) + \Delta q f_{1q}(p_k, q_k), \tag{C13}$$

$$f_2(P, Q) = 0 \approx f_2(p_k, q_k) + \Delta p f_{2p}(p_k, q_k) + \Delta q f_{2q}(p_k, q_k), \tag{C14}$$

where $\Delta p = p_{k+1} - p_k$ and $\Delta q = q_{k+1} - q_k$. Hence Δp and Δq may be found by simultaneous solution of Eqs. (C13) and (C14).

If p_k and q_k were good estimates, $f_1(p_k, q_k)$ would be almost equal to $f_1(P, Q)$, and Δp and Δq would be small. If the estimates were not good, then large values of Δp and Δq would imply that (p_{k+1}, q_{k+1}) is closer to the actual expansion point. Hence we next try

$$p_{k+1} = p_k + (p_{k+1} - p_k) = p_k + \Delta p$$

and

$$q_{k+1} = q_k + (q_{k+1} - q_k) = q_k + \Delta q.$$

We continue in this fashion until Δp and Δq are within preassigned limits.

Although Eqs. (C13) and (C14) are explicitly stated, we have as yet no means of calculating the following partial derivatives:

$$\frac{\partial f_1}{\partial p} = \frac{\partial B_{n-1}}{\partial p},$$

$$\frac{\partial f_1}{\partial q} = \frac{\partial B_{n-1}}{\partial q},$$

$$\frac{\partial f_2}{\partial p} = \frac{\partial B_n}{\partial p} + p\frac{\partial B_{n-1}}{\partial p} + B_{n-1}, \tag{C15}$$

$$\frac{\partial f_2}{\partial q} = \frac{\partial B_n}{\partial q} + p\frac{\partial B_{n-1}}{\partial q}.$$

Equations (C13) and (14) may be simplified in the following manner.

(1) Substituting f_1, f_2, and the derivatives of Eqs. (C15) to give

$$B_{n-1} + \frac{\partial B_{n-1}}{\partial p} \Delta p + \frac{\partial B_{n-1}}{\partial q} \Delta q = 0, \tag{C16}$$

$$B_n + p_k B_{n-1} + \left(\frac{\partial B_n}{\partial p} + p_k \frac{\partial B_{n-1}}{\partial p} + B_{n-1}\right) \Delta p + \left(\frac{\partial B_n}{\partial q} + p_k \frac{\partial B_{n-1}}{\partial q}\right) \Delta q = 0; \tag{C17}$$

(2) subtracting p_k times Eq. (C16) from Eq. (C14) to give

$$B_{n-1} + \frac{\partial B_{n-1}}{\partial p} \Delta p + \frac{\partial B_{n-1}}{\partial q} \Delta q = 0, \tag{C18}$$

$$B_n + \left(\frac{\partial B_n}{\partial p} + B_{n-1}\right) \Delta p + \frac{\partial B_n}{\partial q} \Delta q = 0. \tag{C19}$$

Since we need the derivative of B at n and at $n - 1$, and since n may be any positive integer, it is clear that a recursion relationship for $\partial B_k/\partial p$ and $\partial B_k/\partial q$ must be obtained. The general term was given by Eq. (C4) and differentiating it, we have

$$\frac{\partial B_k}{\partial p} = -B_{k-1} - p \frac{\partial B_{k-1}}{\partial p} - q \frac{\partial B_{k-2}}{\partial p}, \tag{C20}$$

$$\frac{\partial B_k}{\partial q} = -B_{k-2} - p \frac{\partial B_{k-1}}{\partial q} - q \frac{\partial B_{k-2}}{\partial q}. \tag{C21}$$

Since both B_{-1} and B_0 are constant, we have

$$\frac{\partial B_{-1}}{\partial p} = \frac{\partial B_{-1}}{\partial q} = \frac{\partial B_0}{\partial p} = \frac{\partial B_0}{\partial q} = 0. \tag{C22}$$

Although the partial derivatives of Eqs. (C18) and (C19) could be calculated directly from the recursion relationship of Eqs. (C20) and (C21), we choose the following procedure which avoids the awkward notation.

Just as we factored Eq. (C1) to get Eq. (C2), so we may factor $(y^2 + py + q)$ from Eq. (C2) [Note: R and S are zero in the final form of (C2)] to get

$$(y^2 + py + q)(y^{n-4} + C_1 y^{n-5} + \cdots + C_{n-5} y + C_{n-4}) + R^* y + S^*,$$

where $R^* y + S^*$ is the linear remainder analogous to R and S in Eq. (C2). Expanding and equating coefficients of like powers as we did in Eq. (C2), we have for the general recursion relationship,

$$C_k = B_k - pC_{k-1} - qC_{k-2} \quad \text{with} \quad C_{-1} = 0 \quad \text{and} \quad C_0 = 1, \tag{C23}$$

which may be extended for $k = 1, \ldots, n$. Equation (C23) may be altered to give $-C_{k-1} = B_{k-1} + pC_{k-2} + qC_{k-3}$. When this is compared with Eq. (C20), it is implied that

$$\partial B_k/\partial p = -C_{k-1}. \tag{C24}$$

Altering Eq. (C23) results in $-C_{k-2} = -B_{k-2} + pC_{k-3} + qC_{k-4}$, which may be compared with Eq. (C21) to yield

$$\partial B_k / \partial q = -C_{k-2}. \tag{C25}$$

Substituting Eqs. (C24) and (C25) into Eqs. (C18) and (C19), we have

$$C_{n-2}\,\Delta p + C_{n-3}\,\Delta q = B_{n-1}, \tag{C26}$$

$$-(\partial B_n / \partial p + B_{n-1})\,\Delta p + C_{n-2}\,\Delta q = B_n. \tag{C27}$$

Letting $-(\partial B_n / \partial p + B_{n-1}) = \overline{C}_{n-1}$, we have

$$\overline{C}_{n-1} = C_{n-1} - B_{n-1} = -pC_{n-2} - qC_{n-3},$$

from which we obtain the final forms of Bairstow's equations:

$$C_{n-2}\,\Delta p + C_{n-3}\,\Delta q = B_{n-1} \tag{C28}$$

and

$$\overline{C}_{n-1}\,\Delta p + C_{n-2}\,\Delta q = B_n. \tag{C29}$$

Equations (6.17) and (6.18) in Chapter 6 are obtained by solving Eqs. (C28) and (C29) simultaneously.

To recapitulate the computational procedure, we start with an initial assumption for p and q; then we calculate the B-terms from the A-terms using Eq. (C4) expanded for $k = 1, \ldots, n$; next we compute the C-terms from Eq. (C23), which need be extended only for $k = 1, \ldots, (n - 1)$; and finally we calculate $\overline{C}_{n-1} = C_{n-1} - B_{n-1}$. The coefficients C_{n-2}, C_{n-3}, and C_{n-1}, are then substituted along with B_{n-1} and B_n into the two Bairstow equations, (C28) and (C29). These equations can be solved simultaneously for Δp and Δq. If Δp and Δq are smaller in magnitude than some preassigned positive number, say, $|\Delta p| + |\Delta q| < \epsilon$, where ϵ is usually between 0 and 1, then the solution has converged to a p and a q, which yield the desired quadratic factor $(y^2 + py + q)$; if the convergence test is not passed, p and q are modified by the amounts Δp and Δq, and the whole iterative process is repeated.

To compute the two roots associated with the factor $(y^2 + py + q)$ is an easy matter. If we let the two roots be of the form $(D + Ej)$ and $(D - Ej)$, then the quadratic factor would be $[y - (D + Ej)][y - (D - Ej)] = y^2 - 2Dy + (D^2 + E^2)$. Equating coefficients of like powers of y from the quadratic factor $y^2 + py + q$ yields

$$p = -2D, \quad q = (D^2 + E^2) \quad \text{or} \quad D = -p/2, \quad E = \sqrt{q - p^2/4}.$$

If the initial values of p and q are sufficiently close to the true values, then the method will always converge. Since this is Newton's process, convergence, once established, is quite rapid. In cases where the initial values of p and q are merely random guesses, the method may not converge. To protect against this possibility, an upper limit should be placed on the number of iterations.

ANSWERS TO
SELECTED PROBLEMS

CHAPTER 6

1. 1.391 radians
2. a) 0.30677, 4.4939, 7.7254, etc. b) 0.14393, 4.4934, 7.7253, etc.
3. *Case* 1: $t = 2.364$; *Case* 2: $t = 17.48$
4. 2.3650
5. 4.4934
6. 2.2185
7. 0.788146
8. $\alpha_1 = 0.07951$, $\alpha_2 = 0.145475$, $\alpha_3 = 0.221908$, $\alpha_4 = 0.312662$, $\alpha_5 = 0.407639$
9. 0.567143
10. For $c = 0.8$, $t = -0.42548$; for $c = 0.9$, $t = -0.45851$
11. 0.91175
12. 0.4725
13. 0.008826
14. 0.8354
15. $0, \pm 1, \pm 2$
16. $1 \pm j, 2 \pm 3j$
17. (1) $y = 11.954$, $u + v = 79.1$; (2) $u + v = 56$

CHAPTER 7

1. $t = 10$, $h = 9.7199$ for the closed-form solution.
2. $t = 2.5$, $x = 990.992$ from the closed-form solution.
5. $t = 5$ hours, $c = 1.4484$ for the closed-form solution.
6. for $\Delta t = 0.125$: $t = 0.125$, $x = 0.718199$ (Runge-Kutta).

8. $t = 0.001$, $v = 0.021445$ from the closed-form solution.

9. $x = 0.2$, $y = 0.6534$ from the closed-form solution.

11. $x = 3.01$, $v = 3.01376$ for the closed-form solution.

12. For $\Delta r = 0.1$: at $r = 1$, $T = 899.212$, $dT/dr = -1.57$.

13. For $\Delta x = 0.05$ in.: at $x = 0.6$ in., $T = 1014.82$, $dT/dR = -18077.8$.

17. For $t = 0.1$: at $t = 3$, $y = 0.346$, $dy/dt = 0.306$.

19. For $x = 0.1$: at $x = 3$ ft, $y = 124.29$, $dy/dx = 8.456$.

20. For $\Delta t = 0.2$: at $t = 1.2$, $y = 0.0000156$, $y' = 0.00000874$.

21. At $t = 2$, $x = -0.8607$, $dx/dt = -1.2285$.

CHAPTER 8

4. $\theta_a = 2.954$

5. $I_1 = 0.377$, $I_2 = 1.262$, $I_3 = -0.639$, $I_4 = 0.532$, $I_5 = 0.35$, $I_6 = 0.154$, $I_7 = 0.322$, $I_8 = 0.532$, $I_9 = 0.99$, $I_{10} = 0.477$

6. First set: $x = 3$, $y = -18$, $z = 11$, $w = 3$
   Second set: $x = 12.647$, $y = -0.529$, $z = -38.823$

7. DET $= -360$, Inverse $= \begin{bmatrix} -0.1027 & 0.1055 & 0.0638 \\ 0.1888 & 0.0222 & -0.1444 \\ -0.0194 & -0.0611 & 0.1472 \end{bmatrix}$

8. $x = 9.26$, $y = -0.92$, $z = -2.51$, $w = 3.27$

CHAPTER 9

1. $\lambda_1 = 5.05$

2. Eigenvectors:
$$\begin{matrix} -0.080 & 0.339 & -0.937 \\ -0.831 & -0.544 & -0.119 \\ -0.555 & 0.767 & 0.330 \end{matrix}$$

8. a) $[T] = \begin{bmatrix} \sqrt{2}/2 & -\sqrt{2}/2 \\ \sqrt{2}/2 & \sqrt{2}/2 \end{bmatrix}$

   b) $\lambda_1 = 3$, $\lambda_2 = 1$
   c) same as $[T]$

9. $\lambda_1 = 2 + \sqrt{3}$, $\lambda_2 = -1$, $\lambda_3 = 2 - \sqrt{3}$

CHAPTER 10

2. 0.96517, 0.82303, 0.76280, 0.63536, 0.50832, 0.28497, 0.02444

3. 0.773465, 0.959982, 0.997214, 0.975020

CHAPTER 11

1. $Y = 948.184 - 40.18x + 19.3069x^2 - 1.1208x^3$ (based on $x = 1, 2, \ldots, 11$)

2. a) $T = 0.0321 + 0.01346t$
   b) $T = 0.0332 + 0.01237t + 0.000142t^2$
   c) $T = 0.03192 + 0.01457t - 0.000586t^2 + 0.0000612t^3$

3. From normal equations:

$$T = 138.4505 - 0.8437R + 0.5120R^2 - 0.028R^3$$

From Chebyshev polynomials:

$$T = 139.0476 - 1.2933R + 0.6045R^2 - 0.033R^3$$

CHAPTER 12

3. 0.499968
4. 0.7854
5. 1.8518
6. $A_0 = 15.915$, $A_1 = 12.5$, $A_2 = 5.308$, $A_3 = 0.00$, $A_4 = -1.0638$,
   $A_5 = 0.00$, $A_6 = 0.45179$, $A_7 = 0.2118$, $A_8 = 0.1773$, $A_9 = 0.2125$
7. (I) 0.0868567   (II) 0.399876   (III) 0.4867327   (IV) 0.4966294

CHAPTER 14

3. 0.49652
5. Theoretical answer: $\frac{4}{3}\pi$
8. 16 steps: 7.29954
    200 steps: 24.7477

CHAPTER 15

1. a) $x_1 = 1\frac{1}{5}$, $x_2 = 3\frac{1}{5}$, $P = 36$
   b) Six basic solutions
   c) Four feasible basic solutions
3. Produce $6\frac{2}{3}$ product A and $35\frac{5}{9}$ product B (or 7 product A and 36 product B).
5. Number of cakes $A = 0$, number of cakes $B = 12$, profit $= \$36$.

Index

Index